High-Performance Scientific Computing

Michael W. Berry • Kyle A. Gallivan •
Efstratios Gallopoulos • Ananth Grama •
Bernard Philippe • Yousef Saad • Faisal Saied
Editors

High-Performance Scientific Computing

Algorithms and Applications

Springer

Editors
Michael W. Berry
Dept. Electrical Eng. & Computer Science
University of Tennessee
Knoxville, TN, USA

Kyle A. Gallivan
Department of Mathematics
Florida State University
Tallahassee, FL, USA

Efstratios Gallopoulos
Dept. Computer Engineering & Informatics
University of Patras
Patras, Greece

Ananth Grama
Department of Computer Science
Purdue University
West Lafayette, IN, USA

Bernard Philippe
IRISA
INRIA Rennes - Bretagne Atlantique
Rennes, France

Yousef Saad
Dept. of Computer Science & Engineering
University of Minnesota
Minneapolis, MN, USA

Faisal Saied
Department of Computer Science
Purdue University
West Lafayette, IN, USA

ISBN 978-1-4471-5888-2 ISBN 978-1-4471-2437-5 (eBook)
DOI 10.1007/978-1-4471-2437-5
Springer London Dordrecht Heidelberg New York

British Library Cataloguing in Publication Data
A catalogue record for this book is available from the British Library

Printed on acid-free paper

Springer is part of Springer Science+Business Media (www.springer.com)

Preface

This collection is a tribute to the intellectual leadership and legacy of Prof. Ahmed H. Sameh. His significant contributions to the field of Parallel Computing, over his long and distinguished career, have had a profound influence on high performance computing algorithms, applications, and systems. His defining contributions to the field of Computational Science and Engineering, and its associated educational program, resulted in a generation of highly trained researchers and practitioners. His high moral character and fortitude serve as exemplars for many in the community and beyond.

Prof. Sameh did his graduate studies in Civil Engineering at the University of Illinois at Urbana-Champaign (UIUC). Upon completion of his Ph.D. in 1966, he was recruited by Daniel L. Slotnick, Professor and Director of the Illiac IV project, to develop various numerical algorithms. Prof. Sameh joined the Department of Computer Science as a Research Assistant Professor, subsequently becoming a Professor, and along with Profs. Duncan Lawrie, Daniel Gajski and Edward Davidson served as the Associate Director of the Center for Supercomputing Research and Development (CSRD). CSRD was established in 1984 under the leadership of Prof. David J. Kuck to build the University of Illinois Cedar multiprocessor. Prof. Sameh directed the CSRD Algorithms and Applications Group. His visionary, yet practical outlook, in which algorithms were never isolated either from real applications or from architecture and software, resulted in seminal contributions. By 1995 CSRD's main mission had been accomplished, and Prof. Sameh moved to the University of Minnesota as Head of the Computer Science Department and William Norris Chair for Large-Scale Computing. After a brief interlude, back at UIUC, to lead CSRD, during which he was very active in planning the establishment of Computational Science and Engineering as a discipline and an associated graduate program at UIUC, he returned to Minnesota, where he remained until 1997. He moved to Purdue University as the Head and Samuel D. Conte Professor of Computer Science. Prof. Sameh, who is a Fellow of SIAM, ACM and IEEE, was honored with the IEEE 1999 Harry H. Goode Memorial Award "For seminal and influential work in parallel numerical algorithms".

It was at Purdue that over 50 researchers and academic progeny of Prof. Sameh gathered in October 2010 to celebrate his 70th birthday. The occasion was the *Con-*

ference on High Performance Scientific Computing: *Architectures, Algorithms, and Applications* held in his honor. The attendees recalled Prof. Sameh's many academic achievements, including, not only his research but also his efforts in defining the interdisciplinary field of Computational Science and Engineering and his leadership and founding Editor-in-Chief role in the IEEE CS&E Magazine as well as the many doctoral candidates that he has graduated: At UIUC, Jonathan Lermit (1971), John Larson (1978), John Wisniewski (1981), Joseph Grcar (1981), Emmanuel Kamgnia (1983), Chandrika Kamath (1986), Mark Schaefer (1987), Hsin-Chu Chen (1988), Randall Bramley (1988), Gung-Chung Yang (1990), Michael Berry (1990), Felix G. Lou (1992), Bart Semeraro (1992) and Vivek Sarin (1997); Ananth Grama (1996) at the University of Minnesota; and Zhanye Tong (1999), Matt Knepley (2000), Abdelkader Baggag (2003), Murat Manguoglu (2009) and Carl Christian Kjelgaard Mikkelsen (2009) at Purdue.

This volume consists of a survey of Prof. Sameh's contributions to the development high performance computing and sixteen editorially reviewed papers written to commemorate the occasion of his 70th birthday.

Knoxville, USA	Michael W. Berry
Tallahassee, USA	Kyle A. Gallivan
Patras, Greece	Stratis Gallopoulos
West Lafayette, USA	Ananth Grama
Rennes, France	Bernard Philippe
Minneapolis, USA	Yousef Saad
West Lafayette, USA	Faisal Saied

Acknowledgements

We are especially grateful to Profs. Zhiyuan Li, Alex Pothen, and Bob Skeel for many arrangements that made the conference possible. We are also grateful to Dr. Eric Cox, who undertook the heavy load of making many of the local arrangements and Dr. George Kollias and Ms. Eugenia-Maria Kontopoulou for their help in compiling this volume. Finally, we thank Springer and especially Mr. Simon Rees for patiently working with us on this project and Donatas Akmanavičius of VTeX Book Production for great editing work in compiling the volume.

Contents

List of Contributors

P.-A. Absil Department of Mathematical Engineering, ICTEAM Institute, Université catholique de Louvain, Louvain-la-Neuve, Belgium

Jean-Thomas Acquaviva UVSQ/Exascale Computing Research, Versailles, France

Abdelkader Baggag Université Laval, College of Science and Engineering, Quebec City, Canada

Allison H. Baker Lawrence Livermore National Laboratory, Center for Applied Scientific Computing, Livermore, CA, USA

Martin Bečka Mathematical Institute, Department of Informatics, Slovak Academy of Sciences, Bratislava, Slovak Republic

Michael W. Berry Center for Intelligent Systems and Machine Learning (CISML), University of Tennessee, Knoxville, TN, USA

Jean-Christophe Beyler UVSQ/Exascale Computing Research, Versailles, France

Jack Dongarra Department of Electrical Engineering and Computer Science, University of Tennessee, Knoxville, USA

Victor Eijkhout Texas Advanced Computing Center, The University of Texas at Austin, Austin, USA

Robert D. Falgout Lawrence Livermore National Laboratory, Center for Applied Scientific Computing, Livermore, CA, USA

Kyle A. Gallivan Department of Mathematics, Florida State University, Tallahassee, FL, USA

Efstratios Gallopoulos CEID, University of Patras, Rio, Greece

Robert van de Geijn Computer Science Department, The University of Texas at Austin, Austin, USA

Ananth Grama Computer Science Department, Purdue University, West-Lafayette, IN, USA

C.W. Mattias Holm LIACS, Leiden University, Leiden, CA, The Netherlands

William Jalby UVSQ/Exascale Computing Research, Versailles, France

Sami A. Kilic Department of Civil Engineering, Faculty of Engineering, Bogazici University, Bebek, Istanbul, Turkey

Jingu Kim School of Computational Science and Engineering, College of Computing, Georgia Institute of Technology, Atlanta, USA

Tzanio V. Kolev Lawrence Livermore National Laboratory, Center for Applied Scientific Computing, Livermore, CA, USA

David J. Kuck Intel Corporation, Urbana, USA

Jakub Kurzak Department of Electrical Engineering and Computer Science, University of Tennessee, Knoxville, USA

Piotr Luszczek Department of Electrical Engineering and Computer Science, University of Tennessee, Knoxville, USA

Murat Manguoglu Department of Computer Engineering, Middle East Technical University, Ankara, Turkey

Carl Christian Kjelgaard Mikkelsen Department of Computing Science and HPC2N, Umeå University, Umeå, Sweden

Gabriel Okša Mathematical Institute, Department of Informatics, Slovak Academy of Sciences, Bratislava, Slovak Republic

Haesun Park School of Computational Science and Engineering, College of Computing, Georgia Institute of Technology, Atlanta, USA

Bernard Philippe INRIA Research Center Rennes Bretagne Atlantique, Rennes, France

Eric Polizzi Department of Electrical and Computer Engineering, University of Massachusetts, Amherst, MA, USA

Andrey A. Puretskiy Department of Electrical Engineering and Computer Science, University of Tennessee, Knoxville, TN, USA

Chunhong Qi Department of Mathematics, Florida State University, Tallahassee, FL, USA

Yousef Saad Department of Computer Science and Engineering, University of Minnesota, Minneapolis, MN, USA

Faisal Saied Computer Science Department, Purdue University, West-Lafayette, IN, USA

Danny Sorensen Computational and Applied Mathematics, Rice University, Houston, TX, USA

Harmen L.A. van der Spek LIACS, Leiden University, Leiden, CA, The Netherlands

Stanimire Tomov Department of Electrical Engineering and Computer Science, University of Tennessee, Knoxville, USA

Marián Vajteršic Mathematical Institute, Department of Informatics, Slovak Academy of Sciences, Bratislava, Slovak Republic; Department of Computer Sciences, University of Salzburg, Salzburg, Austria

Mu Wang Tsinghua University, Beijing, P.R. China

Xiaoge Wang Tsinghua University, Beijing, P.R. China

Harry A.G. Wijshoff LIACS, Leiden University, Leiden, CA, The Netherlands

David C. Wong Intel Corporation, Urbana, USA

Jianlin Xia Department of Mathematics, Purdue University, West Lafayette, IN, USA

Ulrike Meier Yang Lawrence Livermore National Laboratory, Center for Applied Scientific Computing, Livermore, CA, USA

Chapter 1
Parallel Numerical Computing from Illiac IV to Exascale—The Contributions of Ahmed H. Sameh

Kyle A. Gallivan, Efstratios Gallopoulos, Ananth Grama, Bernard Philippe, Eric Polizzi, Yousef Saad, Faisal Saied, and Danny Sorensen

K.A. Gallivan (✉)
Department of Mathematics, Florida State University, Tallahassee, FL, USA
e-mail: gallivan@math.fsu.edu

E. Gallopoulos
CEID, University of Patras, Rio, Greece
e-mail: stratis@ceid.upatras.gr

A. Grama · F. Saied
Computer Science Department, Purdue University, West-Lafayette, IN, USA

A. Grama
e-mail: grama@cs.purdue.edu

F. Saied
e-mail: fsaied@purdue.edu

B. Philippe
INRIA Research Center Rennes Bretagne Atlantique, Rennes, France
e-mail: Bernard.Philippe@inria.fr

E. Polizzi
Department of Electrical and Computer Engineering, University of Massachusetts, Amherst, MA, USA
e-mail: polizzi@ecs.umass.edu

Y. Saad
Department of Computer Science and Engineering, University of Minnesota, Minneapolis, MN, USA
e-mail: saad@cs.umn.edu

D. Sorensen
Computational and Applied Mathematics, Rice University, Houston, TX, USA
e-mail: sorensen@rice.edu

M.W. Berry et al. (eds.), *High-Performance Scientific Computing*,
DOI 10.1007/978-1-4471-2437-5_1, © Springer-Verlag London Limited 2012

Abstract As exascale computing is looming on the horizon while multicore and GPU's are routinely used, we survey the achievements of Ahmed H. Sameh, a pioneer in parallel matrix algorithms. Studying his contributions since the days of Illiac IV as well as the work that he directed and inspired in the building of the Cedar multiprocessor and his recent research unfolds a useful historical perspective in the field of parallel scientific computing.

1.1 Illiac IV and Cedar Legacies on Parallel Numerical Algorithms

Ahmed Sameh's research on parallel matrix algorithms spans more than four decades. It started in 1966 at the University of Illinois at Urbana-Champaign (UIUC) when he became involved in the Illiac IV project [22] as research assistant while pursuing his Ph.D. in Civil Engineering, following his undergraduate engineering studies at the University of Alexandria in Egypt and a M.Sc. as Fulbright scholar at Georgia Tech. Via his advisor, Professor Alfredo Ang at UIUC, Sameh became a descendent of Nathan Newmark, Hardy Cross, and David Hilbert (see Fig. 1.1).

At the invitation of Daniel Slotnick (also a Hilbert descendant), who was the director of the Illiac IV project Sameh looked at eigenvalue problems. The result of that effort was the first genuinely parallel algorithm (and accompanying Illiac IV assembly code) for the computation of eigenvalues of symmetric matrices [111]; see Sect. 1.2.2. By the time he completed his doctoral thesis (on "Numerical analysis of axisymmetric wave propagation in elastic-plastic layered media") in 1968 [107] (see also [108]) Sameh was deeply involved in the Illiac IV project [6]. It was the most significant parallel computing project in a U.S. academic institution, in fact the first large-scale attempt to build a parallel supercomputer, following the early prototypes of Solomon I at Westinghouse [130]. Not surprisingly, at that time there were very few publications on parallel numerical algorithms, even fewer on parallel matrix computations and practically no implementations since no parallel computers were available. The reasons for abandoning the classical von Neumann architecture and the motivation for the Illiac IV model of parallel computing were outlined in detail in [22]. Quoting from the paper:

> "The turning away from the conventional organization came in the middle 1950s, when the law of diminishing returns began to take effect in the effort to increase the operational speed of a computer. Up until this point the approach was simply to speed up the operation of the electronic circuitry which comprised the four major functional components."

The Illiac IV, modeled after the earlier Solomon I [130], was an SIMD computer initially designed to have 256 processors, though only a quadrant of 64 PEs was finally built.

One can say that the Illiac IV work initiated a new era by bringing fundamental change. Though the hardware side of the project faced difficulties due to the challenges of the technologies adopted, the results and by-products of this work have

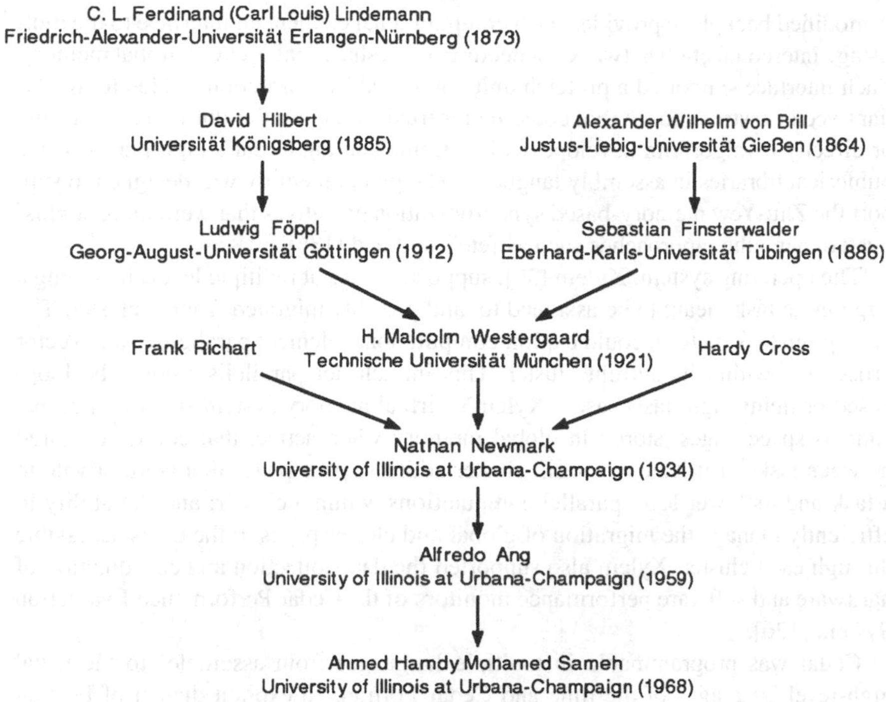

Fig. 1.1 Ahmed H. Sameh's 10 most recent scientific ancestors (from the Mathematics Genealogy project)

been felt for decades. In particular, it had a tremendous vitalizing effect on parallel numerical algorithms [129].

In 1985, about a decade after the Illiac IV was finally deployed, Sameh joined the Center for Supercomputing Research and Development (CSRD) to participate in the development of another groundbreaking Illinois parallel computing system, called Cedar, an effort led by David Kuck.[1] Sameh served as Associate Director of CSRD leading the Algorithms and Applications group until 1991 and was Director of CSRD in 1992.

Cedar was a cluster-based multiprocessor with an hierarchical structure that will seem familiar to those acquainted with today's systems. It comprised multiple clusters, each of which was a tightly coupled hierarchical shared memory multivector processor (an Alliant FX/8) [40, 83]. The computational elements were register-based vector processors that shared an interleaved multi-bank cache. Memory-based synchronization and a dedicated concurrency control bus provided low-overhead synchronization and dynamic scheduling on the eight computational elements.

[1]David J. Kuck was the 2011 recipient of the IEEE Computer Society Computer Pioneer Award "for pioneering parallel architectures including the Illiac IV, the Burroughs BSP, and Cedar; and, for revolutionary parallel compiler technology including Parafrase and KAP.2009".

A modified backplane provided each computational element an interface to the multistage interconnection network connecting to the shared interleaved global memory. Each interface supported a prefetch unit controlled by instructions added to the Alliant vector instruction set that could be inserted by a compiler during restructuring or directly by algorithm developers when implementing Cedar's high-performance numerical libraries in assembly language. The global memory was designed to support the Zhu-Yew memory-based synchronization primitives that were more sophisticated than other approaches such as fetch-and-add [136].

The operating system, Xylem [39], supported tasks at multiple levels including a large grain task meant to be assigned to, and possibly migrated from, a cluster. The large grain task, in turn, could exploit computational element parallelism and vector processing within its current cluster. This intracluster parallelism could be loop-based or lightweight task-based. Xylem's virtual memory system supported global address space pages, stored in global memory when active, that could be shared between tasks or private to a task; cluster address space pages that were private to a task and its lower level parallel computations within a cluster; and the ability to efficiently manage the migration of global and cluster pages to the disks accessible through each cluster. Xylem also supported the data collection and coordination of hardware and software performance monitors of the Cedar Performance Evaluation System [126].

Cedar was programmable in multiple languages from assembler to the usual high-level languages of the time and Cedar Fortran, an explicit dialect of Fortran that supported hierarchical task-based and hierarchical loop-based parallelism as well as combinations of the two. In the hierarchical loop-based approach, loops at the outer level were spread across clusters, loops at the middle level were spread across computational elements within a cluster, and loops at the inner level were vectorized [69]. In addition to explicit parallel coding, a Cedar Fortran restructuring compiler provided directive-based restructuring for parallelism.

The Cedar system was very flexible in the viewpoints that could be adopted when investigating algorithm, architecture, and application interaction—a characteristic that was exploited extensively in the work of the Algorithms and Application group with the strong encouragement of Sameh. Indeed, the significant level of research and its continuing influence was attributable as much to the leadership of Sameh as it was to the state-of-the-art hardware and software architecture of Cedar. Below, we review a few of Sameh's contributions, specifically those related to high performance numerical linear algebra but in closing this section we briefly mention his view of the research strategies and priorities of the Algorithms and Applications group and a few of the resulting multidisciplinary activities.

Sameh based his motivation on the premise that fundamental research and development in architecture (hardware and software), applications and algorithms must be equally responsible for motivating progress in high performance computation and its use in science and engineering. As a result, it was the responsibility of the Algorithms and Applications group to identify critical applications and the algorithms that were vital to their computations and similarly identify algorithms that were vital components of multiple critical applications. The resulting matrix, see Table 1.1,

Table 1.1 Applications vs. Computational Kernels [32]

	1	2	3	4	5	6	7	8	9	10
lattice gauge (QCD)	X	–	–	X	–	–	–	–	X	–
quantum mechanics	–	–	–	X	–	–	–	X	–	X
weather	–	–	–	–	X	X	–	–	–	–
CFD	X	–	X	–	X	X	X	–	–	–
geodesy	X	X	–	–	–	–	–	–	–	–
inverse problems	–	X	–	–	X	–	–	–	–	–
structures	X	–	X	X	–	–	–	–	–	–
device simulation	X	–	X	–	–	X	X	–	X	–
circuit simulation	X	–	X	–	–	–	–	X	–	–
electromagnetics	X	X	X	X	X	X	–	–	–	–

1—linear systems, 2—least squares, 3—nonlinear systems, 4—eigenvalues/SVD's, 5—fast transforms, 6—rapid elliptic solvers, 7—multigrid, 8—stiff ODE, 9—Monte Carlo, 10—integral transforms

identified efforts where fundamental research in algorithms could promote progress in applications and demands for increasingly sophisticated computational capabilities in applications could promote progress in algorithms research.[2] Implicit in this, of course, is an unseen third dimension to the table—system architecture. All application/algorithm interaction was driven by assessing the current capabilities of system architecture in hardware and software to identify good fits and to facilitate critiques leading to improved systems design. Effectively considering these three legs of the triad required a fairly wide range of variation in each hence the value of the flexibility of Cedar and the resulting breadth of Algorithm and Applications group research.

The group interacted with many external application specialists to improve algorithmic approaches, algorithm/architecture/application mixes and application capabilities in a variety of areas including: circuit and device simulation, molecular dynamics, geodesy, computational fluid mechanics, computational structural mechanics, and ocean circulation modeling. In addition to this and a significant amount of algorithm research—a glimpse of which is evident in the rest of this article—members of the group collaborated with the other groups in CSRD and external parties in many areas but in particular in performance evaluation of Cedar and other systems, benchmarking, performance prediction and improvement, problem solving environments and restructuring compilers. These activities included: intense performance evaluation of the Cedar system as it evolved [46, 79]; memory system and compiler-driven characterization and prediction of performance for numerical algorithms [45] benchmarking of performance for sparse matrix computing [105, 106]; the Perfect Club for systematic application-level performance evaluation of supercomputers [15]; data dependence analysis [56]; restructuring of codes exploiting matrix structure such as sparsity [16, 17, 93]; defining the area of problem solving

[2]It is interesting to note that this table is occasionally referenced as the "Sameh table" in the literature; see e.g. [36]. Indeed, it is possible to see this as an early precursor of the "Berkeley Dwarfs" [5].

environments [52]; and algebraically driven restructuring within a problem solving environment [35]. Sameh's leadership of the group was instrumental in making this research successful.

1.2 Algorithms for Dense Matrices

1.2.1 Primitives, Dense and Banded Systems

The evolution of the state-of-the-art in the understanding of the interaction of algorithm and architecture for numerical linear algebra can be traced in Sameh's contributions to the Illiac IV and Cedar projects. In particular, comparing the focus and depth of Sameh's survey in 1977 [115] or Heller's from 1978 [70] to Sameh's survey of 1990 [50] shows the significant improvement in the area. The discussions moved from complexity analysis of simple approaches under unrealistic or idealized architectural assumptions to detailed models and systematic experiments combined with algebraic characterizations of the algorithms that ease the mapping to any target architecture.

The early papers separated the algorithms and analyses into unlimited and limited parallelism versions with the former essentially complexity analyses and the latter more practical approaches. A good example of the unlimited parallelism class is the work of Sameh and Brent [117] on solving dense and banded triangular systems in $0.5\log^2 n + O(\log n)$ time and $\log m \log n + O(\log^2 m)$ time respectively where m is the bandwidth. The results are unlimited parallelism since $n^3/68 + O(n^2)$ and $0.5m^2 n + O(mn)$ processors are required respectively. This renders the dense triangular solver impractical for even moderately sized systems in floating point arithmetic but it does have uses in other situations, e.g., boolean recurrences, and was consider for such alternatives in [135]. For small m the complexity for banded triangular systems is reasonable on array processors but there is a superior limited parallelism approach that is more significant to this discussion.

The key contribution that has lasted in this "product form" algorithm for dense triangular systems is the emphasis on using an algebraic characterization to derive the algorithm and show its relationship to other approaches. The algorithm is simply derived by applying associativity to a factorization of the triangular matrix L to yield a fan-in tree of matrix-matrix and matrix-vector products. The particular factors chosen in this case are the elementary triangular matrices that each correspond to a column or row of L but many others are possible. The product form for $n = 8$ and $L = M_1 M_2 \cdots M_7$ is given by the expression

$$x = (((M_7^{-1} M_6^{-1})(M_5^{-1} M_4^{-1}))((M_3^{-1} M_2^{-1})(M_1^{-1} f)))$$

and the log time is easily seen. It is also clear that a portion of L^{-1} is computed leading to the need for $O(n^3)$ processors and a significant computational redundancy compared to the standard $O(n^2)$ sequential row or column-based algorithm given by

$$x = (M_7^{-1}(M_6^{-1}(M_5^{-1}(M_4^{-1}(M_3^{-1}(M_2^{-1}(M_1^{-1}f))))))).$$

The notion of simple pairwise grouping can be altered to exploit or maintain structure, e.g., sparsity, for a potentially practical algorithm.

In [28] Chen, Kuck and Sameh introduced a limited parallelism algorithm for the solution of banded triangular systems using an idea that is the basis for much of the discussion in the remainder of this chapter. When solving a lower triangular system of equations, i.e., a linear recurrence, that is banded, ideally one would be able to solve p independent systems yielding a speedup of p. This is not possible, in general, but it is the first step in the algorithm and corresponds to a block diagonal transformation on the system, e.g.,

$$D^{-1}Lx = D^{-1}f$$

where $D = \text{diag}(L_1, L_2, \ldots, L_p)$ contains the block diagonal part of L. The system then has a special form as seen in this example with $n = 12$ and $p = 3$:

$$\left(\begin{array}{cccc|cccc|cccc}
1&0&0&0&0&0&0&0&0&0&0&0\\
0&1&0&0&0&0&0&0&0&0&0&0\\
0&0&1&0&0&0&0&0&0&0&0&0\\
0&0&0&1&0&0&0&0&0&0&0&0\\
\hline
0&0&0&a&1&0&0&0&0&0&0&0\\
0&0&0&b&0&1&0&0&0&0&0&0\\
0&0&0&c&0&0&1&0&0&0&0&0\\
0&0&0&d&0&0&0&1&0&0&0&0\\
\hline
0&0&0&0&0&0&0&e&1&0&0&0\\
0&0&0&0&0&0&0&f&0&1&0&0\\
0&0&0&0&0&0&0&g&0&0&1&0\\
0&0&0&0&0&0&0&h&0&0&0&1
\end{array}\right)
\left(\begin{array}{c}
\xi_1\\ \xi_2\\ \xi_3\\ \xi_4\\ \hline \xi_5\\ \xi_6\\ \xi_7\\ \xi_8\\ \hline \xi_9\\ \xi_{10}\\ \xi_{11}\\ \xi_{12}
\end{array}\right)
=
\left(\begin{array}{c}
\gamma_1\\ \gamma_2\\ \gamma_3\\ \gamma_4\\ \hline \gamma_5\\ \gamma_6\\ \gamma_7\\ \gamma_8\\ \hline \gamma_9\\ \gamma_{10}\\ \gamma_{11}\\ \gamma_{12}
\end{array}\right).$$

The equations defined by rows 4, 5, 8, 9, and 12, i.e., those above and below the partitioning lines and the last equation, can then be solved for the corresponding unknowns. If all that is required is the final value, as is sometimes the case with a recurrence, the algorithm terminates, otherwise the rest of the unknowns can be recovered by independent vector operations.

Unlimited and limited parallelism approaches to solving dense linear systems were presented in [120] and [110] respectively. Both were based on orthogonal factorization to ensure stability without pivoting but nonorthogonal factorization versions were also developed and used on several machines including Cedar. The unlimited parallelism factorization algorithm computes the QR factorization based on Givens rotations and yields the classic knight's move pattern in the elements that can be eliminated simultaneously. For example, for $n = 6$ we have

X					
5	X				
4	6	X			
3	5	7	X		
2	4	6	8	X	
1	3	5	7	9	X

The elements can be eliminated simultaneously in the set indicated by the number in their positions. The elimination is done by rotating the row containing the element and the one above it. This algorithm yields $O(n)$ time given $O(n^2)$ processors. While it was considered unlimited parallelism due to the need for $O(n^2)$ it is easily adapted as the basis for practical algorithms on semi-systolic and systolic arrays and was effectively used on a geodesy problem on Cedar. This pattern can also be applied to the pairwise pivoting approach for factorization analyzed by Sorensen [131].

The limited parallelism *QR* algorithm of [110] assumes p processors and consists of an initial set of independent factorizations followed by a series of "waves" that eliminate the remaining elements. If the matrix to be factored A is partitioned into p blocks by grouping consecutive sets of rows, i.e., $A^T = (A_1^T \dots A_p^T)$ then each block can be reduced to upper triangular independently yielding $R^T = (R_1^T \dots R_p^T)$ where R_i is upper triangular.

The first row of R_1 can be used to eliminate the $(1,1)$ element of R_2, then after this modification, it can be passed on to eliminate the $(1,1)$ element of R_3, and the process can be repeated to eliminate all $(1,1)$ elements of the blocks. Note that after each $(1,1)$ element is eliminated from R_i the same processor can eliminate the entire nonzero diagonal to create a triangular matrix of nonzeros with dimension reduced by 1 by a series of Givens rotations. These diagonal eliminations can overlap the elimination of the $(1,1)$ elements and diagonals of later blocks.

The second row of R_1 can then be used in a similar fashion to eliminate the $(2,2)$ elements in R_2 to R_p followed by independent diagonal elimination. After each row of R_1 has been used in this manner all blocks other than the updated R_1 are 0 and the factorization is complete. The algorithm offers many possible patterns of communication and is easily adaptable to shared or distributed memory organizations. It was modified for Cedar as described in [50] and has been reintroduced many times subsequently in parallel dense factorization algorithms.

The divide and conquer algorithm presented above for banded triangular systems was generalized by Sameh and Kuck for solving tridiagonal systems [120]. The method assumes A is nonsingular and that A and A^T are unreduced tridiagonal matrices. The unreduced assumptions are required to adapt the algorithm to be robust when a diagonal block was singular. The assumptions guarantee that a $d \times d$ diagonal block's rank is at least $d - 1$.

Suppose we are to solve a linear system $Ax = b$ where A is tridiagonal, i.e., nonzero elements are restricted to the main diagonal and the first super and sub-diagonal. The cost of solving a tridiagonal system on a scalar machine is $O(n)$ computations using standard Gaussian elimination. The standard algorithm for doing this may, at first, appear to be intrinsically sequential and a major question was whether or not it was possible to solve a tridiagonal system in time less than $O(n)$ computations. A few authors started addressing this problem in the mid 1960s and methods were discovered that required $O(\log(n))$ computations. Two algorithms in this category are worth noting, one is the recursing doubling method by Stone [133], and the second is the cyclic reduction algorithm, first discussed by Hockney [72] in 1965. While both of these algorithms dealt with cases that required no pivoting, Kuck and Sameh presented a method for the general case described above.

Their divide and conquer algorithm [120], consists of five stages. In what follows, p is the number of processors, j is the processor number, and $m = n/p$. The first stage is simply a preparation phase. The system is scaled and partitioned. The second stage consists of using unitary transformations in each processor (e.g., plane rotations) to transform each local tridiagonal system into upper triangular form. If a diagonal block is singular then its (m, m) element will be 0. When such blocks exist, each column of the transformed matrix containing such a 0 element is replaced by its sum with the following column yielding the matrix, $A^{(1)}$. This is a simple nonsingular transformation applied from the right and guarantees that all diagonal blocks are nonsingular and upper triangular.

Stage 3 consists of a backward-elimination to transform each diagonal block into the identity and create dense "spikes" in the columns immediately before and after each diagonal block yielding the matrix, $A^{(2)}$. The matrices, $A^{(1)}$ and $A^{(2)}$, resulting from these (local) transformations have the forms

$$A^{(1)} = \left(\begin{array}{cccc|cccc|cccc}
x&x&x&0&0&0&0&0&0&0&0&0\\
0&x&x&x&0&0&0&0&0&0&0&0\\
0&0&x&x&x&0&0&0&0&0&0&0\\
0&0&0&x&x&0&0&0&0&0&0&0\\
\hline
0&0&0&x&x&x&x&0&0&0&0&0\\
0&0&0&x&0&x&x&x&0&0&0&0\\
0&0&0&x&0&0&x&x&x&0&0&0\\
0&0&0&x&0&0&0&x&x&0&0&0\\
\hline
0&0&0&0&0&0&0&x&x&x&x&0\\
0&0&0&0&0&0&0&x&0&x&x&x\\
0&0&0&0&0&0&0&x&0&0&x&x\\
0&0&0&0&0&0&0&x&0&0&0&x
\end{array}\right), \quad
A^{(2)} = \left(\begin{array}{cccc|cccc|cccc}
1&0&0&0&x&0&0&0&0&0&0&0\\
0&1&0&0&x&0&0&0&0&0&0&0\\
0&0&1&0&x&0&0&0&0&0&0&0\\
0&0&0&1&x&0&0&0&0&0&0&0\\
\hline
0&0&0&x&1&0&0&0&x&0&0&0\\
0&0&0&x&0&1&0&0&x&0&0&0\\
0&0&0&x&0&0&1&0&x&0&0&0\\
0&0&0&x&0&0&0&1&x&0&0&0\\
\hline
0&0&0&0&0&0&0&x&1&0&0&0\\
0&0&0&0&0&0&0&x&0&1&0&0\\
0&0&0&0&0&0&0&x&0&0&1&0\\
0&0&0&0&0&0&0&x&0&0&0&1
\end{array}\right).$$

Applying a permutation on the right to interchange the columns on either side of the vertical partitioning lines moves the spikes into the blocks. The resulting matrix,

$A^{(3)} = A^{(2)}P$, has the form

$$A^{(3)} = A^{(2)}P = \left(\begin{array}{cccc|cccc|cccc} 1&0&0&x&0&0&0&0&0&0&0&0\\ 0&1&0&x&0&0&0&0&0&0&0&0\\ 0&0&1&x&0&0&0&0&0&0&0&0\\ 0&0&0&x&1&0&0&0&0&0&0&0\\ \hline 0&0&0&1&x&0&0&x&0&0&0&0\\ 0&0&0&0&x&1&0&x&0&0&0&0\\ 0&0&0&0&x&0&1&x&0&0&0&0\\ 0&0&0&0&x&0&0&x&1&0&0&0\\ \hline 0&0&0&0&0&0&0&1&x&0&0&0\\ 0&0&0&0&0&0&0&0&x&1&0&0\\ 0&0&0&0&0&0&0&0&x&0&1&0\\ 0&0&0&0&0&0&0&0&x&0&0&1 \end{array}\right).$$

Finally, Kuck and Sameh observe that unknowns $m,\ m+1,\ 2m,\ 2m+1,\dots,\ (p-1)m,\ (p-1)m+1,\ pm$ satisfy an independent tridiagonal system of $2p-1$ equations. In Stage 4, this system is solved. Stage 5 consists on a back-substitution to get the remaining unknowns.

Kuck and Sameh show that a tridiagonal linear system of dimension n can be solved in $11+9\log n$ steps and the evaluation of a square root, using $3n$ processors. Schemes for solving banded systems were later developed based on related divide and conquer ideas, see [37, 88] and the method was adapted and analyzed for Cedar for block tridiagonal systems by Berry and Sameh [13]. The idea was later generalized to yield a class of methods under the name of "Spike" solvers, see, e.g, [100, 101], which are the object of another section of this survey.

One of the most enduring contributions of Sameh and the Algorithm and Applications group is their work on the design and analysis of numerical linear algebra algorithms that efficiently exploit the complicated multilevel memory and parallelism hierarchies of Cedar. These issues have reappeared several times since CSRD as new combinations of the basic computational and memory building blocks are exploited in new implementations of systems. As a result, the practice on Cedar of analyzing these building blocks with various relative contributions to the performance of an architecture created a solid foundation for performance analysis, algorithm design and algorithm modification on many of the systems currently available. Sameh's contribution in [47] and in an expanded form in [50] was significant and crucial. The Algorithm and Applications group combined algorithm characteristics, architecture characteristics, and empirical characterizations into an effective performance modeling and design strategy (see [50] for a summary of contemporary investigations of the influence of memory architecture).

In [47] this approach was used to present a systematic analysis of the performance implications of the BLAS level-3 primitives for numerical linear algebra computation on hierarchical memory machines. The contributions included design techniques for achieving high performance in the critical BLAS level-3 kernels as well as the design and analysis of high performance implementations of the *LU* factorization and the Modified Gram-Schmidt (MGS) algorithm. The performance trends were analyzed in terms of the various blocking parameters available at the kernel and algorithm level, and the resulting predictions were evaluated empirically. Performance improvements such as multi-level blocking in both *LU* and MGS were justified based on the models and verified empirically. The numerical properties of the multilevel block MGS algorithm were subsequently investigated by Jalby and Philippe in [76] and further improvements to the algorithm suggested. The insights gained in this work were the basis for the entire Cedar numerical library and for portions of the performance improving compiler work and problem solving environment work mentioned earlier. The algorithms and insights have been of continuing interest since then (see for example the use of block MGS and a version of the limited parallelism *QR* algorithm of [110] mentioned above in the recent thesis of M. Hoemmen [73]).

1.2.2 Jacobi Sweeps and Sturm Sequences

For diagonalizing a symmetric matrix, the oldest method, introduced by Jacobi in 1846 [75], consists of annihilating successively off-diagonal entries of the matrix via orthogonal similarity transformations. The scheme is organized into sweeps of $n(n-1)/2$ rotations to annihilate every off-diagonal pairs of symmetric entries once. One sweep involves $6\,n^3 + O(n^2)$ operations when symmetry is exploited in the computation. The method was abandoned due to high computational cost but has been revived with the advent of parallelism.

A parallel version of the cyclic Jacobi algorithm was given by Sameh [114]. It is obtained by the simultaneous annihilation of several off-diagonal elements by a given orthogonal matrix U_k, rather than only one rotation as is done in the serial version. For example, let A be of order 8 (see Fig. 1.2) and consider the orthogonal matrix U_k as the direct sum of four independent plane rotations simultaneously determined. An example of such a matrix is

$$U_k = R_k(1,3) \oplus R_k(2,8) \oplus R_k(4,7) \oplus R_k(5,6),$$

where $R_k(i,j)$ is that rotation which annihilates the (i,j) off-diagonal element ($\oplus$ indicates that the rotations are assembled in a single matrix and extended to order n by the identity). Let one sweep be the collection of such orthogonal similarity transformations that annihilate the element in each of the $\frac{1}{2}n(n-1)$ off-diagonal positions (above the main diagonal) only once, then for a matrix of order 8, the first sweep will consist of seven successive orthogonal transformations with each one annihilating distinct groups of maximum 4 elements simultaneously as described in Fig. 1.2.

For symmetric tridiagonal matrices, Sturm sequences are often used when only the part of the spectrum, in an interval $[a,b]$, is sought. Since the parallel computation of a Sturm sequence is poorly efficient it is more beneficial to consider simultaneous computation of Sturm sequences by replacing the traditional bisection

Fig. 1.2 Annihilation scheme as in [114] (First regime)

x	3	6	2	5	1	4	7
	x	2	5	1	4	7	6
		x	1	4	7	3	5
			x	7	3	6	4
				x	6	2	3
					x	5	2
						x	1
							x

of intervals by multisection. This approach has been defined for the Illiac IV [74, 82] in the 1970s and revisited by A. Sameh and his coauthors in [89]. Multisections are efficient only when most of the created sub-intervals contain eigenvalues. Therefore a two-step strategy was proposed: (i) Isolating all the eigenvalues with disjoint intervals, (ii) extracting each eigenvalue from its interval. Multisections are used for step (i) and bisections or other root finders are used for step (ii). This approach proved to be very efficient. When the eigenvectors are needed, they are computed independently by Inverse Iterations. A difficulty could arise if one wishes to compute all the eigenvectors corresponding to a cluster of very poorly separated eigenvalues. Demmel, Dhillon and Ren in [34], discussed the reliability of the Sturm sequence computation in floating point arithmetic where the sequence is no longer monotonic.

1.2.3 Fast Poisson Solvers and Structured Matrices

One exciting research topic in scientific computing "making headlines" in the early 1970s was *Fast Poisson Solvers*, that is, direct numerical methods for the solution of linear systems obtained from the discretization of certain partial differential equations, typically elliptic and separable, defined on rectangular domains, with sequential computational complexity $O(N \log_2 N)$ or less, where N is the number of unknowns (see [51] for an outline of these methods). At the University of Illinois, the Center for Advanced Computation (CAC), an organization engaged in the Illiac IV project, issued a report of what is, to the best of our knowledge, the earliest published investigation on parallel algorithms in this area. That was the Master's thesis of James H. Ericksen [41], presenting an adaptation of the groundbreaking FACR algorithm of Hockney for the Illiac IV. It used the algorithms of Cooley et al. for the FFT [30] and the well-known Thomas algorithm for solving the tridiagonal systems instead of Hockney's cyclic reduction. Ericksen was from the Department of Atmospheric Sciences and a goal of that research was the solution of a CFD problem (the non-dimensional Boussinesq equations in a two-dimensional rectangular region describing Bernard-Rayleigh convection) in the streamfunction-vorticity formulation which required the repeated solution of Poisson's equation. A modified version of FACR, MFACR, that did not contain odd-even reduction was also implemented. This CAC report did not report on the results of an implementation, but provided codes in GLYPNIR (the Illiac IV Algol-like language), partial implementations in the ASK assembly and timing estimates based on the clocks required by the instructions in these codes. The direct method was found to be faster and to require less memory than competing ones based on SOR and ADI. Some results from this study were reported by Wilhelmson in [134] and later extended in [42], where odd-even reduction was also applied to economize in storage for a system that could not be solved in core.

At about the time that the first Erickson report was issued, Bill Buzbee outlined in a widely cited paper [25] the opportunities for parallelism in the simplest version of FACR (like MFACR, which is the Fourier Matrix Decomposition method [26]; see also [51]) "It seldom happens that the application of L processors would yield an L-fold increase in efficiency relative to a single processor, but that is the case with the MD algorithm." Recalling that effort Buzbee noted in [127] "Then, by that point we were getting increasingly interested in parallel computing at Los Alamos, and I saw some opportunities with Hockney's scheme for parallel computing, so I wrote a paper on that."

Given this background, the 1974 Sameh, Chen and Kuck Technical Report [118], entitled *Parallel direct Poisson and biharmonic solvers*, and the follow-up paper [119] appear to have been the first detailed studies of rapid elliptic solvers for a parallel computational model that for many years dominated the analyses of many algorithms. Comparisons with competing iterative algorithms were also provided. The theoretical analysis in these papers together with the practical Illiac IV study in [42] are essential references in the early history of rapid elliptic solvers.

The methods presented in the paper built on earlier work of Kuck and coauthors on the fast solution of triangular systems. They are also based on the fundamental work on the parallel computation of the FFT, the matrix decomposition by Buzbee, Dorr, Golub and Nielson [26] as well as work in [44] on the use of the Toeplitz structure of the matrices occurring when discretizing the Poisson equation. One major result was that the $n^2 \times b^2$ block Toeplitz tridiagonal system resulting from the discretization of the Poisson equation on the unit square with the standard 5-point finite difference approximation can be solved in $T_p = 12 \log n$ steps (omitting terms of $O(1)$) using at most n^2 processors with speedup $O(n^2)$ and efficiency $O(1)$. In order to evaluate the performance of these methods on the parallel computational model mentioned earlier, their performance characteristics were compared to those of SOR and ADI.

It is remarkable that [118, 119] also contained the first published parallel algorithm for the biharmonic equation. That relies on the fact that the coefficient matrix of order n^2 has the form $G + 2FF^\top$, where $F \in \mathbb{R}^{n^2 \times 2n}$ is $F = \mathrm{diag}(E, \dots, E)$, and hence is of rank $2n$ and G is the square of the usual discrete Poisson operator slightly modified. It is worth noting that even recent approaches to the fast solution of the biharmonic equation are based on similar techniques, see e.g. [7]. It was shown in [119] that the biharmonic equation can be solved in $T_p = 6n + \frac{1}{2}\log^2 n + 28.5 \log n$ steps using $O(n^3)$ processors. It was also proved that the equation can be solved in $T_p = 50n \log n + O(n)$ steps when there are only $4n^2$ processors. The complexity can be further reduced if some preprocessing is permitted.

Another significant paper of Sameh on rapid elliptic solvers, entitled *A fast Poisson solver for multiprocessors*, appeared in 1984 [116]. Therein, a parallel matrix-decomposition framework for the standard discrete Poisson equation in two and three dimensions is proposed. A close study of these methods reveals that they can be viewed as special cases of the Spike banded solver method, which emerged, as is evident in this volume, as an important algorithmic kernel in Sameh's work. They are also related to domain decomposition. The three-dimensional version is based on a six phase parallel matrix-decomposition framework that combines independent FFTs in two dimensions and tridiagonal system solutions along the third dimension. The computational models used were a ring of $p < n$ processors for the two dimensional problem and a mesh of n^2 processors consisting of n rings of n processors each for the 3-d problem. In these rings, each processor has immediate access to a small local memory while one processor has access to a much larger memory. It was assumed that each processor was able to perform simultaneously an arithmetic operation as well as to receive one floating-point number and transmit another from and to a neighboring processor. The influence of the work presented in [116] can be seen in papers that appeared much later, e.g. the solvers for the Cray T3E presented in [57].

Sameh's suggestion to E. Gallopoulos to study the "Charge Simulation Method", was also pivotal as it led to interesting novel work by Daeshik Lee, then Ph.D. student at CSRD, who studied the use of these "boundary-integral" type methods to perform non-iterated domain decomposition and then apply suitable rapid elliptic solvers (cf. [53] and [54]). The area of CSM-based methods for solving elliptic equations dramatically expanded later on and is still evolving, especially in the context of meshless methods (see e.g. [43]).

Algorithms for matrix problems with special structure have emerged as an important topic in matrix computations. In our opinion, together with the FFT, rapid elliptic solvers have been the first example of research in this area. Historically then, parallel FFT and rapid elliptic solver algorithms can be considered as the first examples of parallel algorithms for structured matrices. Toeplitz structure, that is, matrices in which individual elements or submatrices (blocks) are constant along diagonals are of great importance in a variety of areas, from signal processing to the solution of partial differential equations. Sameh's contribution in this area is significant and can be considered groundbreaking. In the article *On Certain Parallel Toeplitz Linear System Solvers* [67], coauthored with his Ph.D. student J. Grcar, they described fast parallel algorithms for banded Toeplitz matrices of semi-bandwidth m. Specifically, they described a practical algorithm of parallel complexity $O(\log n)$ for the solution of banded Toeplitz systems useful when the matrix can be embedded in a nonsingular circulant matrix. When this assumption does not hold but the matrix is spd they described an $O(m \log n)$ algorithm under the weaker assumption that all principal minors are nonzero. Under somewhat more restrictive conditions (related to the factorization of the "symbol" of A), a less expensive, $O(\log m \log n)$ algorithm was also described. The numerical behavior of all algorithms was also investigated. This is probably the first published paper proposing parallel algorithms for banded Toeplitz systems, laying groundwork for important subsequent developments by experts from the fast growing structured matrix algorithms community (see for example [18, 20] and the survey [19]).

Structured matrices were also central in the work by Sameh and Hsin-Chu Chen, on linear systems with matrices that satisfy the relation $A = PAP$, where P is some symmetrical signed permutation matrix. These matrices are called reflexive and also said to possess the SAS property. Based on this they showed, in the 1987 paper *Numerical linear algebra algorithms on the Cedar system* and then expanded in Chen's Ph.D. thesis and the follow-up article [27], that for matrices with this structure, it is possible to decompose the original problem into two or more independent subproblems (in what they termed SAS decomposition) via orthogonal transformations, leading to algorithms possessing hierarchical parallelism suitable for a variety of parallel architectures. Actually, the advantage of these techniques was demonstrated over a variety of architectures (such as the Alliant FX/8 vector multiprocessor and the University of Illinois Cedar system).

1.3 Algorithms for Sparse Matrices

1.3.1 Computing Intermediate Eigenvalues

Sameh has had a keen interest in solvers for large eigenvalue problems throughout his career. During the 1970s, the most popular eigenvalue methods for large symmetric matrices were the Simultaneous Iteration method as studied by Rutishauser [102] and the Lanczos method [84]. These two methods are efficient for the computation of extreme eigenvalues, particularly those of largest absolute value.

For computing interior eigenvalues, methods based on spectral transformations were introduced and thoroughly discussed during the 1980s. The most effective approach to computing interior eigenvalues was the Shift-and-Invert technique which enables the computation of those eigenvalues of A that are the nearest to a given $\sigma \in \mathbb{R}$ by applying the Lanczos processes on $(A - \sigma I)^{-1}$ in place of A. If $(A - \sigma I)^{-1} q = q\mu$, then $Aq = q\lambda$ with $\lambda = \sigma + 1/\mu$, and thus the extreme eigenvalues of $(A - \sigma I)^{-1}$ transform to the eigenvalues of A closest to the shift σ. The shift-inverse approach is very effective and is the most commonly used technique for computing interior eigenvalues. However, shift-invert requires an accurate numerical solution of a linear system at each iteration. Sameh and his coauthors were pioneers in considering a well constructed polynomial transformation $p(A)$ in place of $(A - \sigma I)^{-1}$. In [121], they combined a quadratic transformation $B = I - c(A - aI)(A - bI)$ with the method of simultaneous iteration for computing all the eigenvalues of A lying in a given interval $[a, b]$. The scaling factor c is chosen so that $\min(p(c_1), p(c_2)) = -1$ where the interval $[c_1, c_2]$ includes as closely as possible the spectrum of A. They provide an analysis of the rate of convergence based upon Chebyshev polynomials.

Although polynomial transformations usually give rise to a considerably higher number of iterations than Shift-and-Invert transformations, they are still useful because of their ability to cope with extremely large matrices and to run efficiently on modern high performance computing architectures.

As an alternative to both of these approaches, Sameh and his students developed an algorithm that allows preconditioning and inaccurate solves of linear systems so that the efficiency of shift-invert is nearly recovered but no factorizations of the shit-invert matrix are required. This approach is described in the next section.

1.3.2 The Trace Minimization Algorithm

The generalized eigenvalue problem

$$Aq = Bq\lambda,$$

with A, B matrices of order n, q a non-zero vector and λ a scalar provides a significant challenge for large scale iterative methods. For example, methods based upon an Arnoldi or Lanczos process will typically require a spectral transformation

$$(A - \sigma B)^{-1} Bq = q\mu, \quad \text{with } \lambda = \sigma + \frac{1}{\mu} \tag{1.1}$$

to convert the system to a standard eigenvalue problem. The spectral transformation enables rapid convergence to eigenvalues near the shift σ. As mentioned previously, this transformation is highly effective, but its implementation requires a sparse direct factorization of the shifted matrix $A - \sigma B$. When this is not possible due to storage or computational costs, one is forced to turn to an inner–outer approach with an iterative method replacing the sparse direct solution. There are numerous difficulties with such an approach.

The Trace Minimization Algorithm (Trace Min) offers a very different subspace iteration approach that does not require an explicit matrix factorization. Trace Min remains as a unique and important contribution to large scale eigenvalue problems. It addresses the special case of Eq. (1.1) assuming that

$$A = A^T, \quad B = B^T \quad \text{pos.def.}$$

Some important features of Trace Min are

1. There is no need to factor $(A - \sigma B)$.
2. Often there is a sequence of related parametrically dependent eigenvalue problems to solve. Since Trace Min is a type of subspace iteration, the previous basis V may be used to start the iteration of a subsequent problem with new parameter.
3. At each iteration, a linear system must be solved approximately. Preconditioning for this task is possible and natural. In Trace Min, inaccurate solves of these linear systems are readily accommodated.
4. Trace Min is ideal suited for parallel computation.

1.3.2.1 The Trace Min Idea

The foundation of the Trace Min algorithm is the fact that the smallest eigenvalue of the problem must satisfy

$$\lambda_{\min} = \min_{v \neq 0} \frac{v^T A v}{v^T B v}.$$

This is a Raleigh quotient for A in a norm weighted by B. If the eigenvalues of the pair (A, B) are ordered so that $\lambda_1 \leq \lambda_2 \leq \cdots \leq \lambda_n$, then there is a well-known subspace variant of this Raleigh condition:

$$\lambda_1 + \lambda_2 + \cdots + \lambda_k = \min \operatorname{tr}\{V^T A V\}$$
$$\text{s.t.} \quad V^T B V = I_k.$$

Courant–Fischer Theory implies that the optimal V satisfies

$$V^T A V = \Lambda_k, \quad V^T B V = I_k, \quad \Lambda_k = \operatorname{diag}(\lambda_1, \lambda_2, \ldots, \lambda_k).$$

An iteration that amounts to a sequence of local tangent space minimization steps will lead to Trace Min. If V is the current basis with $V^T B V = I_k$ then a local tangent space search is facilitated by noting that

$$(V - \Delta)^T B (V - \Delta) = I_k + \Delta^T B \Delta$$

if $V^T B \Delta = 0$.

The local tangent space minimization subproblem is given by

$$\min\{\operatorname{tr}\{(V - \Delta)^T A (V - \Delta)\}\} \quad \text{s.t.} \quad V^T B \Delta = 0.$$

The symmetry of A may be used to derive the equivalent formulation

$$\min\{\operatorname{tr}(-2\Delta^T A V + \Delta^T A \Delta)\} \quad \text{s.t.} \quad V^T B \Delta = 0.$$

The KKT conditions for this formulation provide a block bordered system of linear equations that are suitable for computation:

$$\begin{bmatrix} A & BV \\ V^T B & 0 \end{bmatrix} \begin{bmatrix} \Delta \\ L \end{bmatrix} = \begin{bmatrix} AV \\ 0 \end{bmatrix}.$$

The correction required to update V is $V - \Delta$ and this must be rescaled so that

$$V_+ = (V - \Delta) S \tag{1.2}$$

with

$$V_+^T B V_+ = I_k \quad \text{and} \quad V_+^T A V_+ = \Lambda_+.$$

This rescaling may be accomplished with the formula

$$(V - \Delta)^T B(V - \Delta) = I_k + \Delta^T B\Delta = U\big(I_k + D^2\big)U^T,$$

where $UD^2U^T = \Delta^T B\Delta$ is the eigensystem of the symmetric positive definite matrix $\Delta^T B\Delta$. The scaling matrix is given by $S = U(I_k + D^2)^{-1/2}W$, with

$$W\Lambda W^T = \hat{V}^T A\hat{V}$$

is the eigensystem of $\hat{V}^T A\hat{V}$ with $\hat{V} = U(I_k + D^2)^{-1/2}$. Alternatively, one may take $S = L^{-T}W$ where $LL^T = I_k + \Delta^T B\Delta$ is the Cholesky factorization.

It is easily seen that

$$\mathrm{tr}\big\{-2\Delta^T AV + \Delta^T A\Delta\big\} = \mathrm{tr}\big\{(V-\Delta)^T A(V-\Delta)\big\} - \mathrm{tr}\big\{V^T AV\big\}$$

is a strictly convex function of Δ and hence the problem

$$\min\big\{\mathrm{tr}\big\{\big(-2\Delta^T AV + \Delta^T A\Delta\big)\big\}\big\} \quad \text{s.t.} \quad V^T B\Delta = 0$$

has a strictly convex objective subject to convex constraints. Moreover, at the optimal Δ,

$$\mathrm{tr}\big\{(V-\Delta)^T A(V-\Delta)\big\} < \mathrm{tr}\big\{V^T AV\big\}$$

(unless $\Delta = 0$ and V optimal). It is also possible to show that the scaled V_+ in Eq. (1.2) also satisfies

$$\mathrm{tr}\big\{V_+^T AV_+\big\} < \mathrm{tr}\big\{V^T AV\big\}$$

so that rescaling still gives descent at each iteration.

Since the optimal Δ must satisfy the KKT equations

$$\begin{bmatrix} A & BV \\ V^T B & 0 \end{bmatrix}\begin{bmatrix} \Delta \\ Ł \end{bmatrix} = \begin{bmatrix} AV \\ 0 \end{bmatrix}, \tag{1.3}$$

the algorithm will be fully specified with a procedure to solve this system. To this end, let

$$BV = QR = [Q_1, Q_2]\begin{bmatrix} \hat{R} \\ 0 \end{bmatrix}$$

be the long form QR-factorization. Then Eq. (1.3) will be equivalent to

$$\begin{bmatrix} Q^T AQ & R \\ R^T & 0 \end{bmatrix}\begin{bmatrix} G \\ L \end{bmatrix} = \begin{bmatrix} F \\ 0 \end{bmatrix}. \tag{1.4}$$

Equation (1.4) may be further partitioned into

$$\begin{bmatrix} A_{11} & A_{12} & \hat{R} \\ A_{21} & A_{22} & 0 \\ R^T & 0 & 0 \end{bmatrix}\begin{bmatrix} G_1 \\ G_2 \\ L \end{bmatrix} = \begin{bmatrix} F_1 \\ F_2 \\ 0 \end{bmatrix} \tag{1.5}$$

where $A_{ij} = Q_i^T A Q_j$, $G_i = Q_i^T \Delta$, and $F_i := Q_i^T AV$ for $i, j \in \{1,2\}$. The last block of these equations will imply $G_1 = 0$ since BV is full rank and $\hat{R}$ must be nonsingular. Thus

$$\Delta = QG = Q_1 G_1 + Q_2 G_2 = Q_2 G_2.$$

Hence, Eq. (1.5) reduces to the auxiliary system

(i) Solve $Q_2^T A Q_2 G_2 = Q_2^T AV$ for G_2.
(ii) Put $\Delta = Q_2 G_2$.

Unfortunately, Q_2 is size $n \times (n-k)$ so that A_{22} is an order $n-k$ matrix with $k \ll n$ (recall n is huge and k is small). Most likely, it will not even be possible to compute Q_2 for large n. However, there is an effective remedy to this problem. Since $Q_2 Q_2^T = I - Q_1 Q_1^T$ and $Q_2 Q_2^T A Q_2 Q_2^T G = Q_2 Q_2^T AV$, the following system is equivalent:

$$\left(I - Q_1 Q_1^T\right) A \left(I - Q_1 Q_1^T\right) \Delta = \left(I - Q_1 Q_1^T\right) AV, \tag{1.6}$$

which derives from the fact $\Delta = Q_2 G_2$ and $(I - Q_1 Q_1^T) Q_2 = Q_2$.

Note that Eq. (1.6) is a consistent symmetric positive semi-definite system and hence may be solved via the preconditioned conjugate gradient method (PCG). In PCG we only need matrix-vector products of the form $w = (I - Q_1 Q_1^T) A (I - Q_1 Q_1^T) v$ which, of course, may be implemented in the form

1. $z_1 = Q_1^T v$,
2. $y_1 = v - Q_1 z_1$,
3. $q = A y_1$,
4. $z_2 = Q_1^T q$,
5. $y_2 = v - Q_1 z_2$.

Global and Rapid Convergence: Using relations to Rutishauser's simultaneous iteration for eigenvalues of $A^{-1}B$ Sameh and Wisniewski were able to prove

Theorem 1.1 *Assume* $\lambda_1 \le \lambda_2 \le \cdots \le \lambda_k < \lambda_{k+1}$ *with corresponding generalized eigenvectors* v_i. *Let* $v_i^{(j)}$ *be the* i*th column of* $V = V^{(j)}$ *at the* j*th iteration. Then*

(i) $v_i^{(j)} \to v_i$, *at rate asymptotic to* $\lambda_i / \lambda_{k+1}$
(ii) $(v_i^{(j)} - v_i)^T A (v_i^{(j)} - v_i)$ *is reduced asymptotically by factor* $(\lambda_i / \lambda_{k+1})^2$.

Putting all this together provides the Trace Min algorithm which is shown as Algorithm 1.1. Many more computational details and insights plus several convincing numerical experiments are presented in the Sameh and Wisniewski paper.

Trace Min is one of the very few methods for solving the generalized eigenvalue problem without factoring a matrix. Another such method called the Jacobi–Davidson method was published by Van der Vorst and Sleijpen [128]. It is interesting to note that Trace Min preceded Jacobi–Davidson by a decade. Moreover, from the derivation given above, it is readily seen that these two methods have a great deal in common.

Algorithm 1.1: The Tracemin algorithm

Algorithm: $[V, D] = \text{Tracemin}((A, B, k, tol))$

Data: A an $n \times n$ spd matrix; B an $n \times n$ spd matrix; k a positive integer $(k \leq n)$; tol requested accuracy tolerance;

Result: V an $n \times k$ B-orthogonal matrix; D a $k \times k$ positive diagonal matrix with $AV = BVD$;

$[V, R] = \text{qr}(\text{randn}(n, k), 0)$;

$[U, S] = \text{eig}(V^T BV)$; $S = \text{diag}(\text{sqrt}(\text{diag}(S)))$;

$V = VUS^{-1}$;

$Resid = 10 \cdot tol$;

while *(Resid > tol)* **do**

- $[Q, R] = \text{qr}(BV, 0)$;
- `/*` Solve $(I - QQ^T)A(I - QQ^T)Z = AV$ with pcg--Preconditioned Conjugate Gradient `*/`
- $Z = \text{pcg}(A, Q, AV)$;
- $V = V - Z$;
- $[U, S] = \text{eig}(V^T BV)$; $S = \text{diag}(\text{sqrt}(\text{diag}(S)))$;
- $V = VUS^{-1}$;
- $H = V^T AV$;
- $W = AV$; $H = V' * W$; $H = .5 * (H + H^T)$;
- $[Y, D] = \text{eig}(H)$;
- $[s, p] = \text{sort}(\text{diag}(D))$; $V = V * Y(:, p)$; $D = D(p, p)$;
- $Resid = \|AV - BVD\|$

1.3.2.2 Trace Minimization and Davidson

We have seen that Trace Min is a subspace iteration approach that is a considerable advance over the existing simultaneous iteration that preceded it. However, the subspace iteration approach languished within the numerical analysis and numerical linear algebra communities which tended to favor Krylov subspace approaches such as Arnoldi and Lanczos. However, the rigid structure of Krylov spaces did not lend itself well to modifications that would accelerate convergence other than the previously mentioned shift-invert transformation.

Quite a different approach emerged from the computational chemistry community in the form of Davidson's method. For various reasons, the chemists preferred this over the Lanczos method. The Davidson method can be viewed as a modification of Newton's method applied to the KKT system that arises from treating the symmetric eigenvalue problem as a constrained optimization problem involving the Rayleigh quotient. From our previous discussion of Trace Min, there is an obvious connection. The Jacobi–Davidson method is a related approach that is now viewed as a significant advance over the original Davidson method. However, much of technology in Jacobi–Davidson had already been developed in Trace Min.

These approaches essentially take the viewpoint that the eigenvalue problem is a nonlinear system of equations and attempt to find a good way to correct a given approximate eigenpair $(\tilde{\lambda}, \tilde{u})$, by enriching the most recent subspace of approximants with Newton-like directions. In practice, this means that we need to solve the correction equation, i.e., the equation which updates the current approximate eigenvector, in a subspace that is orthogonal to the most current approximate eigenvectors.

Let us consider that, at a given iteration, the current approximate eigenpair $(\tilde{\lambda}, \tilde{u})$ is a Ritz pair obtained from the current subspace spanned by the orthonormal columns of $V \in \mathbb{R}^{n\times k}$ ($k \ll n$ and $V^T V = I_k$). Denoting $\tilde{u} = Vw$ and assuming that $\|\tilde{u}\|_2 = 1$ reveals that the Ritz value $\tilde{\lambda} = \tilde{u}^T A\tilde{u}$ is the Rayleigh quotient of $\tilde{u}$ and w is the corresponding eigenvector of the reduced problem $(V^T AV)w = \tilde{\lambda} w$. The residual $r = A\tilde{u} - \tilde{\lambda}\tilde{u}$ satisfies $V^T r = 0$. One can think of the problem as that of solving $(A - (\tilde{\lambda} + \delta)I)(\tilde{u} + z) = 0$, but since there are $n+1$ unknowns, a constraint must be added, for example, $\|\tilde{u} + z\|_2 = 1$,

$$\begin{cases} ((A - \tilde{\lambda} I) - \delta I)(\tilde{u} + z) = 0, \\ (\tilde{u} + z)^T(\tilde{u} + z) = 1\,. \end{cases} \tag{1.7}$$

Ignoring second order terms, this yields the system of equations

$$\big(A - \tilde{\lambda} I\big)z - Vw\,\delta = -r, \tag{1.8}$$

$$-w^T V^T z = 0. \tag{1.9}$$

Equation (1.8) is a linear system of rank $n - k$: for any solution z_0 of the system all the vectors $z_0 + Vw$ are also solutions. Since only non-redundant information must be appended to V to define the next subspace, Eq. (1.9) is replaced by

$$V^T z = 0 \tag{1.10}$$

to reach a full rank for the global system. By invoking the orthogonal projector $P = I - VV^T$, and observing that $P\tilde{u} = 0$ and $Pr = r$, it yields,

$$P\big(A - \tilde{\lambda} I\big)Pz = -r, \tag{1.11}$$

$$Pz = z. \tag{1.12}$$

Note that the correction δ to $\tilde{\lambda}$ can be ignored since the new approximate eigenvalue will just be defined as the new Rayleigh quotient. So we are left with the Eq. (1.11). We look now at several attempts that have been considered to solve approximately this equation.

In 1982, Sameh and Wisjniewski derived that system (see Eq. (2.20) in [123]) except that the entire derivation of the method is written in the context of the generalized eigenvalue problem.

In 1986, Morgan and Scott generalized the Davidson method, which was already known in chemistry [97]. They defined an approximate equation which may be written:

$$(M - \tilde{\lambda} I)z = -r, \tag{1.13}$$

where M is a preconditioner (i.e. an approximation of A). Later in 1994, Crouzeix et al. proved the convergence of the method in [31].

In 1996, Sleijpen and van der Vorst rederived Eq. (1.11) in [128], for their famous Jacobi–Davidson method. In 2000, Sameh and Tong revisited Trace Min in [113], and discussed some comparisons with a block version of Jacobi–Davidson. The main difference comes from the strategy for selecting the shifts in the correction step.

From this discussion, it is clear that the eigenvalue solvers based on the Trace Min or the Jacobi–Davidson method should behave similarly. One of the differences concerns the sequence of subspaces that are constructed in each method: the subspaces are of increasing dimension for the latter but of fixed dimension for the former. However, both of these methods can give rise to a large range of versions which may be commonly derived from them.

1.3.3 Algorithms for Large Scale SVD

In the late 1980s, research of Michael Berry supervised by Sameh at CSRD led to a host of algorithms for computing the singular value decomposition on multiprocessors. After initial work on the dense problem [14]. Berry in his Ph.D. thesis considered Lanczos and block Lanczos, subspace iteration and trace minimization algorithms for approximating one or more of the largest or smallest singular values and vectors of unstructured sparse matrices and their implementations on Alliant and Cray multiprocessors [8, 9]. One novelty of these works is their emphasis on information retrieval (IR), especially the Latent Semantic Indexing model that had just appeared in the literature. This work played a major role in the development of algorithms for large scale SVD computations and their use in IR; cf. [10–12]. It was also key in publicizing the topic of IR to the linear algebra and scientific computing communities.

1.3.4 Iterative Methods for Linear Systems of Equations

When it comes to solving large sparse linear systems of equations, iterative methods have a definite advantage over direct methods in that they are easy to parallelize. In addition, their memory requirements are generally quite modest. Sameh and coworkers considered a number of parallel iterative methods, see, e.g., [23, 24, 77, 78, 103, 104].

A particular scheme which is sketched here is one based on variants of row-projection methods. A row-projection method, such as the Kaczmarz method uses a row of the matrix to define a search direction for reducing a certain objective function (error norm or residual norm). For example, if r is the current residual vector $r = b - Ax$, and if e_i is the ith column of the identity, the Kaczmarz update

$$x := x + \frac{e_i^T r}{\|A^T e_i\|_2^2} A^T e_i \tag{1.14}$$

simply performs a relaxation step (Gauss–Seidel) for solving the normal equation $AA^T y = b$, where the unknown x is set to $x = A^T y$. It can also be viewed as a projection method along the direction $A^T e_i$ for solving $Ax = b$, by minimizing the next error norm. Regardless of the viewpoint taken, when Eq. (1.14) is executed cyclically from $i = 1$ through $i = n$, we would essentially accomplish a Gauss–Seidel sweep for solving $AA^T y = b$. Generally this scheme is sensitive to the condition number of A, so block schemes were sought by Sameh and coworkers in an effort to improve convergence rates on the one hand and improve parallelism at the same time. In particular, an important idea discussed in [23, 24, 77, 78] is one in which rows are grouped in blocks so that a block SOR scheme would lead to parallel steps. This means that rows which have no overlap must be identified and grouped together.

Consider a generalization of the scheme Eq. (1.14) in which the vector e_i is replaced by a block V_i of p columns of e_j's. Then since we are performing a projection step in the space span$(A^T V_i)$ the scheme in Eq. (1.14) will be replaced by a step like

$$x := x + (A^T V_i) u_i \tag{1.15}$$

where u_i is now a vector determined by the requirement that the new residual be orthogonal to V_i, giving

$$V_i^T \big(r - A(A^T V_i) u_i\big) = 0 \quad \rightarrow \quad u_i = \big[(A^T V_i)^T (A^T V_i)\big]^{-1} V_i^T r.$$

The $p \times p$ matrix $S_i = (A^T V_i)^T (A^T V_i)$ can have an advantageous structure for parallel environments provided a good selection of the sets of rows is made. For example, S_i can be diagonal if the rows $A^T e_j$ are orthogonal to each other for the columns e_js of the associated V_i. This is the basis of the contribution by Kamath and Sameh [78]. Later the idea was further refined to improve convergence properties and give specific partition vectors for 3-D elliptic Partial Differential Equations [23, 24]. This class of methods can be quite effective for problems that are very highly indefinite since other methods will most likely fail in this particular situation.

In [103] and later in [104] a combination of Chebyshev iteration and block Krylov methods was exploited. The Chebyshev iteration has clear advantages in a parallel computing environment as it is easy to parallelize and has no inner products. The block-Krylov method can be used to extract eigenvalue information and, at the same time, to perform a projection step to speed-up convergence. The method has great appeal even today, though implementation can be difficult and this can discourage potential users.

One of the most successful alternatives to direct solution techniques for solving symmetric linear systems is the preconditioned conjugate gradient algorithm and at the time when Cedar was being built it was imperative to study this approach in detail. Sameh and coauthors published several papers on this. One of these papers, the article [94] by Meier and Sameh, considered in detail the performance of a few different schemes of the preconditioned Conjugate Gradient algorithm in a parallel environment. The issue was revisited a few years later in collaboration with Gupta and Kumar [68]. For solving general sparse linear systems, the authors of [48, 49] explore 'hybrid methods'. The idea is to use a direct solver and drop small terms when computing the factorization. The authors show that a hybrid method of this type is often better than the corresponding direct and pure iterative methods, or methods based on level-of-fill preconditioners.

Another interesting contribution was related to the very complex application of particulate flow [124]. This application brings together many challenges. First, the problem itself is quite difficult to solve because the particles must satisfy physical constraints and this leads to the use of Arbitrary Lagrangian Eulerian (ALE) formulations. When finite elements are used the problem must be remeshed as the time discretization progresses and with the remeshing a re-partitioning must also be applied. Finally, standard preconditioners encounter serious difficulties. Sameh and his team contributed several key ideas to this project. One idea [124] is a projection type method to perform the simulation. The simulation was performed matrix-free on a space constrained to be incompressible in the discrete space. A multilevel preconditioner is also devised by Sarin and Sameh [122] to build a basis of the space of divergence-free functions. The algorithm showed good scalability and efficiency for particle benchmarks on the SGI Origin 2000.

The above overview of Sameh's contributions to parallel iterative methods has deliberately put an emphasis on work done around the Cedar project and before. Sameh has continued to make contributions to parallel iterative methods and a few of the papers in this volume discuss his more recent work.

1.3.5 The Spike Algorithm

Sparse linear systems $Ax = b$ can often be reordered to produce either banded systems or low-rank perturbations of banded systems in which the width of the band is but a small fraction of the size of the overall system. In other instances, banded systems can act as effective preconditioners to general sparse systems which are solved via iterative methods. Existing algorithms and software using direct methods for banded matrices are commonly based on the LU factorization that represents a matrix A as a product of lower and upper triangular matrices i.e. $A = LU$. Consequently, solving $Ax = b$ can be achieved by solutions of two triangular systems $Lg = b$ and $Ux = g$. In contrast to the LU factorization, the Spike algorithm, introduced by Sameh in the late 1970s [115, 120], relies on a DS factorization of the

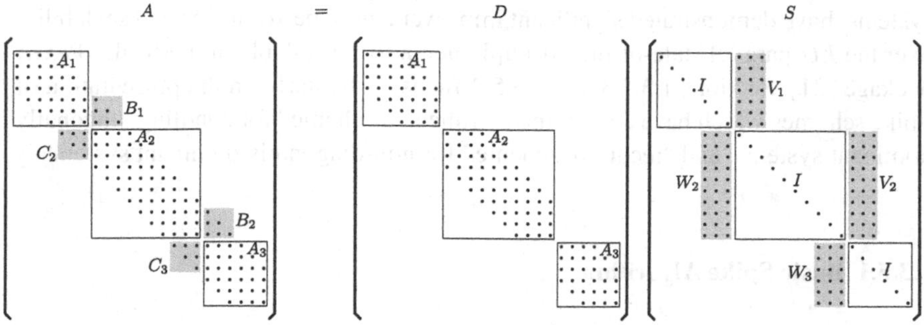

Fig. 1.3 Example of DS factorization of the banded matrix A for the case of three partitions. The diagonal blocks in D are supposed non-singular, and their size much larger than the size the off-diagonal blocks B and C (the system is said to be narrow banded)

banded matrix A where D is a block-diagonal matrix, and S has the structure of an identity matrix with some extra "spikes" (S is called the spike matrix). Assuming a direct partitioning of the banded matrix A in the context of parallel processing, the resulting DS factorization procedure is illustrated in Fig. 1.3.

As a result, solving $Ax = b$ can be achieved by solving for a modified right hand side $Dg = b$ which consists of decoupled block diagonal systems, and the spike system $Sx = g$ which is also decoupled to a large extent except for a reduced system that can be extracted near the interfaces of each of the identity blocks. The Spike algorithm is then similar to a domain decomposition technique that allows performing independent calculations on each subdomain or partition of the linear system, while the interface problem leads to a reduced linear system of much smaller size than that of the original one. Multiple arithmetic operations can indeed be processed simultaneously in parallel such as the factorization of each partition of the diagonal matrix D (using for example a LU factorization), the generation of the spike matrix S, or the retrieval of the entire solution once the reduced system solved. All the communication operations are then concentrated in solving the reduced system. The Spike algorithm is then ideally suited for achieving linear scalability in parallel implementation since it naturally leads to low communication cost. In addition and in comparison to other divide-and-conquer approach which enforces the LU factorization paradigm [4, 29], Spike naturally minimizes memory references (no reordering needed for performing the DS factorization on banded systems) as well as the arithmetic cost for obtaining the reduced system (the reduced system is directly extracted from the spike matrix and not generated via Schur complement for example). Since its first publications several enhancements and variants of the Spike algorithm have been proposed by Sameh and coauthors in [13, 37, 58, 88, 91, 92, 99, 100, 109, 112]. Spike can be cast as a hybrid and polyalgorithm that uses many different strategies for solving large banded linear systems in parallel and can be used either as a direct scheme or a preconditioned iterative scheme. In the following, Sect. 1.3.5.1 briefly summarizes the basic Spike algorithm, while Sect. 1.3.5.2 presents the polyalgorithm nature of Spike. From all the different possible options for Spike, two highly efficient direct methods recently introduced in [100, 101] for solving dense banded systems, have demonstrated significant improvement in performance and scalability over the LU parallel state-of-the-art implementations available in the ScaLAPACK package [21]. Sections 1.3.5.3 and 1.3.5.4 focus particularly on the presentation of Spike schemes which have been named "truncated scheme" for handling diagonally dominant systems, and "recursive scheme" for non-diagonally dominant systems.

1.3.5.1 Basic Spike Algorithm

As illustrated in Fig. 1.3, a $(n \times n)$ banded matrix A can be partitioned into a block tridiagonal form $\{C_j, A_j, B_j\}$ where A_j is the $(n_j \times n_j)$ diagonal block j, and B_j (i.e. C_j) is the $(ku \times ku)$ (i.e. $(kl \times kl)$) right block (i.e. left block). Using p partitions, it comes that n_j is roughly equal to n/p. In order to ease the description of the Spike algorithm but without loss of generality, the size off-diagonal blocks are both supposed equal to m $(kl = ku = m)$. The size of the bandwidth is then defined by $b_d = 2m + 1$ where $b_d \ll n_j$. Each partition j $(j = 1, \ldots, p)$, can be associated to one processor or one node allowing multilevel of parallelism. Using the DS factorization, the obtained spike matrix S has a block tridiagonal form $\{W_j, I_j, V_j\}$, where I_j is the $(n_j \times n_j)$ identity matrix, V_j and W_j are the $(n_j \times m)$ right and left spikes. The spikes V_j and W_j are solutions of the following linear systems:

$$A_j V_j = \begin{bmatrix} 0 \\ \vdots \\ 0 \\ B_j \end{bmatrix}, \quad \text{and} \quad A_j W_j = \begin{bmatrix} C_j \\ 0 \\ \vdots \\ 0 \end{bmatrix} \tag{1.16}$$

respectively for $j = 1, \ldots, p-1$ and $j = 2, \ldots, p$.

Solving the system $Ax = b$ now consists of two steps:

$$\text{(a) solve} \quad Dg = b, \tag{1.17}$$

$$\text{(b) solve} \quad Sx = g. \tag{1.18}$$

The solution of the linear system $Dg = b$ in Step (a), yields the modified right-hand side g needed for Step (b). In case of assigning one partition to each processor, Step (a) is performed with perfect parallelism. To solve $Sx = g$ in Step (b), one should observe that the problem can be reduced further by solving a system of much smaller size which consists of the m rows of S immediately above and below each partitioning line. Indeed, the spikes V_j and W_j can also be partitioned as follows:

$$V_j = \begin{bmatrix} V_j^{(t)} \\ V_j' \\ V_j^{(b)} \end{bmatrix} \quad \text{and} \quad W_j = \begin{bmatrix} W_j^{(t)} \\ W_j' \\ W_j^{(b)} \end{bmatrix} \tag{1.19}$$

where $V_j^{(t)}$, V_j', $V_j^{(b)}$, and $W_j^{(t)}$, W_j', $W_j^{(b)}$, are the top m, the middle $n_j - 2m$ and the bottom m rows of V_j and W_j, respectively. Here,

$$V_j^{(b)} = [0 \quad I_m]V_j; \quad W_j^{(t)} = [I_m \quad 0]W_j, \tag{1.20}$$

and

$$V_j^{(t)} = [I_m \quad 0]V_j; \quad W_j^{(b)} = [0 \quad I_m]W_j. \tag{1.21}$$

Similarly, if x_j and g_j are the jth partitions of x and g, it comes

$$x_j = \begin{bmatrix} x_j^{(t)} \\ x_j' \\ x_j^{(b)} \end{bmatrix} \quad \text{and} \quad g_j = \begin{bmatrix} g_j^{(t)} \\ g_j' \\ g_j^{(b)} \end{bmatrix}. \tag{1.22}$$

It is then possible to extract from a block tridiagonal reduced linear system of size $2(p-1)m$ which involves only the top and bottom elements of V_j, W_j, x_j and g_j.

As example, the reduced system obtained for the case of four partitions ($p = 4$) is given by

$$\begin{bmatrix} I_m & V_1^{(b)} & & & & \\ W_2^{(t)} & I_m & & V_2^{(t)} & & \\ W_2^{(b)} & & I_m & V_2^{(b)} & & \\ & W_3^{(t)} & I_m & & V_3^{(t)} & \\ & W_3^{(b)} & & I_m & V_3^{(b)} & \\ & & & W_4^{(t)} & I_m & \end{bmatrix} \begin{bmatrix} x_1^{(b)} \\ x_2^{(t)} \\ x_2^{(b)} \\ x_3^{(t)} \\ x_3^{(b)} \\ x_4^{(t)} \end{bmatrix} = \begin{bmatrix} g_1^{(b)} \\ g_2^{(t)} \\ g_2^{(b)} \\ g_3^{(t)} \\ g_3^{(b)} \\ g_4^{(t)} \end{bmatrix}. \tag{1.23}$$

Finally once the solution of the reduced system is obtained, the global solution x can be reconstructed from $x_k^{(b)}$ $(k = 1, \ldots, p-1)$ and $x_k^{(t)}$ $(k = 2, \ldots, p)$ either by computing

$$\begin{cases} x_1' = g_1' - V_1' x_2^{(t)}, \\ x_j' = g_j' - V_j' x_{j+1}^{(t)} - W_j' x_{j-1}^{(b)}, \quad j = 2, \ldots, p-1, \\ x_p' = g_p' - W_j' x_{p-1}^{(b)}, \end{cases} \tag{1.24}$$

or by solving

$$\begin{cases} A_1 x_1 = b_1 - \left[\begin{smallmatrix} 0 \\ I_m \end{smallmatrix}\right] B_j x_2^{(t)}, \\ A_j x_j = b_j - \left[\begin{smallmatrix} 0 \\ I_m \end{smallmatrix}\right] B_j x_{j+1}^{(t)} - \left[\begin{smallmatrix} I_m \\ 0 \end{smallmatrix}\right] C_j x_{j-1}^{(b)}, \quad j = 2, \ldots, p-1, \\ A_p x_p = b_p - \left[\begin{smallmatrix} I_m \\ 0 \end{smallmatrix}\right] C_j x_{p-1}^{(b)}. \end{cases} \tag{1.25}$$

1.3.5.2 Spike: A Hybrid and Polyalgorithm

Multiple options are available for efficient parallel implementation of the Spike algorithm depending on the properties of the linear system as well as the architecture of the parallel platform. More specifically, the following stages of the Spike algorithm can be handled in several ways resulting in a polyalgorithm.

Factorization of the Diagonal Blocks A_j Depending on the sparsity pattern of the matrix and the size of the bandwidth, these diagonal blocks could be considered either dense or sparse within the band.

- For the dense banded case, different strategies based on the LU decomposition of each A_i can be applied. Those include variants such as LU with pivoting, LU without any pivoting but diagonal boosting, as well as a combination of LU and UL decompositions, either with or without pivoting. In order to minimize memory references, it is indeed advantageous to factorize the diagonal blocks A_j using LU without any pivoting but adding a diagonal boosting if a "zero-pivot" is detected. Hence the original A matrix is not exactly the product DS and rather takes the form $A = DS + R$, where R represents the correction which, even if non-zero, is by design small in some sense. Outer iterations via Krylov subspace schemes or iterative refinement, would then be necessary to obtain sufficient accuracy as Spike would act on $M = DS$ (i.e. the approximate Spike decomposition for M is used as effective preconditioner).
- For the sparse banded case, it is common to use a sparse direct linear system solver to reorder and then factorize the diagonal blocks. However, solving the various linear systems for A_j can also be achieved using an iterative solver with preconditioner.
- In order to address the case where the block diagonal A_j are nearly singular (i.e. ill-conditioned), and when even the LU decomposition with partial pivoting is expected to fail, a Spike-balance scheme has also been proposed by Golub, Sameh and Sarin in [58].

Finally, each partition in the decomposition can be associated with one or several processors (one node), enabling multilevel parallelism.

Computation of the spikes If the spikes V_j and W_j are determined entirely, Eq. (1.24) can be used to retrieve the entire solution once the reduced system of Eq. (1.23) solved. Since Eq. (1.25) can also be used to retrieve the solution, the spikes may not be computed entirely but only for the top and bottom $(m \times m)$ blocks of V_j and W_j needed to form the reduced system. The spike tips $W_j^{(t)}$, $W_j^{(b)}$, $V_j^{(t)}$, and $V_j^{(b)}$, can be respectively defined by $(I_m\ 0)A_j^{-1}\binom{I_m}{0}C_j$, $(0\ I_m)A_j^{-1}\binom{I_m}{0}C_j$, $(I_m\ 0)A_j^{-1}\binom{I_m}{0}B_j$, $(0\ I_m)A_j^{-1}\binom{I_m}{0}B_j$. It is then easy to show that computing the spike tips is also equivalent to obtaining the four $(m \times m)$ corners of A_j^{-1} as illustrated in Fig. 1.4

Finally, it is important to note that the determination of the top and bottom spikes is also not explicitly needed for computing the actions of the multiplications with

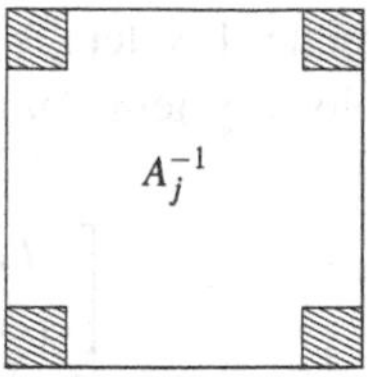

Fig. 1.4 Representation of the four corners of the inverse of the diagonal block A_j, which can be used to compute the spike tips

$W_j^{(t)}$, $W_j^{(b)}$, $V_j^{(t)}$ and $V_j^{(b)}$. These latter can then be performed "on-the-fly" in the case where the bandwidth becomes too large (i.e. systems with large sparse bandwidth).

Solution Scheme for the Reduced System The reduced system of Eq. (1.23) can be solved either iteratively or directly.

- Iterative Krylov-subspace based methods for solving the reduced system conferred to Spike its hybrid nature. Most often, they can be used along with a block Jacobi preconditioner (i.e. diagonal blocks of the reduced system) if the bottom of the V_j spikes and the top of the W_j spikes are computed explicitly. In turn, the matrix-vector multiplication operations of the iterative technique, can be done explicitly or implicitly (i.e. "on-the-fly").
- For large number of partitions, solving the reduced system using iterative methods with or without preconditioner, may result in high interprocessor communication cost. Direct methods such as the truncated and the recursive schemes, have then been recently introduced in [100, 101] to enhance robustness and scalability for solving the reduced system. The truncated Spike scheme represents an optimized version of the Spike algorithm with enhanced use of parallelism for handling diagonally dominant systems. The recursive Spike scheme can be used more generally for solving non-diagonally dominant systems and consists of successive iterations of the Spike algorithm from solving the reduced system.

1.3.5.3 The Truncated Spike Scheme for Diagonally Dominant Systems

Diagonally dominant systems may arise from several science and engineering applications, and are defined if the degree of diagonally dominance, d_d, of the matrix A is greater than one; where d_d is given by $\min\{|A_{i,i}|/\sum_{j\neq i}|A_{i,j}|\}$. If this property is satisfied, one can show that the magnitude of the elements of the right spikes V_j, would decay from bottom to top, while the elements of the left spikes W_j would decay in magnitude from top to bottom [33, 96]. Since the size of the diagonal blocks A_j is assumed much larger than the size m of the blocks B_j and C_j, the bottom blocks of the left spikes $W_j^{(b)}$ and the top blocks of the right spikes $V_j^{(t)}$ can be approximately set equal to zero. More generally, the truncated Spike scheme described here is valid as long as the two $(m \times m)$ blocks at the top right and bottom left corners of A_j^{-1} in Fig. 1.4 are approximately equal to zero (which is always satisfied if $d_d > 1$ and the matrix is narrow banded). It follows that the resulting "truncated" reduced system is simply block diagonal composed by $p-1$ independent $2m \times 2m$ block systems (p represents the number of partitions) of this form:

$$\begin{bmatrix} I_m & V_j^{(b)} \\ W_{j+1}^{(t)} & I_m \end{bmatrix} \begin{bmatrix} x_j^{(b)} \\ x_{j+1}^{(t)} \end{bmatrix} = \begin{bmatrix} g_j^{(b)} \\ g_{j+1}^{(t)} \end{bmatrix}, \quad j = 1, \ldots, p-1. \tag{1.26}$$

The reduced linear systems are then decoupled and can be solved in parallel.

Within the framework of the truncated scheme, two other major contributions have also been proposed for improving computing performance of the factorization and solve stages: (i) a LU/UL strategy [100], and (ii) a new unconventional partitioning scheme [95]. The LU/UL strategy can be used to avoid computing (generating) the entire spikes in order to obtain the tips V_j^b and W_{j+1}^t ($j = 1, \ldots, p-1$). For a number of partitions greater than two, however, each middle partition $j = 2, \ldots, p-1$ has now to perform both LU and UL factorizations. In order to decrease the number of arithmetic operations, a new parallel distribution of the system matrix can be considered, which involves using fewer partitions p than number of processors k (where $p = (k+2)/2$). As compared to a sequential **LU** algorithm, the speed-up for the factorization stage of this Spike scheme, named the TA-scheme, is then expected ideally to be equal to the new number of partitions, i.e. 2× on two processors, 3× on four, 5× on eight, etc.

1.3.5.4 The Recursive Spike Scheme for Non-diagonally Dominant Systems

In contrast to diagonally dominant systems, the tips V_j^t and W_{j+1}^b ($j=1, \ldots, p-1$) cannot be set equal to zero if the system is non-diagonally dominant. Additionally, the probable appearance of "zero-pivot" in the LU and UL factorization stages necessitates the use of a diagonal boosting strategy along with outer-refinement steps. For larger number of partitions, a recursive scheme has been proposed for solving the reduced system which results in better balance between the costs of computation and communication as compared to iterative methods. In contrast to a cyclic reduction approach often used in LU parallel banded schemes [4], the recursive scheme comes very naturally as it consists of successive iterations of the Spike algorithm for solving the reduced system.

The recursive scheme assumes that the original number of (conventional) partitions for the reduced system is given by $p = 2^d$ ($d > 1$). The bottom and top blocks of the V_j and W_j spikes are then computed explicitly to form the reduced system (i.e. all four corners of A_j^{-1} in Fig. 1.4 are computed). In practice, a modified version of the reduced system is preferred which also includes the top block V_1^t and bottom block W_p^b. For the case $p = 4$, the reduced system in Eq. (1.23) can also take the form of the following "reduced spike system" $\hat{S}\hat{x} = \hat{g}$:

$$
\left(\begin{array}{cc|cc|cc|cc}
I_m & & V_1^{(t)} & & & & & \\
 & I_m & V_1^{(b)} & & & & & \\
\hline
 & W_2^{(t)} & I_m & & V_2^{(t)} & & & \\
 & W_2^{(b)} & & I_m & V_2^{(b)} & & & \\
\hline
 & & & W_3^{(t)} & I_m & & V_3^{(t)} & \\
 & & & W_3^{(b)} & & I_m & V_3^{(b)} & \\
\hline
 & & & & & W_4^{(t)} & I_m & \\
 & & & & & W_4^{(b)} & & I_m
\end{array}\right)
\begin{bmatrix} x_1^{(t)} \\ x_1^{(b)} \\ \hline x_2^{(t)} \\ x_2^{(b)} \\ \hline x_3^{(t)} \\ x_3^{(b)} \\ \hline x_4^{(t)} \\ x_4^{(b)} \end{bmatrix}
=
\begin{bmatrix} g_1^{(t)} \\ g_1^{(b)} \\ \hline g_2^{(t)} \\ g_2^{(b)} \\ \hline g_3^{(t)} \\ g_3^{(b)} \\ \hline g_4^{(t)} \\ g_4^{(b)} \end{bmatrix}. \tag{1.27}
$$

This reduced spike system contains p partitions with p diagonal identity blocks. The system can be easily redistributed in parallel using only $p/2$ partitions which are factorized by Spike recursively up until obtaining two partitions only. For the case $p = 4$ where only one recursion can happen, the new spike matrix is obtained from the DS factorization of Eq. (1.27) which takes the form

$$
\left(\begin{array}{cccc|cccc}
I_m & & V_1^{(t)} & & & & & \\
 & I_m & V_1^{(b)} & & & & & \\
 & W_2^{(t)} & I_m & & & & & \\
 & W_2^{(b)} & & I_m & & & & \\
\hline
 & & & & I_m & & V_3^{(t)} & \\
 & & & & & I_m & V_3^{(b)} & \\
 & & & & & W_4^{(t)} & I_m & \\
 & & & & & W_4^{(b)} & & I_m
\end{array}\right)
$$

$$
\times \left(\begin{array}{cccc|cccc}
I_m & & & & V_{1,2}^{(t)} & & & \\
 & I_m & & & V_{1,2}' & & & \\
 & & I_m & & V_{1,2}'' & & & \\
 & & & I_m & V_{1,2}^{(b)} & & & \\
\hline
 & & & W_{3,4}^{(t)} & I_m & & & \\
 & & & W_{3,4}' & & I_m & & \\
 & & & W_{3,4}'' & & & I_m & \\
 & & & W_{3,4}^{(b)} & & & & I_m
\end{array}\right). \tag{1.28}
$$

Here, the spikes $V_{1,2}$ and $W_{3,4}$ are associated to the partitions named $\{1,2\}$ and $\{3,4\}$ of the new spike matrix. It can be shown [100] that the main computational kernel for obtaining the new spikes consists of a succession of $2m \times 2m$ linear system similar to the one presented in Eq. (1.26). Since the Spike algorithm is applied on the reduced system and then does not generate long vectors V and W, those can be computed explicitly (for example the middle of spikes V', V'', W' and W'' in Eq. (1.28) would explicitly be generated). From the recursion of DS factorization, the right hand side can be modified recursively as well by solving block diagonal systems in parallel using relation similar to Eq. (1.24). Finally, the entire solution $\hat{x}$ can be retrieved explicitly using Eq. (1.24) from the final two partitions reduced spike system (as the one in Eq. (1.28) for the case $p = 4$).

1.3.5.5 Spike: Current and Future Directions

Since the publications of the first Spike algorithm in the late 1970s [115, 120], many variations and new schemes have been proposed and implemented. In recent years, a comprehensive MPI-Fortran 90 Spike package for distributed memory architecture has been developed which includes, in particular, all the different family of Spike algorithms: recursive, truncated and on-the-fly schemes. These Spike solvers rely on a hierarchy of computational modules, starting with the data locality-rich BLAS level-3, up to the blocked LAPACK [3] algorithms for handling dense banded systems, or up to the direct sparse solver PARDISO [125] for handling sparse banded systems, with Spike being on the outermost level of the hierarchy. The Spike package also includes new primitives for banded matrices that make efficient use of BLAS level-3 routines. Those include: banded triangular solvers with multiple right-hand sides, banded matrix-matrix multiplications, and *LU*, *UL* factorizations with diagonal boosting strategy. Using this new Spike implementation for solving the dense banded systems, both truncated and recursive Spike solvers exhibit the same degree of accuracy as compared to the corresponding LAPACK computational routines, and with significant improvements in performance and scalability as compared to the ScaLAPACK ones for large number of processors on high-end computing platforms [100]. It is also important to note that the capabilities and domain applicability of the Spike-PARDISO scheme have recently been significantly enhanced by Manguoglu, Sameh and Schenk in [92]. In addition, it has been shown that Spike can effectively be used to enhance the parallel scalability of iterative solvers using banded preconditioners for solving general sparse systems [90].

In addition, the large number of options/decision schemes available for Spike created the need for the automatic generation of a sophisticated runtime decision tree "Spike-ADAPT" that has been proposed by Kuck and developed at Intel. This adaptive layer indicates the most appropriate version of the Spike algorithm capable of achieving the highest performance for solving banded systems that are dense within the band. The relevant linear system parameters in this case are: system size, number of nodes/processors to be used, bandwidth of the linear system, and degree of diagonal dominance. Spike and Spike-ADAPT have been regrouped into one library package and released to the public in June 2008 on the Intel experimental website [132].

Finally, the emergence of multicore computing platforms these recent years have brought new emphasis on parallel algorithms and linear system solvers in particular, for achieving better net speed-up over the corresponding best sequential algorithms starting from a small number of processors/cores. Typical speed-up performance results for narrow banded matrix are presented and discussed in Fig. 1.5 for the Spike

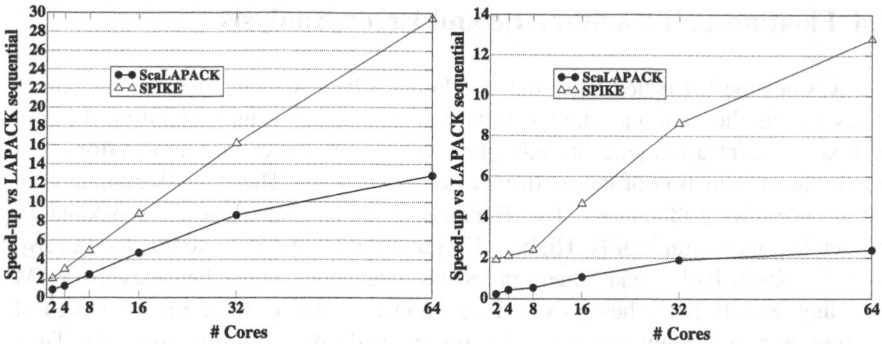

Fig. 1.5 In these experiments, a banded linear system of size $n = 1{,}000{,}000$ and bandwidth 321 ($m = 160$) with only one right hand side, is considered. On the *left* a diagonally dominant system and the Spike-TA scheme are considered, while on the *right* the experiment are performed using a non-diagonally dominant system along with the Spike recursive scheme. Both Spike and ScaLAPACK are running on a Linux Intel Nehalem cluster X5550 featuring eight nodes with infiniband, eight cores per node running at 2.66 GHz, and with 48 Gb total memory per node. The computational modules run using real double precision Intel MKL BLAS, LAPACK and ScaLAPACK. The total time taken by MKL-LAPACK to solve the banded system is 10 s on one core. The accuracy results on the residuals obtained by both LAPACK, Spike, and ScaLAPACK, not reported here, are comparable (accuracy machine). For the non-diagonally dominant case on the right, outer-iterative refinements have not been performed for Spike as no zero-pivot has been detected in the factorization stage. In case of boosting, however, iterative refinements often represent only a very small fraction of the total Spike time

and ScaLAPACK algorithms versus the corresponding best sequential routines in LAPACK on recent Intel Nehalem cluster architecture from 2 to 64 cores. These overall performance and scalability results are very similar to the ones obtained on high-end computing architectures [100] (the loss of linear scalability going from 32 to 64 cores for the non-diagonally dominant case can be attributed to limitations of the multicore architecture). In addition, one can note that Spike aims at achieving linear scalability from small number of processors/cores since $2\times$ speed-up are obtained from only two cores. These performances have recently motivated the development of a shared-memory version of the Spike package [95] to offer a high efficient alternative parallel strategy to the LAPACK-BLAS-threaded *LU* model for solving banded linear systems on current and emerging multicore architectures.

After more than thirty years of innovations in parallel architectures, parallel programming models and numerical libraries, it is remarkable that the Spike algorithm continues to produce such sustainable scalability and ideal speed-up performances. Spike finds its strength from a deceptively simple algorithm which leads to a rich and complex variety of numerical schemes ideally suited for addressing the challenges in modern large-scale applications on a wide range of parallel computing platforms. Over the years, Spike has been a valuable tool to many applications in computational science and engineering including nanoelectronics, oil reservoirs modeling, structural mechanics, fluid-structure interactions, graphic animation, etc. It can be expected that new functionalities and domain of applicability for Spike will continue to flourish in the future.

1.4 Floating-Point Arithmetic and Error Analysis

Today, since the IEEE floating-point standard has been widely adopted, one sometimes forgets the pain and many efforts that have gone in understanding the role of floating-point arithmetic, its interplay with architectures and applications and finally the culmination of these efforts into the standard. These are documented in extensive on-line publications of W. Kahan (http://www.cs.berkeley.edu/~wkahan/), the seminal monograph of N. Higham [71] and the comprehensive handbook [98]. In 1975, Kuck, Parker and Sameh proposed a new rounding scheme called ROM rounding [80]. In [81], they provided a systematic study of the error properties of floating-point arithmetic using ROM rounding and other rounding methods. They also explained and evaluated the fundamental role of extra bits (guard digits and the sticky bit) in floating-point arithmetic. Error analysis was the subject of the Ph.D. work of John Larson directed by Sameh [85]. In [87], they extended work by Bauer on the propagation of relative errors and obtained a system of equations relating roundoff errors and their effects. They proved that this could be done in linear rather than quadratic time and storage with respect to the length of the straight-line program for the algorithm under consideration. These ideas were also applied in [86] to automate forward and backward error analyses as well as a mixture of the two.

1.5 Contributions to n-Body Methods, Fast Multipole Methods, Boundary Integral Solvers, and Their Applications

Sameh recognized early on the power of hierarchical approximations and multipole methods. Working with his colleagues at the University of Minnesota (in particular, Vipin Kumar) and his students at that time (Ananth Grama and Vivek Sarin), he made a number of key contributions on parallel algorithms, error control, linear system solvers and preconditioners (with suitable Green's function kernels), and applications in boundary element solvers. More recently, this work was also applied to a class of molecular dynamics methods, called reactive molecular dynamics, for charge equilibration.

1.5.1 Multipole-Based Hierarchical Approximation Techniques

The problem of simulating the motion of a large set of bodies arises in a variety of domains such as astrophysics, electrostatics, molecular dynamics, fluid dynamics, and high energy physics. The all-to-all nature of interaction between various bodies renders this problem extremely computation-intensive. Techniques based on hierarchical approximations have effectively reduced the complexity of this problem. Coupled with parallel processing, these techniques hold the promise of large scale n-body simulations.

Sameh and his group developed parallel formulations of multipole-based Barnes–Hut methods that are highly scalable to large machine configurations and yield excellent performance for highly unstructured particle distributions. The key difference between these new schemes and other schemes is the use of function shipping, as opposed to more traditional data-shipping. In function shipping, a particle needing to interact with a subdomain at a remote processor is shipped to the remote processor (and the result of the interaction—typically the force or potential is returned). This is in contrast to other approaches where the data associated with the remote subdomain are shipped to the local processor on demand and the computation performed locally. A direct consequence of this is that as accuracy of the simulation is increased by increasing multipole degree, the efficiency of the scheme increases. This is not the case with competing approaches. Furthermore, these schemes have the advantage that all communication is sender initiated and the associated communication overhead is very low. Combined with an effective load balancing scheme that distributes the domain across processors, Sameh and his colleagues demonstrated very high efficiencies and raw performance large configurations of the Cray T3D (and earlier versions on a TMC CM5, both of which were state-of-the-art machines at the time of the research). These evaluations were based on astrophysical simulations of a variety of Gaussian and Plummer galaxies. Detailed results of this study are published in [59, 63].

1.5.2 Multipole Based Dense Linear System Solvers and Preconditioners

An important application of hierarchical multipole methods such as Barnes–Hut and FMM is in solving dense linear systems arising from boundary element modeling of a variety of phenomena in electromagnetics. Boundary element modeling is useful in domains where the Sommerfeld radiation conditions must be satisfied at infinity. Consequently, a finite element modeling of such domains would have to use accurate absorbing boundary conditions (ABCs) for satisfying these radiation conditions. The main drawback of boundary element methods is that the linear systems resulting from them are dense, consequently, having significant compute and memory requirements.

Multipole methods can be used to effectively reduce the complexity of the underlying matrix-vector product from $O(n^2)$ to $O(n \log n)$ and its memory requirement to $O(n)$. This algorithmic speedup from approximation can be combined with parallelism to yield very fast dense solvers. Sameh's group was among the first to develop efficient parallel formulations of dense iterative solvers based on hierarchical approximations for solving potential integral equations of the first kind. They studied the impact of various parameters on the accuracy and performance of the parallel solver and demonstrated that their parallel formulation incurs minimal parallel processing overhead and scales up to a large number of processors. They also proposed two preconditioning techniques for accelerating the convergence of the iterative solver. These techniques are based on an inner–outer scheme and a block diagonal scheme based on a truncated Green's function. Detailed experimental results on moderate configurations of a Cray T3D demonstrated excellent raw performance and scalability. This was among the first parallel dense solver-preconditioner toolkits based on multipole methods [60–62]. This work also provides the basis for a number of more recent results that require solution to dense linear systems with electrostatic $(1/r)$ kernels in atomistic modeling.

1.5.3 Improving Error Bounds for Hierarchical Approximation Techniques

Multipole-based hierarchical approximation techniques relied on fixed-degree multipole approximations of clusters of charges (or masses). Sameh et al. presented analysis and experiments to illustrate that fixed-degree multipole approximations can lead to large aggregate errors. They developed an alternate strategy based on careful selection of the multipole degree that leads to asymptotically lower errors, while incurring minimal computation overhead. First, they estimate the error associated with each particle-cluster interaction and the aggregate error for each particle. They then describe a technique for computing the multipole degree of each interaction with a view to reducing aggregate error, and establish the computational complexity of the new method. Numerical experiments demonstrate significantly enhanced error properties of the new method, at the expense of marginal increase in computation [64–66].

1.5.4 Algorithms for Atomistic Modeling

Sameh's results on linear system solvers have found significant recent applications in atomistic modeling of large reactive systems. Modeling atomic and molecular systems requires computation-intensive quantum mechanical methods such as, but not limited to, density functional theory (DFT). These methods have been successful in predicting various properties of chemical systems at atomistic detail. Due to the inherent nonlocality of quantum mechanics, the scalability of these methods ranges from $O(N^3)$ to $O(N^7)$ for an N-atom system, depending on the method used and approximations involved. This significantly limits the size of simulated systems to a few thousands of atoms, even on large scale parallel platforms. On the other hand, classical approximations of quantum systems, although computationally (relatively) easy to implement, yield simpler models that lack essential chemical properties such as reactivity and charge transfer. The recent work of van Duin et al. [38] overcomes the limitations of non-reactive classical molecular dynamics (MD) approximations by carefully incorporating limited nonlocality (to mimic quantum behavior) through an empirical bond order potential. This reactive classical MD method, called ReaxFF, achieves essential quantum properties, while retaining the computational simplicity of classical molecular dynamics, to a large extent. Implementation of reactive force fields presents significant algorithmic challenges. Since these methods model bond breaking and formation, efficient implementations must rely on complex dynamic data structures. Charge transfer in these methods is accomplished by minimizing electrostatic energy through charge equilibration. This requires the solution of large linear systems (10^8 degrees of freedom and beyond) with shielded electrostatic kernels at each time-step. Individual time-steps are themselves typically in the range of tenths of femtoseconds, requiring optimizations within and across time-steps to scale simulations to nanoseconds and beyond, where interesting phenomena may be observed.

Based on Sameh's work, the group of Grama et al. [2] developed the sPuReMD (serial Purdue Reactive Molecular Dynamics) program, a unique reactive classical MD code. At the heart of this code is a Krylov subspace solver relies on a preconditioner based on incomplete LU factorization (ILUT), specially designed for the application. sPuReMD has been validated for performance and accuracy on a variety of systems, ranging from hydrocarbons and biophysical systems to nanoscale devices (Si-Ge nanorods) and explosives (RDX). Parallel versions of PuReMD have now been integrated into commonly used MD package LAMMPS and are widely used [1].

The breadth and depth of Sameh's contributions in this area have served as one of the bases for a large class of simulations on current high performance computing platforms.

1.6 Computational Science and Engineering

Sameh was among the earliest proponents of the discipline of Computational Science Engineering (CSE). A position paper by George Cybenko, David Kuck and Ahmed Sameh was written in 1990 that described a proposal to create a CSE program at the University of Illinois at Urbana-Champaign. This paper was widely circulated on the UIUC campus and gave a crucial impulse to the idea of a CS&E program.

The importance of this proposal was recognized by the IEEE, where the Computer Society's Technical Segment Committee (TSC) on CSE under the chairmanship of John Riganati proposed in April 1992 setting up an IEEE magazine devoted to CSE. The formal approval came in February 1993. The October 1993 issue of the IEEE Computer Magazine was devoted to Computational Science and Engineering, with Sameh and John Riganati as Guest Editors. This issue served as a prototype of the new magazine. Ted Lewis, the Editor-in-Chief of the Computer magazine at the time, was a key supporter at IEEE. In their introduction the Guest Editors write

> The term "computational science" was first used by Ken Wilson (awarded the Nobel Prize for his work in physics) to refer to those activities in science and engineering that exploit

> computing as their main tool. Indeed, in the past 10 years, computing has become an essential tool for advancing various science and engineering disciplines. Moreover, with the advent of parallel computers, it has become clear that parallel computing has become an integral part of many of aspects of these disciplines An understanding of parallel or distributed (heterogeneous) computer systems often motivates new solution strategies not previously envisioned when only uniprocessors were used. Such interaction between the applications, algorithms, architectures, and underlying system software characterizes computational science and engineering (CSE).

The first issue of the IEEE Computational Science and Engineering Magazine appeared in Spring 1994, with Sameh as Editor-in-Chief and Francis Sullivan as Associate Editor-in-Chief. Contributors to the first issue included many distinguished researchers who were important leaders in various areas of CSE: S. Kamal Abdali, Bill Buzbee, George Cybenko, David P. Dobkin, David K. Ferry, Efstratios (Stratis) Gallopoulos, Eric Gross, Karl Hess, Charles H. Holland, Malvin H. Kalos, Jacob V. Maizel, Jr., Edmund K. Miller, Ahmed K. Noor, William H. Press, John R. Rice, John P. Riganati, Bruce Shriver, Tayfun Tezduyar, Donald G. Truhlar, Adan Wheeler, Paul Woodward, Richard N. Draper, William Jalby, David K. Kahaner, Alvin K. Thaler, Robert G. Voigt, and Harry A.G. Wijshoff. This trend of very high quality contributions was continued in subsequent issues of the magazine. In keeping with Sameh's vision, the magazine developed into an important forum for articles on interdisciplinary CSE research that reaches a very broad audience. The paper [55] by Gallopoulos and Sameh expands on the notion of CSE, its relation with Computer Science, and the key role of problem solving environments [52]. "Doing" CSE was described as research and development activities surrounding the traversal of data and information across concentric layers of various "models" (especially the discrete, arithmetic, and computational) and investigating questions of concept (How many steps can be automated?) design (What is the system that will make this possible?), implementation (What are the enabling technologies and how can a system be implemented?) and progress (How does the system adapt to evolving requirements? How is it evaluated?) It is worth noting that this view also makes natural the notion of "information loss" as model layers are traversed. Discretization and roundoff errors, computer performance obfuscation by pseudocode and programming language can be put and studied in this framework.

The last issue of IEEE Computational Science and Engineering with Sameh as Editor-in-Chief was the Winter 1995 issue. In 1999, the CSE magazine was merged with the publication Computers in Physics, with the new name Computing in Science and Engineering (CISE).

References

1. Aktulga, H.M., Fogarty, J.C., Pandit, S.A., Grama, A.Y.: Parallel reactive molecular dynamics: Numerical methods and algorithmic techniques. Parallel Comput. (2011). doi:10.1016/j.parco.2011.08.005
2. Aktulga, H., Pandit, S., van Duin, A., Grama, A.: Reactive molecular dynamics: Numerical methods and algorithmic techniques. SIAM J. Sci. Comput. (2011, to appear)

3. Anderson, E., Bai, Z., Bischof, C., Blackford, S., Demmel, J., Dongarra, J., Croz, J.D., Greenbaum, A., Hammerling, S., McKenney, A., Sorensen, D.: LAPACK Users' Guide, 3rd edn. SIAM, Philadelphia (1999)
4. Arbenz, P., Cleary, A., Dongarra, J., Hegland, M.: A comparison of parallel solvers for diagonally dominant and general narrow-banded linear systems. Parallel Dist. Comp. Pract. **2**, 385–400 (1999)
5. Asanovic, K., Bodik, R., Catanzaro, B.C., Gebis, J.J., Husbands, P., Keutzer, K., Patterson, D.A., Plishker, W.L., Shalf, J., Williams, S.W., Yelick, K.A.: The landscape of parallel comput. research: A view from Berkeley. Tech. Rep. UCB/EECS-2006-183, EECS Department. University of California, Berkeley (2006). http://www.eecs.berkeley.edu/Pubs/TechRpts/2006/EECS-2006-183.html
6. Barnes, G., Brown, R., Kato, M., Kuck, D., Slotnick, D., Stokes, R.: The ILLIAC IV computer. IEEE Trans. Comput. **17**, 746–757 (1968). http://doi.ieeecomputersociety.org/10.1109/TC.1968.229158
7. Ben-Artzi, M., Croisille, J.P., Fishelov, D.: A fast direct solver for the biharmonic problem in a rectangular grid. SIAM J. Sci. Comput. **31**(1), 303–333 (2008)
8. Berry, M.: Multiprocessor sparse SVD algorithms and applications. Ph.D. thesis, University of Illinois at Urbana-Champaign (1991)
9. Berry, M.: Large scale singular value decomposition. Int. J. Supercomput. Appl. **6**, 13–49 (1992)
10. Berry, M., Drmac, Z., Jessup, E.: Matrices, vector spaces, and information retrieval. SIAM Rev. **41**, 335–362 (1998)
11. Berry, M., Dumais, S., O'Brien, G.: Using linear algebra for intelligent information retrieval. SIAM Rev. **37**, 573–595 (1995)
12. Berry, M., Mezher, D., Philippe, B., Sameh, A.: Parallel algorithms for the singular value decomposition. In: Kontoghiorghes, E. (ed.) Handbook of Parallel Computing and Statistics, pp. 117–164. Chapman & Hall/CRC, Boca Raton (2006)
13. Berry, M., Sameh, A.: Multiprocessor schemes for solving block tridiagonal linear systems. Int. J. Supercomput. Appl. **2**(3), 37–57 (1988)
14. Berry, M., Sameh, A.: An overview of parallel algorithms for the singular value and symmetric eigenvalue problems. J. Comput. Appl. Math. **27**(1–2), 191–213 (1989). doi:10.1016/0377-0427(89)90366-X. http://www.sciencedirect.com/science/article/pii/037704278990366X. Special Issue on Parallel Algorithms for Numerical Linear Algebra
15. Berry, M., et al.: The Perfect club benchmarks: Effective performance evaluation of supercomputers. Int. J. High Perform. Comput. Appl. **3**(3), 5–40 (1989)
16. Bik, A.J.C., Wijshoff, H.A.G.: Compilation techniques for sparse matrix computations. In: Proc. Int'l. Conf. Supercomp, pp. 416–424 (1993)
17. Bik, A.J.C., Wijshoff, H.A.G.: Advanced compiler optimizations for sparse computations. J. Parallel Distrib. Comput. **31**(1), 14–24 (1995)
18. Bini, D.: Parallel solution of certain Toeplitz linear systems. SIAM J. Comput. **13**, 268–276 (1984)
19. Bini, D.: Matrix structures in parallel matrix computations. Calcolo **25**(1–2), 37–51 (1988)
20. Bini, D., Capovani, M.: Fast parallel and sequential computations and spectral properties concerning band Toeplitz matrices. Calcolo **20**, 177–189 (1983)
21. Blackford, L., Choi, J., Cleary, A., D'Azevedo, E., Demmel, J., Dhillon, I., Dongarra, J., Hammarling, S., Henry, G., Petitet, A., Stanley, K., Walker, D., Whaley, R.: ScaLAPACK User's Guide. SIAM, Philadelphia (1997). See also www.netlib.org/scalapack
22. Bouknight, W., Denenberg, S., McIntyre, D., Randall, J., Sameh, A., Slotnick, D.: The ILLIAC IV system. Proc. IEEE **60**(4), 369–388 (1972)
23. Bramley, R., Chen, H.C., Meier, U., Sameh, A.: On some parallel preconditioned CG schemes. In: Axelsson, O. (ed.) Preconditioned Conjugate Gradient Methods. Lecture Notes in Mathematics. Springer, Berlin (1990)
24. Bramley, R., Sameh, A.: Row projection methods for large nonsymmetric linear systems. SIAM J. Sci. Stat. Comput. **13**, 168–193 (1992)

25. Buzbee, B.: A fast Poisson solver amenable to parallel computation. IEEE Trans. Comput. **C-22**(8), 793–796 (1973)
26. Buzbee, B., Golub, G., Nielson, C.: On direct methods for solving Poisson's equation. SIAM J. Numer. Anal. **7**(4), 627–656 (1970)
27. Chen, H.C., Sameh, A.: A matrix decomposition method for orthotropic elasticity problems. SIAM J. Matrix Anal. Appl. **10**(1), 39–64 (1989)
28. Chen, S.C., Kuck, D.J., Sameh, A.H.: Practical band triangular system solvers. ACM Trans. Math. Softw. **4**(3), 270–277 (1978)
29. Cleary, A., Dongarra, J.: Implementation in ScaLAPACK of divide and conquer algorithms for banded and tridiagonal linear systems. Tech. Rep. UT-CS-97-358, University of Tennessee Computer Science Technical Report (1997)
30. Cooley, J., Lewis, P., Welch, P.: The fast Fourier transform algorithm: Programming considerations in the calculation of sine, cosine, and Laplace transforms. J. Sound Vib. **12**(2), 315–337 (1970)
31. Crouzeix, M., Philippe, B., Sadkane, M.: The Davidson method. SIAM J. Sci. Stat. Comput. **15**, 62–76 (1984)
32. Davidson, E., Kuck, D., Lawrie, D., Sameh, A.: Supercomputing tradeoffs and the Cedar system. In: Wilhelmson, R. (ed.) High-Speed Computing, Scientific Applications and Algorithm Design. University of Illinois Press, Champaign (1986)
33. Demko, S., Moss, W., Smith, P.: Decay rates for inverses of band matrices. Math. Comput. **43**(168), 491–499 (1984)
34. Demmel, J., Dhillon, I., Ren, H.: On the correctness of some bisection-like parallel eigenvalue algorithms in floating point arithmetic. Electron. Trans. Numer. Anal. **3**, 116–149 (1995)
35. DeRose, L., Gallivan, K., Gallopoulos, E., Marsolf, B., Padua, D.: FALCON: A MATLAB interactive restructuring compiler. In: Huang, C.H., et al. (eds.) Languages and Compilers for Parallel Comp. Lecture Notes in Computer Science, vol. 1033, pp. 269–288. Springer, Berlin (1996)
36. Dongarra, J.: Getting the performance out of high performance computing. Presentation (2003). http://www.netlib.org/utk/people/JackDongarra/SLIDES/scidac-napa-0303.pdf. DOE SciDAC Review
37. Dongarra, J., Sameh, A.: On some parallel banded system solvers. Parallel Comput. **1**(3–4), 223–236 (1984)
38. van Duin, A.C.T., Dasgupta, S., Lorant, F. III: ReaxFF: A reactive force field for hydrocarbons. J. Phys. Chem. A **105**, 9396–9409 (2001)
39. Emrath, P.: XYLEM: An operating system for the Cedar multiprocessor. IEEE Softw. **2**(4), 30–37 (1985)
40. Emrath, P., Padua, D., Yew, P.C.: Cedar architecture and its software. In: Architecture Track, Proc. of the Twenty-Second Annual Hawaii Intl. Conf. on System Sciences, vol. 1, pp. 306–315 (1989)
41. Ericksen, J.: Iterative and direct methods for solving Poisson's equation and their adaptability to ILLIAC IV. Tech. Rep. UIUCDCS-R-72-574, University of Illinois at Urbana-Champaign, Department of Computer Science (Dec. 1972)
42. Ericksen, J., Wilhelmson, R.B.: Implementation of a convective problem requiring auxiliary storage. ACM Trans. Math. Softw. **2**, 187–195 (1976)
43. Fairweather, G., Karageorghis, A., Martin, P.: The method of fundamental solutions for scattering and radiation problems. Eng. Anal. Bound. Elem. **27**(7), 759–769 (2003). Special issue on Acoustics
44. Fischer, D., Golub, G., Hald, O., Leiva, C., Widlund, O.: On Fourier–Toeplitz methods for separable elliptic problems. Math. Comput. **28**(126), 349–368 (1974)
45. Gallivan, K., Jalby, W., Malony, A., Wijshoff, H.: Performance prediction for parallel numerical algorithms. Int. J. High Speed Comput. **3**(1), 31–62 (1991)
46. Gallivan, K., Jalby, W., Malony, A., Yew, P.C.: Performance analysis on the Cedar system. In: Martin, J.L. (ed.) Performance Evaluation of Supercomp.s, pp. 109–142. Elsevier, Amsterdam (1988)

47. Gallivan, K., Jalby, W., Meier, U., Sameh, A.H.: Impact of hierarchical memory systems on linear algebra algorithm design. Int. J. Supercomput. Appl. **2**(1), 12–48 (1988)
48. Gallivan, K., Sameh, A., Zlatev, Z.: A parallel hybrid sparse linear system solver. Comp. Sys. Eng. **1**(2–4) (1990)
49. Gallivan, K., Sameh, A., Zlatev, Z.: Solving general sparse linear systems using conjugate gradient-type methods. Proc. Int. Conf. Supercomput. **18**, 132–139 (1990) doi:http://doi.acm.org/10.1145/255129.255149
50. Gallivan, K.A., Plemmons, R.J., Sameh, A.H.: Parallel numerical algorithms for dense linear algebra computations. SIAM Rev. **32**(1), 54–135 (1990)
51. Gallopoulos, E.: Rapid elliptic solvers. In: Padua, D. (ed.) Encyclopedia of Parallel Comput. Springer, Berlin (2011)
52. Gallopoulos, E., Houstis, E., Rice, J.: Computer as thinker/doer: Problem solving environments for CSE. IEEE Comput. Sci. Eng. **1**(2), 11–23 (1994)
53. Gallopoulos, E., Lee, D.: Boundary integral domain decomposition on hierarchical memory multiprocessors. In: Proc. 1988 ACM Int'l. Conf. Supercomp, pp. 488–499 (1988)
54. Gallopoulos, E., Sameh, A.: Solving elliptic equations on the Cedar multiprocessor. In: Wright, M.H. (ed.) Aspects of Computation on Asynchronous Parallel Processors, pp. 1–12. Elsevier, Amsterdam (1989)
55. Gallopoulos, E., Sameh, A.: CSE: content and product. IEEE Comput. Sci. Eng. Mag. **4**, 39–43 (1997)
56. Gannon, D., Jalby, W., Gallivan, K.: Strategies for cache and local memory management by global program transformation. J. Parallel Distrib. Comput. **5**(5), 587–616 (1988)
57. Giraud, L.: Parallel distributed FFT-based solvers for 3-D Poisson problems in meso-scale atmospheric simulations. Int. J. High Perform. Comput. Appl. **15**(1), 36–46 (2001)
58. Golub, G., Sameh, A., Sarin, V.: Parallel balance scheme for banded linear systems. Numer. Linear Algebra Appl. **8**(5), 297–316 (2001)
59. Grama, A., Kumar, V., Sameh, A.: Scalable parallel formulations of the Barnes–Hut algorithm for n-Body simulations. In: Proceedings of the Supercomputing Conference, Washington, DC, p. 8 (1994)
60. Grama, A., Kumar, V., Sameh, A.: Parallel matrix-vector product using approximate hierarchical methods. In: Proceedings of the Supercomputing Conference, San Diego, CA, p. 8 (1995)
61. Grama, A., Kumar, V., Sameh, A.: Parallel hierarchical solvers and preconditioners for boundary element methods. In: Proceedings of the Supercomputing Conference, Pittsburgh, PA, p. 8 (1996). Proc. on CD and online at http://www.supercomp.org/sc96/proceedings/
62. Grama, A., Kumar, V., Sameh, A.: Parallel hierarchical solvers and preconditioners for boundary element methods. SIAM J. Sci. Comput. **20**(1), 337–358 (1998)
63. Grama, A., Kumar, V., Sameh, A.: Scalable parallel formulations of the Barnes–Hut method for n-Body simulations. Parallel Comput. **24**(5–6), 797–822 (1998)
64. Grama, A., Sarin, V., Sameh, A.: Analyzing the error bounds of multipole-based treecodes. In: Proceedings of the Supercomputing Conference, Orlando, FL, p. 10 (1998). Proc. on CD or online at http://www.supercomp.org/sc98/papers/index.html
65. Grama, A., Sarin, V., Sameh, A.: Improving error bounds for multipole-based treecodes. In: Proceedings of 5th International Conference on High Performance Computing, Chennai, India, p. 8 (1998). Proc. on CD and online at http://www.hipc.org/hipc98/adpgm98.html
66. Grama, A., Sarin, V., Sameh, A.: Improving error bounds for multipole-based treecodes. SIAM J. Sci. Comput. **21**(5), 1790–1803 (2000)
67. Grcar, J., Sameh, A.: On certain parallel Toeplitz linear system solvers. SIAM J. Sci. Stat. Comput. **2**(2), 238–256 (1981)
68. Gupta, A., Kumar, V., Sameh, A.H.: Performance and scalability of preconditioned conjugate gradient methods on parallel computers. IEEE Trans. Parallel Distrib. Syst. **6**(5), 455–469 (1995)
69. Guzzi, M., Padua, D., Hoeflinger, J., Lawrie, D.: Cedar Fortran and other vector and parallel Fortran dialects. In: Proc. Supercomp. 1988, vol. 1, pp. 114–121 (1988)

70. Heller, D.: A survey of parallel algorithms in numerical linear algebra. SIAM Rev. **20**(4), 740–777 (1978)
71. Higham, N.: Accuracy and Stability of Numerical Algorithms, 2nd edn. SIAM, Philadelphia (2002)
72. Hockney, R.: A fast direct solution of Poisson's equation using Fourier analysis. J. Assoc. Comput. Mach. **12**, 95–113 (1965)
73. Hoemmen, M.: Communication-avoiding Krylov subspace methods. Ph.D. thesis, University of California at Berkeley (2010)
74. Huang, H.M.: A parallel algorithm for symmetric tridiagonal eigenvalue problems. CAC Document 109, Center for Advanced Computation, Univ. Illinois at Urbana-Champaign (1974)
75. Jacobi, C.: Über ein leichtes verfahren die in der theorie der säculärstörungen vorkommenden gleichungen numerisch aufzulösen. Crelle's J. für reine und angewandte Mathematik **30**, 51–94 (1846)
76. Jalby, W., Philippe, B.: Stability analysis and improvement of the block Gram-Schmidt algorithm. SIAM J. Sci. Stat. Comput. **12**(5), 1058–1073 (1991)
77. Kamath, C., Sameh, A.: The preconditioned conjugate gradient algorithm on a multiprocessor. In: Fifth IMACS International Symp. on Computer Methods for Partial Differential Equations, pp. 210–217 IMACS (1984)
78. Kamath, C., Sameh, A.: A projection method for solving nonsymmetric linear systems on multiprocessors. Parallel Comput. **9**, 291–312 (1989)
79. Kuck, D., Davidson, E., Lawrie, D., Sameh, A., Zhu, C.Q., Veidenbaum, A., Konicek, J., Yew, P., Gallivan, K., Jalby, W., Wijshoff, H., Bramley, R., Yang, U.M., Emrath, P., Padua, D., Eigenmann, R., Hoeflinger, J., Jaxon, G., Li, Z., Murphy, T., Andrews, J., Turner, S.: The Cedar system and an initial performance study. In: Proc. of the 20th ACM/IEEE Intl. Symposium on Computer Architecture, pp. 213–223. ACM, New York (1993)
80. Kuck, D., Parker, D. Jr., Sameh, A.: ROM rounding: A new rounding scheme. In: Proc. 3d IEEE Symp. Comput. Arith, pp. 67–72 (1975)
81. Kuck, D., Parker, D. Jr., Sameh, A.: Analysis of rounding methods in floating-point arithmetic. IEEE Trans. Comput. **C-26**(7), 643–650 (1977)
82. Kuck, D., Sameh, A.: Parallel computation of eigenvalues of real matrices. In: IFIP Congress 1971, vol. 2, pp. 1266–1272 (1972)
83. Kuck, D.J., Davidson, E.S., Lawrie, D.H., Sameh, A.H.: Parallel supercomputing today and the Cedar approach. Science **231**, 967–974 (1987)
84. Lanczos, C.: An iteration method for the solution of the eigenvalue problem of linear differential and integral operators. J. Res. Natl. Bur. Stand. **45**, 255–282 (1950)
85. Larson, J.L.: Automatic error analysis for serial and parallel algorithms. In: Kuck, D., Sameh, A., Gajski, D. (eds.) High Speed Computers and Algorithm Organization, pp. 457–459. Academic Press, New York (1976)
86. Larson, J.L., Sameh, A.: Algorithms for roundoff error analysis—a relative error approach. Computer **24**(4), 275–297 (1980)
87. Larson, J.L., Sameh, A.: Efficient calculation of the effects of roundoff errors. ACM Trans. Math. Softw. **4**(3), 228–236 (1978)
88. Lawrie, D.H., Sameh, A.H.: The computation and communication complexity of a parallel banded system solver. ACM Trans. Math. Softw. **10**(2), 185–195 (1984)
89. Lo, S., Philippe, B., Sameh, A.: A multiprocessor algorithm for the symmetric tridiagonal eigenvalue problem. SIAM J. Sci. Stat. Comput. **8**, s155–s165 (1987)
90. Manguoglu, M., Koyuturk, M., Sameh, A., Grama, A.: Weighted matrix ordering and parallel banded preconditioners for iterative linear system solvers. SIAM J. Sci. Comput. **32**(3), 1201–1216 (2010)
91. Manguoglu, M., Saied, F., Sameh, A., Grama, A.: Performance models for the SPIKE banded linear system solver. Sci. Program. **19**(1), 13–25 (2011)
92. Manguoglu, M., Sameh, A.H., Schenk, O.: PSPIKE: A parallel hybrid sparse linear system solver. In: Proc. 15th Int'l. Euro-Par Conf. on Parallel Proc., Euro-Par '09, pp. 797–808. Springer, Berlin (2009)

93. Marsolf, B.A., Gallivan, K.A., Wijshoff, H.A.G.: The utilization of matrix structure to generate optimized code from MATLAB programs. Int. J. Parallel Program. **27**(2), 73–96 (1999)
94. Meier, U., Sameh, A.: The behavior of conjugate gradient methods on a multivector processor. J. Comput. Appl. Math. **24**, 13–32 (1988)
95. Mendiratta, K., Polizzi, E.: A threaded SPIKE algorithm for solving general banded systems. Paralel Comput. **37**(12), 733–741 (2011)
96. Mikkelsen, C.C.K., Manguoglu, M.: Analysis of the truncated SPIKE algorithm. SIAM J. Matrix Anal. Appl. **30**, 1500–1519 (2008)
97. Morgan, R., Scott, D.S.: Generalizations of Davidson's method for computing eigenvalues of sparse symmetric matrices. SIAM J. Sci. Stat. Comput. **7**, 817–825 (1986)
98. Muller, J.M., Brisebarre, N., de Dinechin, F., Jeannerod, C.P., Lefèvre, V., Melquiond, G., Revol, N., Stehlé, D., Torres, S.: Handbook of Floating-Point Arithmetic. Birkhäuser, Boston (2010)
99. Naumov, M., Manguoglu, M., Sameh, A.: A tearing-based hybrid parallel sparse linear system solver. J. Comput. Appl. Math. **234**, 3025–3038 (2010)
100. Polizzi, E., Sameh, A.H.: A parallel hybrid banded system solver: The SPIKE algorithm. Parallel Comput. **32**(2), 177–194 (2006)
101. Polizzi, E., Sameh, A.H.: SPIKE: A parallel environment for solving banded linear systems. Comput. Fluids **36**(1), 113–120 (2007)
102. Rutishauser, H.: Simultaneous iteration method for symmetric matrices. Numer. Math. **13**, 204–223 (1970)
103. Saad, Y., Sameh, A.: Iterative methods for the solution of elliptic difference equations on multiprocessors. In: Proc. CONPAR'81, Lecture Notes in Computer Science, pp. 395–413. Springer, Berlin (1981)
104. Saad, Y., Sameh, A., Saylor, P.: Solving elliptic difference equations on a linear array of processors. SIAM J. Sci. Stat. Comput. **6**(4), 1049–1063 (1985)
105. Saad, Y., Wijshoff, H.A.G.: Performance study of some supercomputers using a sparse matrix benchmark. In: Proc. SIAM Conf. Parallel Proc. Sci. Comp, pp. 451–453 (1989)
106. Saad, Y., Wijshoff, H.A.G.: SPARK: a benchmark package for sparse computations. In: Proc. Intl. Conf. Supercomp, pp. 239–253 (1990)
107. Sameh, A.: Numerical analysis of axisymmetric wave propagation in elastic-plastic layered media. Ph.D. thesis, Dept. of Civil Engineering, University of Illinois at Urbana-Champaign (1968)
108. Sameh, A.: A discrete-variable approach for elastic-plastic wave motions in layered solids. J. Comput. Phys. **8**, 342–368 (1971)
109. Sameh, A.: On two numerical algorithms for multiprocessors. In: Proc. NATO Advanced Research Workshop on High-Speed Comp. Series F: Computer and Systems Sciences, p. 18 (1983)
110. Sameh, A.: On some parallel algorithms on a ring of processors. Comput. Phys. Commun. **37**, 159–166 (1985)
111. Sameh, A., Han, L.: Eigenvalue problems. Tech. rep., ILLIAC IV Document 127, Dept. of Computer Science, University of Illinois, Urbana (1968)
112. Sameh, A., Sarin, V.: Hybrid parallel linear solvers. Int. J. Comput. Fluid Dyn. **12**, 213–223 (1999)
113. Sameh, A., Tong, Z.: The trace minimization method for the symmetric generalized eigenvalue problem. J. Comput. Appl. Math. **123**, 155–175 (2000)
114. Sameh, A.H.: On Jacobi and Jacobi-like algorithms for a parallel computer. Math. Comput. **25**, 579–590 (1971)
115. Sameh, A.H.: Numerical parallel algorithms–a survey. In: Kuck, D., Lawrie, D., Sameh, A. (eds.) High Speed Computer and Algorithm Organization, pp. 207–228. Academic Press, San Diego (1977)
116. Sameh, A.H.: A fast Poisson solver for multiprocessors. In: Birkhoff, G., Schoenstadt, A. (eds.) Elliptic Problem Solvers II, pp. 175–186. Academic Press, San Diego (1984)
117. Sameh, A.H., Brent, R.P.: Solving triangular systems on a parallel computer. SIAM J. Numer. Anal. **14**(6), 1101–1113 (1977)

118. Sameh, A.H., Chen, S.C., Kuck, D.J.: Parallel direct Poisson and biharmonic solvers. Tech. Rep. 684, Dept. Computer Science, University of Illinois (1974)
119. Sameh, A.H., Chen, S.C., Kuck, D.J.: Parallel Poisson and biharmonic solvers. Computer **17**, 219–230 (1976)
120. Sameh, A.H., Kuck, D.J.: On stable parallel linear system solvers. J. Assoc. Comput. Mach. **25**(1), 81–91 (1978)
121. Sameh, A.H., Lermit, J., Noh, K.: On the intermediate eigenvalues of symmetric sparse matrices. BIT Numer. Math. **15**, 185–191 (1975)
122. Sameh, A.H., Sarin, V.: Large scale simulation of particulate flows. In: Proceedings 13th International Parallel Processing Symposium/10th Symposium on Parallel and Distributed Processing (IPPS/SPDP '99), 12–16 April 1999, San Juan, Puerto Rico, pp. 660–667. IEEE Computer Society, Los Alamitos (1999)
123. Sameh, A.H., Wisniewski, J.A.: A trace minimization algorithm for the generalized eigenvalue problem. SIAM J. Numer. Anal. **19**(6), 1243–1259 (1982)
124. Sarin, V., Kneppley, M., Sameh, A.H.: Parallel simulation of particulate flows. In: Ferreira, A., Rolim, J.D.P., Simon, H.D., Teng, S.H. (eds.) Proceedings of Solving Irregularly Structured Problems in Parallel, 5th International Symposium, IRREGULAR '98, Berkeley, California, USA, August 9–11, 1998. Lecture Notes in Computer Science, vol. 1457, pp. 226–237. Springer, Berlin (1998)
125. Schenk, O., Gärtner, K.: Solving unsymmetric sparse systems of linear equations with PARDISO. Future Gener. Comput. Syst., **20**(3), 475–487 (2004)
126. Sharma, S., Malony, A., Berry, M., Sinvhal-Sharma, P.: Run-time monitoring of concurrent programs on the Cedar multiprocessor. In: Proc. Supercomp. 1990, pp. 784–793 (1990)
127. SIAM Oral Histories: The history of numerical analysis and scientific computing: An interview with Bill Buzbee (2005). Conducted by Thomas Haigh and accessible from http://history.siam.org/pdfs2/Buzbee_returned_SIAM_copy.pdf
128. Sleijpen, G.L.G., van der Vorst, H.A.: A Jacobi–Davidson iteration method for linear eigenvalue problems. SIAM J. Matrix Anal. Appl. **17**, 401–425 (1996)
129. Slotnick, D., Sameh, A.: Numerical calculation and computer design. Comput. Math. Appl. **3**, 201–210 (1978)
130. Slotnick, D.L., Borck, W.C., McReynolds, R.C.: The SOLOMON computer. In: Proc. Fall Joint Computer Conference, vol. 22, pp. 97–107 AFIPS (1962)
131. Sorensen, D.: Analysis of pairwise pivoting in Gaussian elimination. IEEE Trans. Comput. **C-34**(3), 274–278 (1985)
132. SPIKE. A distributed memory version of the SPIKE package. Obtained from http://software.intel.com/en-us/articles/intel-adaptive-spike-based-solver/
133. Stone, H.S.: An efficient parallel algorithm for the solution of a tridiagonal linear system of equations. J. ACM **20**(1), 27–38 (1973)
134. Wilhelmson, R.: Solving partial differential equations using ILLIAC IV. In: Colton, D., Gilbert, R. (eds.) Constructive and Computational Methods for Differential and Integral Equations. Lecture Notes in Mathematics, vol. 430, pp. 453–476. Springer, Berlin (1974). doi:10.1007/BFb0066281
135. Wisniewski, J.A., Sameh, A.H.: Parallel algorithms for network routing problems and recurrences. SIAM J. Algebr. Discrete Methods **3**(3), 379–394 (1982)
136. Zhu, C.Q., Yew, P.C.: A scheme to enforce data dependences on large multiprocessor systems. IEEE Trans. Softw. Eng. **SE-13**(6), 726–739 (1987)

Chapter 2
Computational Capacity-Based Codesign of Computer Systems

David J. Kuck

Abstract This paper proposes a fast, novel approach for the HW/SW codesign of computer systems based on a computational capacity model. System node bandwidths and bandwidths used by the SW load underlie three sets of linear equations: a model system representing a load running on a computer, a design equation and objective function with goals as inputs, and a capacity sensitivity equation. These are augmented with nonlinear techniques to analyze multirate HW nodes as well as to synthesize system nodes when codesign goals exceed feasible engineering HW choices. Solving the equations rapidly finds the optimal costs of a broad class of architectures for a given computational load. The performance of each component can be determined globally and for each computational phase. The ideas are developed theoretically and illustrated by numerical examples plus results produced by a prototype CAPE tool implementation.

2.1 Introduction

System performance is dominated by the performance of individual system components, and balanced component use in a computation. Designers of computer systems and system HW/SW components must face potential system performance instabilities due to component nonlinear performance behavior and imbalanced use. Instability appears in two forms: a given *system* yields widely varying performances over program types, or a given *program* runs at widely varying performances across (similar) system types. Both are common phenomena.

Traditionally, systems that are more stable and productive arise from application-specialization of system components and architectures. Bandwidth metrics can be used to characterize computations by their dominant constituent phases. Matching system HW/SW components to dominant phases for given sets of computations can increase stability. This can be achieved for any program by decreasing that program's performance deviation from the stable value expected for a target system architecture.

D.J. Kuck (✉)
Intel Corporation, Urbana, USA
e-mail: david.kuck@intel.com

M.W. Berry et al. (eds.), *High-Performance Scientific Computing*,
DOI 10.1007/978-1-4471-2437-5_2,

This leads to specialization in the marketplace—from embedded processors to GPUs and HPC systems. As computer usage broadens, future designs will continue moving toward more-specialized chips and SW that can exploit them effectively in key markets. Computer system designers and application SW providers need tools and analyses to help them design highly productive systems. Success will depend upon developing methods that can approximate the analysis of future-oriented applications to drive the HW/SW co-design of products.

Significant changes are needed in several areas to meet these needs:
Future-oriented application workloads must be used in computer system design. These may include libraries, reference platform application implementations, or whole-application prototypes, plus the full range of data and usage scenarios.
Leading application development tools must drive the HW design to get top performance and avoid regressions. Tools, compilers, and libraries must be available in advance to allow SW developers to provide the applications above.
System and SW codesign process must allow architects to know deliverable design performance in future markets. Beyond the above, this requires a fast, accurate, mixed fidelity, and comprehensive codesign process.

The performance of a communication network, traffic system, or conference room depends on intrinsic physical characteristics as well as type of load and external ambient factors. For each, peak performance can be defined, but in these examples, noise, weather, and event type, respectively, affect usable *capacity*, in practice. This notion, and the term *capacity*, have been used in several ways relative to computer systems. This paper gives the term a precise, comprehensive meaning.

The paper's contributions are a linear capacity-based codesign process and model (Sects. 2.2, 2.3), methods of formulating and optimizing codesign equations (Sects. 2.4, 2.5), initial numerical codesign results (Sect. 2.6), and an overview of multirate nodes and nonlinearities (Sect. 2.7). Computer HW bandwidth, system architecture, and applications load are all captured in the methodology described. SW performance tuning is reviewed (Sect. 2.8).

2.1.1 Background: Performance Basics

Four main contributors determine computer system performance for a given computation: hardware, architecture, system software, and the application code (including data sets) being run, expressed as in formula (2.1):

$$perf(computation(hw, arch, sw, code)) \tag{2.1}$$

Performance can be expressed as the time consumed or as a rate delivered in running a computation. We will discuss several performance metrics; specific engineering details dictate which are used in a particular codesign effort.

- *Computational capacity* C [4, 5] is a fractional-use metric, ranging between 0 and a maximum value (defined here), and it spans one or more HW components. It is a joint property of the HW and the computation being performed. For simplicity, we refer to it as *capacity*. At the *computer control level*, capacity is expressed as Eq. (2.2).

$$C\ [\text{instructions}/sec] = \frac{\text{clock frequency [clock cycles}/sec]}{CPI\ [\text{clock cycles/instruction}]} \tag{2.2}$$

When operations are a more natural metric than *instructions*, we use *operations* O, defined as processing a fixed number of bits. On other occasions, e.g. when data paths are under discussion, capacity measured as BW used in $[bits/sec]$ or $[words/sec]$ is most natural and this will be used throughout the paper.

- *Bandwidth (BW)* is the standard rate metric for HW design. We use BW to describe individual HW nodes in codesign problems, and as a surrogate for *cost*. HW nodes have a path width measured in [bits], and a delay measured in [sec]. BW is expressed in Eq. (2.3).

$$B = \frac{\text{path width } [bits]}{\text{delay } [sec]} \tag{2.3}$$

- *BW used* by a single HW node is written B^u [bits/sec] as one capacity definition. Also, B^u [ops or inst/sec] is required by specific codesign problems. Using Eq. (2.2), assuming $CPI_{\min} = 1$, i.e. one instruction issued per clock,

$$B = B^u_{\max} = \frac{\text{clock frequency}}{CPI_{\min}} = \text{clock frequency } [inst/sec] \tag{2.4}$$

- *Time used* in running a computation as defined with performance [sec] units as Eq. (2.5), is particularly useful to compare several computer systems running one *code*. In the form $O = T^u C$, constant load O is forced through the knothole of usable B in used-time T^u. Thus O is a pivot point for time and capacity. Expressions of T are *time-domain* (Eq. (2.5)) and of B or C are *bandwidth-* or *capacity-domain* (Eqs. (2.2), (2.3)) performance views, respectively.

$$T^u = \frac{O}{C} \tag{2.5}$$

$$C\ [\text{operations or bits/sec}] = \frac{O}{T^u} \tag{2.6}$$

- *Efficiency* is the ratio of delivered performance of one or more HW components to its best possible performance, for a given computation defined in the BW domain. Efficiency Eq. (2.7) is usually discussed globally, but Eq. (2.8) expresses a single node x. Their numerators are used as *perf* functions, so Eqs. (2.7) and (2.8) are *perf/cost* surrogates.

$$E = \frac{C}{B} \leq 1 \tag{2.7}$$

$$E_x = \frac{B^u_x}{B_x} \tag{2.8}$$

- *Bandwidth* wasted is an important diagnostic variable, and can be expressed as Eq. (2.9). Combining Eqs. (2.5), (2.7) and (2.9) for a uniprocessor gives Eq. (2.10), where $T^{\min}$ is the time the computation requires with $B^{\text{waste}} = 0$. Equation (2.10) shows that to minimize execution time, for constant O and B, B^{waste} must be minimized.

$$B^{\text{waste}} = B - C = B - EB = B(1 - E) \tag{2.9}$$

$$T^u = \frac{O}{EB} = \frac{T^{\min}}{E} = \frac{O}{B - B^{\text{waste}}} \tag{2.10}$$

- *Balance* refers to a component pair working together perfectly on a computation, and *imbalance* to one component slowing the other. Imbalanced computation is sometimes referred to as *component-saturated* or *component-bound* computation. Balance is closely related to capacity (see Definition 2.1, Sect. 2.3.1).

Generally, capacity and BW-domain analysis are useful for machine design insight, and time-domain analysis is useful for comparing computational performance. Throughout, we assume fixed O across changes to architectures and compilers—this is an idealization, especially for parallelism which often leads to redundancy.

2.1.2 Performance Background Summary

Increasing the system clock frequency may boost a computation's performance directly (unless e.g. latency limits BW), for a fixed architecture. For a constant clock, architecture, BW, SW, or a code's data set affects performance. An important premise for the following is that an initial computer system design is in place for detailed analysis and improvement. The paper discusses methods for improving various bottlenecked parts of a given HW/SW system.

2.2 Codesign Process

Codesign has been used by computer designers with many meanings. We have two symmetrical reasons for using the term: our overall *goal* is to select HW component speeds, compiler and application structures, and to synthesize specific architectural components. Also, we *drive* the process using comprehensive measurements of applications running on existing architectures. We use global HW and SW data together, to improve both HW and SW in the codesign process.

The goals of codesign using global architecture and SW measurements are to avoid design instabilities and regressions by exploring all important aspects of the design space in advance. The codesign process begins with an existing architecture or design, and existing/proposed applications, to produce *optimal* designs. However, when major performance increments are required to meet design goals, architectural synthesis steps are introduced (Sect. 2.7).

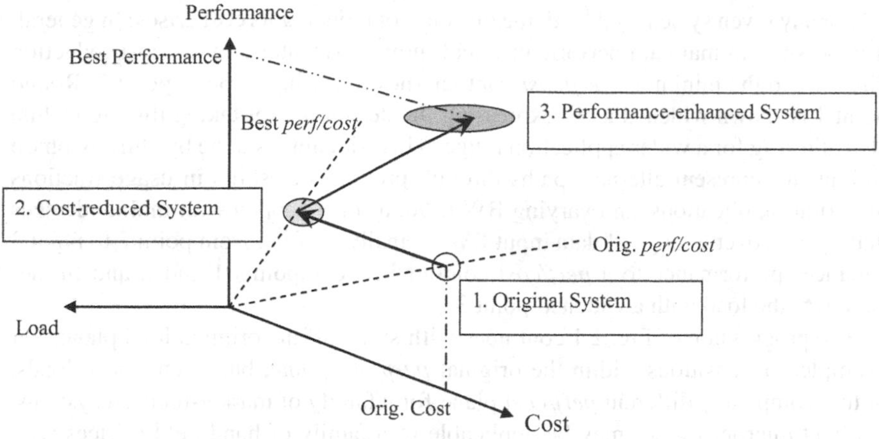

Fig. 2.1 Load normalized to cost-performance plane

2.2.1 Three Dimensions of Codesign

Performance, cost and load are the three dimensions considered in capacity-based codesign. The *codesign process* can accept each as input, and produce each as output; parameters not specified as input are computed as process output.

Performance: As input, a designer chooses specific overall goals or bounds on the performance of specific nodes. The codesign process computes the performance for nodes with unspecified goals.

Cost: A designer can choose specific overall cost goals or bounds for particular nodes; the process computes costs for all other nodes. Cost units can be defined flexibly (to include power, $, etc. as above); here total BW is a cost surrogate.

Load: This is determined by expected market usage of each application and may be hard to specify. The codesign process can reflect variation in BW used per code type (codelet, see Sect. 2.2.2.1) and weights of percent codelet usage. As output, the process can describe performance and cost ranges, when driven by inputs of load uncertainty.

These three dimensions are considered simultaneously here, not incrementally as in current practice. The codesign process we describe cannot be fully automated. Ultimately, designer interaction is integral to the codesign process, e.g. feasible engineering choices for HW nodes and BW values. The key contribution of this approach, however, is to automate far more of the design process than is currently possible without the capacity-based model.

The three codesign dimensions are shown in Fig. 2.1, which illustrates several important points in the space. Suppose an existing design lies at point 1 in Fig. 2.1. We position the origin of the performance/cost plane in codesign 3-space at the SW load's initial measurement. If the load is constant, cost-performance tradeoffs can be made in the plane shown. Two important codesign transitions within that plane are shown as points 2 and 3.

For any given system and load, the question of reducing its cost arises. In general, it is possible to maintain performance and move horizontally in the cost-reduction direction to the minimum *perf/cost* contour shown passing through point 2. Round point 1 is shown widened a bit, expressing the desirability of taking this step while also allowing for a wider application range. This widening is done by shifting phase weights to represent alternate paths through programs or shifts in usage fractions of various applications, and varying BW used, as may happen with mobile devices that operate over a range of data input BWs. Finally, shifting from point 2 to point 3 enhances performance to a *perf/cost* contour between points 1 and 2 and further enhances the load with a widened point 3.

The progression of Fig. 2.1 continues with shifts off the original load plane. For example, to transitions within the original *perf/cost plane*, but to enhanced loads, or to a completely different *perf/cost* plane for a family of market-focused systems. A client microprocessor may be applicable to a family of hand-held devices with new application loads, but only after re-engineering by changing design BW goals as well as load characteristics—i.e. using new sets of codelets in the codesign process.

2.2.2 *Models*

The models used for codesign must cover the complete range of HW and SW uses. Each codesign problem considers a system of linked HW nodes (a HW connection exists from each node to at least one other system node); similarly we consider directly or indirectly linked SW units. A single application is assumed to be control/data dependence linked. To include multiprogrammed systems, applications that share HW concurrently can be regarded as indirectly linked.

2.2.2.1 SW Models

SW *load* on a system will be represented by a collection of computational phases that exhibit steady-state B^u on each HW node. Transient B^u behavior is captured by phase transitions. We represent phases by canonical patterns called *codelets*. The two codelet parameters of interest are B^u and weights representing percent-used of total computation time. Ideally, this method will be used to represent every important computation in the design space of interest, in contrast to the benchmark suites frequently used as simplifications. We define the key SW terms:

Codelet: The term *codelet* is used to represent a small, parameterized segment of code that has properties useful in codesign. Ideal codelets will be discussed here, but the codesign process can proceed using only approximations of the ideal, as will be discussed later. The characteristics defining codelets and their use are:

1. Each codelet has approximately *steady-state* B^u, constant B^u provides ideal codesign results

2. Adequate *dynamic coverage* of important applications
3. Maximum *codelet reuse* across applications—minimal number of codelet classes stored in repository
4. Codelets are *compiler recognizable* and achieve *optimized performance* via compilation—codelet size is small enough to analyze automatically, but large enough for maximal system performance.

Phase: A *phase* is a sequence of executed instructions with relatively *uniform* B^u throughout each phase executed for each modeled system component. In practice, a phase will always be measured dynamically in 100s to 1000s of instructions, as B^u *uniformity* cannot be observed at the finest granularity. A single phase may represent a single simple algorithm in a monoprogrammed system, or several simple algorithms in a multiprogrammed system.

Computation: A *computation* is a sequence of phases. Any idle time encountered in a running computer is ignored.

SW Modeling Objective: The objective of *SW modeling* is to map phases discovered in real application computations to codelets in a codelet repository used with a range of data sets per codelet.

The codesign process does not depend critically on finding ideal codelets, but the accuracy and optimality of results will erode as the load is represented by cruder approximations of ideal codelets. For example, a special-purpose design for a particular algorithm whose computation consists of a single phase simplifies the codesign process; highly efficient designs are possible in this case (Sect. 2.4.4). When computations include more phases/codelets, *phase transitions* from one phase to another tend to cause performance sensitivity and can lead to instabilities (Sect. 2.6.2).

2.2.2.2 Computer System HW Models

HW is represented by *nodes* with peak BW values; this can be extended beyond BW to include power consumption, physical area, $ cost, and so on. *Fidelity* [2] in a given system refers to the node resolution on which a designer chooses to focus, e.g. at a high level one could choose a microprocessor, memory, bus, and network as nodes, and combine these four nodes with a large set of nodes representing individual disks and other I/O devices and controllers. The nodes can be chosen at any fidelity and nodes of various fidelities may be used in one model.

We represent a computer system *architecture* by a *graph* consisting of nodes denoting computer system HW components chosen at any fidelity level, connected by *arcs*, denoting only graph connectivity, i.e. arcs have infinite BW, zero-delay and bidirectional capability. Multiple arcs incident to a node are multiplexed/demultiplexed to and from the node. The following are several types of *nodes*.

Linear nodes: Linear nodes correspond to most low-level system components (e.g. register or arithmetic op). As in Eq. (2.3), their path width is measured in bits and delay is measured in time. Node x has *linear behavior* if its BW is constant and its capacity is a linear function of other node capacities (Eqs. (2.29a), (2.29b)).

Nonlinear nodes: As the modeling abstraction level rises, to vector processors, caches or multicore chips, modeling is harder. A *nonlinear* node has a BW that is a nonlinear function of its linear sub-nodes' capacities. Nonlinear node BW depends on the computation being done, as well as the point at which performance is measured, see [3].

Latency: Since *latency* is delay, the denominator of Eq. (2.3) can be used as a latency surrogate. HW components with BW and latency design issues are represented as two nodes. A memory unit with BW determined by read/write (r/w) cycle time, whose latency depends on wire length (e.g. off-chip delay), is represented by a memory BW supernode (Sect. 2.7) containing two linear nodes: latency and r/w BW. The total delay in Eq. (2.3) is $t_{\mathrm{lat}} + t_{\mathrm{r/w}}$. For word size w, we have Eqs. (2.11), and (2.12). B_{mem} is not a linear function of its constituent BWs, latency or r/w time. In a plot of C_{mem} vs. $C_{\mathrm{r/w}}$, for a given $B_{\mathrm{r/w}}$, varying latency (e.g. disk rotation time) defines a C_{mem} family of load-based B_{mem} values.

$$B_{\mathrm{mem}} = \frac{w}{t_{\mathrm{lat}} + t_{\mathrm{r/w}}} \tag{2.11}$$

$$B_{\mathrm{mem}}^{-1} = B_{\mathrm{lat}}^{-1} + B_{\mathrm{r/w}}^{-1} \tag{2.12}$$

2.2.3 Model Philosophy and Codesign Realities

Several realities separate linear analysis and linear performance response from the real design world. These range from the nonlinearities of HW nodes, through SW variations, to measurement issues, as discussed throughout the paper.

The flow of performance data ranges from continuous (server workloads) to irregularly discrete (laptop use: computation bursts commingled with idle time), and data measurement qualities range from nearly ideal (simulator probes) to crude (Microsoft process-level tools). But if we regard the modeling process as a linear scaling of input data, then even crude approximations can be effectively scaled, if an inverse process exists to carry computed design results faithfully back to the input space. In general, *virtual* HW nodes may be used, if based on appropriate measurement and analysis [2]. For example, a page fault node can be used if there is a way of translating results to and from memory system design (i.e. by virtualizing and devirtualizing the instruction stream measurements).

Figure 2.2 outlines the process of system modeling and obtaining a capacity-based design. The two parameters needed for linear modeling are BW and capacity for each linear node in the model. The details of determining node BW may vary (theoretical peak, microbenchmarking, etc.) but as long as we interpret the results similarly, the method doesn't depend on specific choices. Also, capacity must first be measured and then interpreted for resulting designs.

Mapping is straightforward for linear nodes, following Sects. 2.3, 2.4, and 2.5, and the results are easily mapped back to real hardware. Nonlinear nodes can be dealt with using nonlinear models.

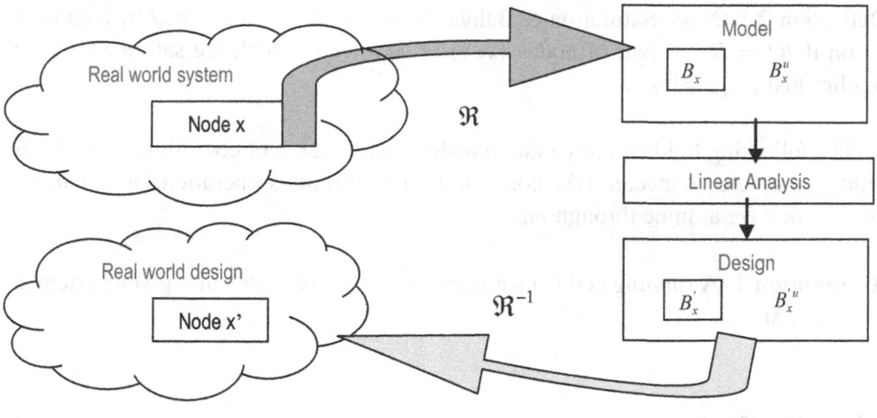

Fig. 2.2 Capacity codesign process

2.3 Linear Computational Capacity Theory

Capacity analysis can be defined on a per computation basis across a system: some simple examples follow.

2.3.1 General Equations for Single Phase Computations

2.3.1.1 Capacity Definitions

For a given system, we define the *computational capacity* of a node pair $\langle x, y \rangle$ as its effective processing bandwidth (BW), either x- or y-*bound*. The physical definitions of Sect. 2.1.1 will not be detailed further, as specific engineering abstraction and measurement of the real world is required for each codesign problem. The BWs of system nodes x and y are represented by B_x and B_y, respectively, according to some abstraction of the system. B_x and B_y represent amounts of those BWs actually used in a computation, by nodes x and y, respectively, giving Eqs. (2.13).

$$0 < B_x^u \leq B_x \quad \text{and} \quad 0 < B_y^u \leq B_y \tag{2.13}$$

We define the *x-y BW ratio* in Eq. (2.14) and the *x-y used-BW ratio* for a given steady-state computation in Eq. (2.15)

$$\alpha_{x,y} = \frac{B_y}{B_x} \tag{2.14}$$

$$\mu_{x,y} = \frac{B_y^u}{B_x^u} = \frac{1}{\mu_{y,x}} \tag{2.15}$$

Definition 2.1 (Node Saturation & Balance) Any *node x is saturated* by a computation if $B_x^u = B_x$. A pair of nodes $\langle x, y\rangle$ is *balanced* if both are saturated, which implies that $\mu_{x,y} = \alpha_{x,y}$.

The following holds for any system under a wide range of conditions (e.g. nodes reflect complete connected HW configuration, continuous operation, no deadlock, etc.), which we assume throughout.

Assumption 1 A running codelet saturates one or more nodes of a given system at each time step.

2.3.1.2 Two Node Systems

Without loss of generality, we analyze the two node system from the point of view of node x.

Definition 2.2 (Computational Capacity with Saturated Node) The *computational capacity of node x* is defined as Eq. (2.16), so from the above we have Eq. (2.17).

$$C_x = B_x^u \tag{2.16}$$

$$C_x = \begin{cases} B_x & \text{if } B_x^u = B_x \quad \text{node } x \text{ saturated} \\ B_x^u & \text{if } B_x^u < B_x \quad \text{node } x \text{ unsaturated} \end{cases} \tag{2.17}$$

In terms of the activity on node y, the second case can be rewritten assuming that node y is saturated, as Eq. (2.18), which we define as the *computational capacity of node x relative to saturated node y*, Eq. (2.19). Because B_x and B_y are defined as non-zero (Eq. (2.13)), capacity is defined only for a pair of nodes that are both actually used a computation. Summarizing, for a pair of nodes, at least one of which is saturated,

$$B_x^u = \left(\frac{\alpha_{x,y}}{\mu_{x,y}}\right) B_x = \mu_{y,x} B_y, \quad \text{for } B_y^u = B_y \tag{2.18}$$

$$C_x = \mu_{y,x} B_y = B_x^u \tag{2.19}$$

Saturated Node Capacity

$$C_x = \begin{cases} B_x & \text{if } B_x^u = B_x \text{ and } (B_y^u = B_y \text{ or } B_y^u < B_y), \text{ i.e. } \alpha_{x,y} \geq \mu_{x,y} \\ \alpha_{x,y} B_x / \mu_{x,y} = \mu_{y,x} B_y & \text{if } B_x^u < B_x \text{ and } B_y^u = B_y, \text{ i.e. } \alpha_{x,y} < \mu_{x,y} \end{cases} \tag{2.20}$$

2.3.1.3 Greater than Two Node Systems

Systems of more than two nodes may lead to neither node x nor node y being saturated, unlike the above discussion. This can happen because a system need only have a single saturated node (Assumption 1). Analysis of a multinode system can be built from node-pair analysis, and if one of a pair is saturated, the analysis proceeds as above. Otherwise, a pair of nodes x and y, neither of which is saturated, gives Eq. (2.21). In this case, it is impossible to bound the $\alpha_{x,y}/\mu_{x,y}$ ratio relative to 1 as for the saturated node case in Eq. (2.20). Following Definition 2.2, and since node x is

$$B_x^u < B_x, \quad \text{and} \quad B_y^u < B_y \tag{2.21}$$

$$C_x = B_x^u \quad \text{for } B_x^u < B_x \tag{2.22}$$

unsaturated (Eq. (2.21)) we write Eq. (2.22). Furthermore, expanding Eq. (2.15),

$$B_x^u = \frac{B_x^u}{B_y^u} B_y^u = \mu_{y,x} \quad \text{for } B_y^u < B_y \tag{2.23}$$

Combining Eqs. (2.22) and (2.23), analogously to Eq. (2.19), we have

Definition 2.3 (Computational Capacity with No Saturated Node) The *computational capacity of unsaturated node x relative to unsaturated node* y is defined as

$$C_x = B_x^u = \mu_{y,x}\, B_y^u \quad \text{for neither node saturated.} \tag{2.24}$$

Combining (2.22) and (2.24) we have Eq. (2.25).

Unsaturated Node Capacity

$$C_x = \begin{cases} B_x^u & \text{if } B_x^u < B_x \\ \mu_{y,x} B_y^u & \text{if } B_x^u < B_x \text{ and } B_y^u < B_y \end{cases} \tag{2.25}$$

Comparing Eq. (2.20) and (2.25) we see that Bs in Eq. (2.20) become B^us in Eq. (2.25), and the condition $B_y = B_y$ in Eq. (2.20) becomes $B_y < B_y$ in Eq. (2.25). Finally, as nodes x and y saturate, Eqs. (2.20) and (2.25) become identical.

2.3.1.4 General Two-Node Capacity Rule

Combining Eqs. (2.20), (2.22), and (2.25), the general rules are summarized in Table 2.1 (Rule 4 is appended for completeness). At this point the relation between B^u and C can be clarified. Definition 2.2 sets them equal, and in subsequent sections we will generalize the notion of capacity to nodes with multiple connections. However, we use both terms to distinguish contexts. B^u refers to the empirical measurements that are used to define μ (Eq. (2.15)), while C terms are variables used

Table 2.1 Two-node capacity rules

Capacity Equations	Conditions B_x	B_y	Rule
$C_x = B_x = \mu_{y,x} B_y$	saturated	don't care	1
$C_x = \mu_{y,x} B_y$	unsaturated	saturated	2
$C_x = \mu_{y,x} C_y$	unsaturated	unsaturated	3
$C_y = \mu_{x,y} C_x$	unsaturated	unsaturated	4

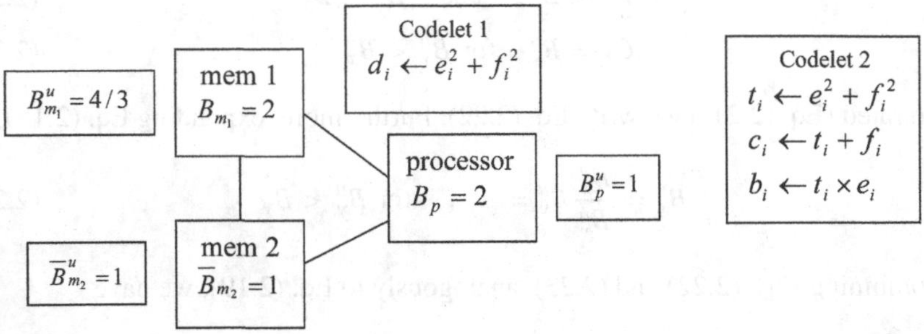

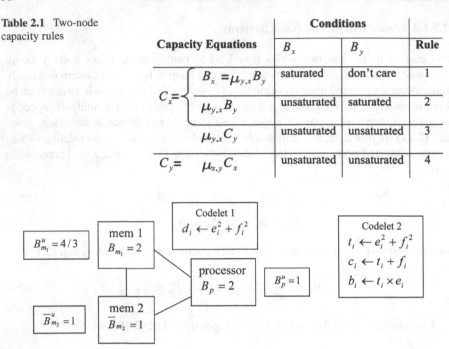

Fig. 2.3 Processor and heterogeneous memory system

to define capacities, (i.e. BW used) in new designs. Finally, μ values are held *invariant* throughout this paper (cf. Sect. 2.5.3). In other words, while B^u values may change when a given program is run on various machines, once a μ value is computed for a computation on any machine, that SW property remains invariant across all machines.

Definition 2.4 (Relative Saturation and Saturation State) The *n-node relative saturation vector* is $\underline{\sigma} = (\sigma_1, \ldots, \sigma_n)$ where $0 < \sigma_i = C_i/B_i = E_i \leq 1$. If neither node of the pair $\langle x, y \rangle$ is saturated and $\sigma_x > \sigma_y$, node x is *relatively saturated* to node y, $\sigma_{x,y} = \sigma_y/\sigma_x < 1$. $\underline{\sigma}^s$ is called the *saturation state vector*, where $\sigma_i = 1$ if node i is saturated, and 0 otherwise. $\underline{\sigma}^s$ varies with a design's B values.

2.3.2 Example: Single Processor-Heterogeneous Memory

To illustrate the use of the theory of Sect. 2.3.1, consider the simple machine model in Fig. 2.3, with two memory units and one processor. This could be an abstraction of a system with a register set, memory and arithmetic unit, for example. The nodes are marked with BW values, and the codelets shown are assumed to execute in a loop indexed by i. Architectural assumptions play an important part here, so we sketch those used in this example.

Counting clocks in a cycle-level execution diagram for this model can give B^u values by simply counting cycles used in a periodic instruction pattern. Assume that data are initially stored in mem 2. Codelet 1 execution starts by fetching e_1 and f_1 from mem2, and writing them one clock later, respectively, in mem1. When both arguments are in mem1, processing begins with two multiplies followed by an add. The d_1 result is then written back directly to mem2, completing the first iteration of the codelet execution. Other iterations are overlapped, so eventually a steady state execution pattern emerges in a 3-clock cycle.

In practice, these numbers would be collected from running computations on a real system using hardware performance monitoring tools, or could be collected from a simulator for an emerging design. B^u boxes correspond to the system load, codelet 1. The overbar on $\overline{B^u_{m_2}}$ indicates mem2 saturation. If this system were improved by increasing B_{m_2}, mem1 would saturate before the processor because $\sigma_{m_1} > \sigma_p$. Codelet 2 will be discussed in Sect. 2.6.1.

2.4 Single Phase Codesign Equations

2.4.1 Capacity Equation Generation

For a single codelet, equations may be generated for each arc in a graph to capture all BW and capacity information, but this is not necessary. The $(n - 1)$ arcs in a spanning tree for an n-node graph, is the minimum needed to capture each node in relation to some other node, but no equations need to be written for arcs that close cycles in the graph.

2.4.1.1 Capacity Equation Generation Algorithm

This algorithm suffices to capture the capacity equations for any graph.

Step 1. Start with any node and build a spanning tree for the entire graph from that root.

Step 2. Write capacity equations for each arc in the spanning tree using Table 2.1. As saturation patterns for a new design are unknown, codesign model equations (Sect. 2.4.3) use only Rule 3 in either orientation. (*Initial equations* can capture the original system and computation by using all three rules, for 2, 1, and 0 nodes saturated, respectively.)

Step 3. This generates $n - 1$ equations for a graph of n nodes, as there is one equation per arc in the spanning tree. If we think of capacity as a nodal relation on a spanning tree, *transitive relations* may be formed between non-adjacent nodes in the graph. This idea and Definition 2.4 are useful in analyzing the sensitivity of solutions (as in the example of Sect. 2.3.2) or in expediting specialized solutions for specified nodes.

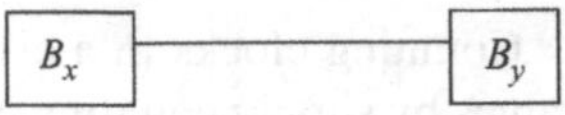

Fig. 2.4 Two-node graph

2.4.2 Codesign Equations

We assume that μ values are constant, while B and C values are variables, i.e. that we are designing HW (B values) to suit given SW (μ) needs. In Sect. 2.5.3, the design of SW in terms of HW will be discussed briefly.

Physical constraints as well as capacity equations are needed to form a complete design equation set. This section contains examples of linear capacity and physical equations that describe a computer system and computation. The nodes can represent any fidelity, but at a high level one may choose multirate nodes whose behaviors vary, depending on their load. Multirate and nonlinear nodes will be discussed further in Sect. 2.7. For this section, it suffices to assume that nodes are chosen at a level that allows linear performance equations to hold.

Physical Constraints It is generally true that for any node x, $0 < B_x^u \leq B_x$. This follows from obvious physical considerations, and because capacity is defined only for nodes where $B^u > 0$ (Eq. (2.13)). To formulate general equations, we represent unknown B^u by C variables, so we rewrite $0 < C_x \leq B_x$ as two inequalities which must be satisfied in all solutions:

$$B_x - C_x \geq 0 \tag{2.26}$$

$$C_x > 0 \tag{2.27}$$

2.4.3 Single Phase Models and Characteristic Equation

We introduce the form of the general design equations by starting with a single phase running on simple systems.

Two Node Systems Figure 2.4, the two node graph case, with saturated node x, has the initial capacity

$$\mu_{xy}\, B_x - C_y = 0 \tag{2.28}$$

Equation (2.28) using Rule 2 of Table 2.1 (rewriting $C_{yx} = C_y$). To cover all possible BW values in system designs executing this computation, Fig. 2.4 is represented by either of Eqs. (2.29a), (2.29b) (Rule 3 Table 2.1) as in general, given a constant μ value, either node x or node y may be saturated in a solution for particular B values (Eq. (2.29a) reduces to Eq. (2.28) if node x is saturated). For any two node system with $B_x = B_y$ and $B_x^u = B_y^u$, node x will saturate in a *new design* with $B_x \ll B_y$, while node y will saturate if $B_x \gg B_y$. Equation (2.29a) for constant μ_{xy} defines x as a *linear node* relative to other nodes y. For $B_x > B_y$, Fig. 2.5 shows the

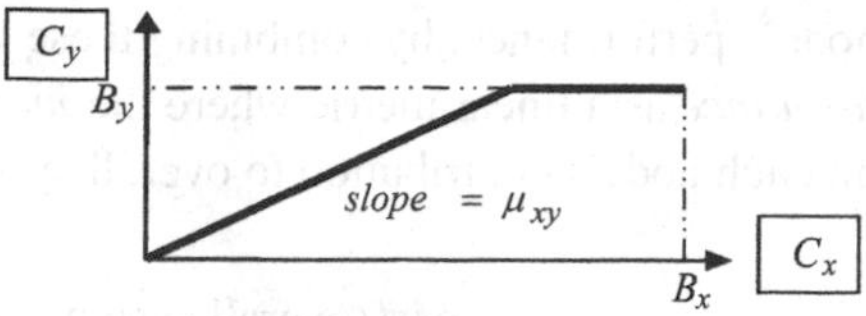

Fig. 2.5 Linear node-pair capacity

behavior of Fig. 2.4, beginning with the slope of Eq. (2.29a) (Rules 3, 4; Table 2.1), then $C_y = B_y$ when B_y saturates (Rule 1; horizontal break), and becomes undefined for $C_x > B_x$ (no Rule). Since $B_x > B_y$, Rule 3 applies until $C_x > B_x$. A spanning tree has $n - 1$ nodes, so Fig. 2.4 has one capacity equation. Following Eqs. (2.26) and (2.27), the *physical equations* are Eqs. (2.30) and (2.31).

$$C_y = \mu_{xy} C_x \tag{2.29a}$$

or

$$C_x = \mu_{yx}\, C_y \tag{2.29b}$$

$$B_x - C_x \geq 0, \quad \text{and} \quad B_y - C_y \geq 0 \tag{2.30}$$

$$C_x > 0, \quad \text{and} \quad C_y > 0 \tag{2.31}$$

We combine these in the *single-phase model system*, Eq. (2.32) as the product of computational parameter matrix M containing parameters μ, 0 and ± 1, and *design vector d*, partitioned into *b* and *c*. *Capacity vector c* corresponds to the performance of a given design, measured in *B*, and *bandwidth vector b* represents the cost of obtaining that performance, measured in *B*. The positions of equality and inequality signs in Eq. (2.32) denote numbers of equalities (starting at =) and inequalities (starting at $\leq$), and 0 is a zero column.

$$M\underline{d} = M\left[\frac{\underline{b}}{\underline{c}}\right] = \begin{bmatrix} 0 & 0 & \mu_{xy} & -1 \\ 1 & 0 & -1 & 0 \\ 0 & 1 & 0 & -1 \\ 0 & 0 & 1 & 0 \\ 0 & 0 & 0 & 1 \end{bmatrix} \begin{bmatrix} B_x \\ B_y \\ C_x \\ C_y \end{bmatrix} \begin{matrix} =0 \\ \geq \underline{0} \\ \\ >\underline{0} \\ \\ \end{matrix} \tag{2.32}$$

$$M = \begin{bmatrix} M_{11} & M_{12} \\ M_{21} & M_{22} \\ M_{31} & M_{32} \end{bmatrix} = \begin{bmatrix} 0 & M \\ I & -I \\ 0 & I \end{bmatrix} \tag{2.33}$$

General Systems The above discussion easily generalizes to a system of n nodes. M is a $(3n - 1) \times 2n$ matrix representing the computation. The $2n$ columns correspond to a B and C per node. The $3n - 1$ rows include $n - 1$ capacity equation rows plus $2n$ for physical inequalities, while $\underline{b}$ and $\underline{c}$ are n-element vectors representing bandwidth and capacity, respectively. We can partition M as Eq. (2.33), where M is a matrix of μ, 0 and -1 values, I is the identity, and 0 the null matrix. M_{11} and M_{12} are $(n - 1) \times n$, and the other M_{ij} are $n \times n$ matrices. As each C_i represents one node's performance, by combining these we can summarize the overall *system performance* as a linear metric where the w_i represent design-importance or emphases on each node's contribution to overall system performance.

$$perf(\text{overall system}) = C_{\text{system}} = \sum_{\text{nodes}} w_i C_i \tag{2.34}$$

From Eq. (2.32), we derive the *single-phase characteristic equation* of an n-node computation, Eq. (2.35), where the inequalities of Eqs. (2.30) and (2.31) have been augmented with one slack variable per physical equation (indicated by primes) to obtain a $(3n-1) \times 4n$ underdetermined system of equations. Identical graph topologies arising from distinct architectures lead to equations of the same nonzero patterns, but represent distinct architectural behavior via distinct μ values per position.

$$M'\underline{d'} = M'\left[\underline{b'}, \underline{c'}\right]^T = 0 \tag{2.35}$$

The characteristic equation contains complete information about performance and cost for any HW system running the single-phase computation used to generate it. The specifics of each computation are represented by μ values. The solution will have $k \geq 1$ saturated nodes. Myriad real HW systems are described by one characteristic equation, in general. Section 2.5.6, gives codesign optimizations for selecting a few practical candidate system designs.

2.4.4 Observations

Obs. SP1: For a single phase computation, it is always possible to design an n-node system with all nodes saturated.
Obs. SP2: Any one node's performance can be set to an arbitrary goal (Sect. 2.5.2) while maintaining Obs. SP1.
Obs. SP3: In any single-phase computation with some unsaturated nodes, changing the BW of saturated nodes changes system-wide performance; for any unsaturated node x, changing B_x such that $B_x < B_x$ does not affect performance.

Observations SP1 and SP2 show how effectively the codesign process can be carried out for single phase computations. They provide a heuristic justification for the many demonstrations since the beginning of computing history that HW specialized to a single algorithm can be far more cost-effective than general purpose systems. In the future, massively multicore chips could allocate substantial real estate to an extensive set of algorithm-level processors. By Obs. SP3, unsaturated node BWs can float until they all reach saturation in the form of Obs. SP1.

2.5 Multiphase Codesign Equations

2.5.1 Multiphase Model and Characteristic Equations

Adding multiple computational phases does not affect the vector $\underline{b}$ in Eq. (2.32), as the machine BW is defined by one set of nodes used in all phases. However, using the designed HW, each phase generally produces distinct performance characteristics. Thus, if a computation has m phases, the $\underline{c}$ vector becomes m times larger than for the single-phase case. We represent the collection of phase performance vectors in Eq. (2.36), where m phases are represented in Eq. (2.37).

$$M\underline{d} = M[\,\underline{b}, \underline{c}^{\text{loc}}]^T \tag{2.36}$$

$$\underline{c}^{\text{loc}} = [\,\underline{b_1}, \,\ldots\,, \underline{c_m}]^T \tag{2.37}$$

As an example, expanding on Sect. 2.4.3, we show a 2-node *two-phase model system* in Eq. (2.38). Assume that in the second phase the saturation is reversed from Fig. 2.4; y is saturated and x is not. Per phase, this gives one capacity equation; third subscripts denote phase numbers. In the M_{phy} partition of M (Eq. (2.38)), rows 3–6 are physical equations of the form $B - C \geq 0$ (Eq. (2.30)); the next four rows are an identity matrix corresponding to Eq. (2.31) for the two phases. There are n elements in $\underline{b}$ and in each $\underline{c}$ vector, for a total of $n(m+1)$ columns in M and rows in $\underline{d}$.

$$M\underline{d} = \begin{bmatrix} M_{C^{\text{loc}}} \\ M_{\text{phy}} \\ M_{C^{\text{glob}}} \end{bmatrix} \begin{bmatrix} \underline{b} \\ \underline{c}^{\text{loc}} \\ \underline{c}^{\text{glob}} \end{bmatrix}$$

$$= \begin{bmatrix} 0 & 0 & \mu_{xy,1} & -1 & 0 & 0 & 0 & 0 \\ 0 & 0 & 0 & 0 & -1 & \mu_{yx,2} & 0 & 0 \\ 1 & & -1 & & & & 0 & 0 \\ & 1 & & & & -1 & 0 & 0 \\ 1 & & & & -1 & & 0 & 0 \\ & 1 & & -1 & & & 0 & 0 \\ & & 1 & & & & 0 & 0 \\ & & & 1 & & & 0 & 0 \\ & & & & 1 & & 0 & 0 \\ & & & & & 1 & 0 & 0 \\ 0 & 0 & \phi_1 & 0 & \phi_2 & 0 & -1 & 0 \\ 0 & 0 & 0 & \phi_1 & 0 & \phi_2 & 0 & -1 \end{bmatrix} \begin{bmatrix} B_x \\ B_y \\ C_{x,1} \\ C_{y,1} \\ C_{x,2} \\ C_{y,2} \\ C_x^{\text{glob}} \\ C_y^{\text{glob}} \end{bmatrix} \begin{matrix} = 0 \\ \\ \geq \underline{0} \\ \\ \\ \\ \geq \underline{0} \\ \\ \\ \\ = \underline{0} \\ \\ \end{matrix} \tag{2.38}$$

In general, we will write a block of equations as above for each of m phases, so from Eq. (2.32), M is an $m(3n-1) \times (m+1)n$ matrix. Some nodes may not be used in a particular phase, e.g. no disk accesses are made, so, as capacity C_{xy} is undefined for unused node y (Eq. (2.13)), it may be dropped from the equation set for that phase. In solving such systems, we can drop nodes for phases whose use approaches the machine's zero value. This leads to a linear system where each phase is reduced in size to represent those nodes active per phase. Generalizing Sect. 2.4.3 with slack variables leads to an underdetermined system, as each phase is underdetermined.

Global component performance in the multiphase case is determined by contributions from each phase. Each phase has a weight ϕ_j, $1 \leq j \leq m$, defined by some combination of the running time of an application segment, the importance of an application containing the segment, etc. The *multiphase global performance* of a HW node x is Eq. (2.39). The last two global performance rows of the model system Eq. (2.38) correspond to Eq. (2.39), for each node, x and y. For an n-node system, this adds n more rows and columns to Eq. (2.38), so M is an $[m(3n-1)+n] \times [(m+2)n]$ matrix. Equation (2.38) yields all BW and performance values for a computation on a computer system, given the μ ratios and ϕ weights.

$$\textit{global perf}(\text{node } x) = C_x^{\text{glob}} = \sum_{\text{phases } j} \phi_j C_{x,j} \tag{2.39}$$

The ϕ_j phase-weights are functions of node BW in general, because running times of individual phases may vary relative to each other based on specific node BWs. Boosting the BW during the design process of node x, unsaturated in phase k, can reduce the running time of phases for nodes that were B_x-bound. In principle, we should readjust the $\phi_j(x)$ values for all nodes (or the most sensitive nodes). As the sizes of changes may be small, and ϕ_j also depends on other qualitative weights, we avoid the complexity of varying ϕ_j, which could be done iteratively.

For multiphase systems, we expand the characteristic equation (Eq. (2.35)) to include global performance equations, and augmenting Eq. (2.38) with $2mn$ slack variables yields Eq. (2.40), an underdetermined *multiphase characteristic equation*, where M is a $[(3m+1)n-m] \times (3m+2)n$ matrix.

$$M'\underline{d}' = M'\left[\underline{b}', \underline{c}'^{\text{loc}}, \underline{c}'^{\text{glob}}\right]^T = 0 \tag{2.40}$$

2.5.2 Codesign Equation

For system design, we rewrite Eq. (2.38) as the *multiphase codesign equation*, Eq. (2.41). Chosen coefficients of $\underline{b}$ and $\underline{c}^{\text{glob}}$ are removed from M and $\underline{d}$, and corresponding positions are set to constant goal values in $\underline{g}$ with appropriate sign changes. The codesign equation can be used to set any BW, capacity or weight goals a designer chooses to target. This reduces the number of columns of M and the size of $\underline{d}$ by g, the number of elements moved to $\underline{g}$. A codesign study may select $\leq n$ global performance equations of the form Eq. (2.39), so $[m(3n-1)+p] \times [(m+2)n-g]$, $p \leq n$ is the size of M. In a codesign study, the $\underline{g}$ values may be chosen to sweep out regions of the overall design space to find optimal designs. See Sect. 2.6.

$$\boxed{M\underline{d} = M\begin{bmatrix}\underline{b}\\ \underline{c}^{\text{loc}}\\ \underline{c}^{\text{glob}}\end{bmatrix} \begin{matrix}=\\ \geq\\ \geq\end{matrix} \begin{bmatrix}\underline{b}^{\text{goal}}\\ 0\\ \underline{c}^{\text{goal}}\end{bmatrix} = \underline{g}} \tag{2.41}$$

2.5.3 Software Design Equations

Although it will not be explored in any detail here, the model Eq. (2.38) has a second interpretation. This paper assumes BWs are variables and μ values are constants. Inverting these relationships, assume a fixed HW system, and instead of Eq. (2.38), write Eq. (2.42), where B is a matrix of BW values and $\underline{\mu}$ is a vector of measured or unknown μ values. Setting capacity and μ *goals* allows the computation of C and μ values, by reasoning analogous to the ideas of this paper. A key difference of SW tuning for given HW, from HW design is that via $\underline{c}^{\text{loc}}$, each phase can be tuned to achieve desired C and μ values, allowing more degrees of freedom than choosing node BW values. It can be viewed as a way of tuning codelets to specific applications. Substantial tuning work may be involved, but performance targets are guaranteed if it succeeds. Some performance tuning methods related to the equations of this paper (ranging from basic capacity to sensitivity analysis) are outlined in Sect. 2.8.

$$B\underline{d} = B\left[\underline{\mu}, \underline{c}^{\text{loc}}, \underline{c}^{\text{glob}}\right]^T \tag{2.42}$$

2.5.4 Multiphase Performance Observations

Obs. MP1: Generally, no system can have all nodes remain saturated throughout a multiphase computation.

Obs. MP2: For a given computation on a well-designed system, each phase saturates some node(s); and each node will be saturated by at least one phase.

Obs. MP3: Linearly scaling all linear model BWs by factor a scales all phase and global capacities by a factor of a.

These observations raise a question. Since we cannot achieve saturation of all nodes throughout a multiphase computation, how should a good design be defined? As discussed earlier, the *benefits* of a design are represented by capacities, and the *costs* of obtaining capacities are BWs. A desirable goal is minimizing *BW wasted per node*, $B_i^{\text{waste}} = (B_i - C_i)$, summed across all nodes. It is exactly the achievement of all-node saturation that is made by optimized single phase computations, Obs. SP1. From Eq. (2.10) minimizing BW waste is equivalent to minimizing time, overall.

2.5.5 Overall System Optimization

In the multiphase case, using Eq. (2.39) the overall system *performance* corresponding to Eq. (2.34) is Eq. (2.43). For n nodes and m phases, the overall system *cost* is Eq. (2.44), so the objective of minimizing total *wasted BW*, B^{waste}, Eq. (2.45), is the difference between Eqs. (2.43) and (2.44). In a codesign problem, certain BW values and performance goals may be chosen a priori, using Eq. (2.45) to minimize the remaining unknown *cost* (BW) values.

$$\textit{perf} \text{ (overall system)} = C_{\text{system}} = \sum_{\text{nodes}} w_i \sum_{\text{phases}} \phi_j C_{i,j} \tag{2.43}$$

$$\textit{cost} \text{ (overall system)} = B_{\text{system}} = \sum_{\text{nodes}} B_i \tag{2.44}$$

The system codesign problem formulation thus becomes the optimization problem of minimizing an objective function (Eq. (2.45)), subject to a set of linear constraints (Eq. (2.41)), a *linear programming* formulation of system codesign. Evaluating simplex solutions at many design points can lead to nonlinear surfaces in codesign space (see Sect. 2.6).

$$\boxed{\min B_{\text{system}}^{\text{waste}} = \min \sum_{i=1}^{n} \left(B_i - w_i \sum_{j=1}^{m} \phi_j C_{i,j} \right)} \tag{2.45}$$

2.5.6 Sensitivity Analysis

Several sensitivities are of interest in codesign studies.

Definition 2.4 (continued) The *n-node m-phase* generalization of relative saturation state (Definition 2.4, Sect. 2.3.1) is the $n \times m$ *saturation matrix* $\sum = [\underline{\sigma}_1, \ldots, \underline{\sigma}_m] = [C_{i,j}/B_i] = [E_{i,j}] = [\sigma_{i,j}]$, where $\underline{\sigma}_j^T$, $1 \le j \le m$ has the form of Definition 2.4. Each set of Σ values defines a computation's *saturation state* Σ^s, where $\sigma_{i,j} = 1$ for saturated node i in phase j, and 0 otherwise. Σ^s varies with phases as well as the B values chosen in each design.

The collective performance- or cost-sensitivity of a given computer system and computation set can be analyzed by examining the saturation matrix. We define *capacity sensitivity*, relating all nodes i to any particular node y in phase j, as Eq. (2.46). Similarly, *bandwidth sensitivity* relating all nodes i to any particular node y in phase j is defined as Eq. (2.47). Using Definition 2.4 for $y = i$, the relation between Eqs. (2.46) and (2.47) is $\Sigma'_B = E \otimes \Sigma'_C$; $\otimes$ is the Hadamard product.

$$\boxed{\Sigma'_{Cy} = \left[\frac{\partial \sigma_{i,j}}{\partial C_{i,j}} \right] = \left[\frac{\mu_{yi,j}}{B_i} \right] = \left[\sigma'_{Cyi,j} \right]} \tag{2.46}$$

$$\Sigma'_{By} = \left[\frac{\partial \sigma_{i,j}}{\partial B_i} \right] = \left[-\mu_{yi,j} C_{y,j} / B_i^2 \right] = \left[\sigma'_{Byi,j} \right] \tag{2.47}$$

Σ'_B offers a view of the most effective BW changes to reduce a design's relative saturation (or efficiency) sensitivity, across all node BWs in the codesign process.

Σ'_C provides a similar effect in reducing the sensitivity of a key node across the capacity sensitivities of all other nodes. Both could be phase-weighted in practice.

This section shows that capacity sensitivity, Eq. (2.46), a function of the model system, joins the model system and codesign optimization equations as a third key element in understanding the codesign problem.

2.6 Using the Codesign Model

Next we explore some codesign issues by sweeping through parts of the design space for noncontrived, simple 3-node, 2-phase examples. Realistic numbers of nodes and phases would make the picture more complex, but follow similar patterns. Sections 2.6.1 and 2.6.2 concern cost and performance sensitivity, respectively, and performance sensitivity is broken into several cases in Sect. 2.6.2. We define *performance stability* as the ratio of maximum to minimum performance over a collection of computations; when an empirical threshold is exceeded, a system is said to be *unstable*. Potential sources of performance sensitivity and nonlinearity that are discussed here and can lead to instability include variations in saturation state by discrete choices of node BW values, architecture changes, and variations in phase weights. A prototype codesign tool CAPE was used to produce the figures shown.

2.6.1 Cost Reduction

Computer manufacturers build only a discrete set of computer systems, and these are made from a discrete set of subsystems, each of which is built from a discrete set of components. For example, most sizes and speeds of memory along a continuum are not feasible design choices for real memory systems—designers use what is reasonable to fabricate in volume. A codesign tool must be constrained to choose among these engineering options. Linear analyses over ranges of discrete constraint choices create nonlinear codesign surfaces that approximate design realities.

Consider the 3-nodes of Fig. 2.3, plus a second phase codelet 2, which yields $B^u_{m_2} = 17/18$, $\overline{B}^u_{m_1} = 36/18 = 2$, and $B^u_p = 22/18 = 11/9$. In contrast to phase 1, mem1 is saturated here, so $\mu_{m_1,m_2,2} = B^u_{m_2}/B_{m_1} = 0.944\ 2 = 0.472$, and $\mu_{m_2,p,2} = B^u_p/B^u_{m_2} = 1.22/0.944 = 1.29$. Assuming that $\phi_1 = \phi_2 = 0.5$, we can find minimum cost solutions among discrete BW values, satisfying engineering codesign constraints. We define system performance in terms of the processor as $C^{\text{glob}}_p = (C_{p,1} + C_{p,2})/2 = 1.11$. Varying processor and memory BW with a step size of 0.1 to simulate engineering constraints, while maintaining original performance, Table 2.2 shows several cost-reduced solutions relative to the original $B_{\text{system}} = 5$ (Eq. (2.44)). These range from $B_{\text{system}} = 4.1$, an 18% cost reduction, to $B_{\text{system}} = 4.3$. The range of BW options covered may have important engineering consequences. For example, the mem1 and proc BWs are reduced from the original design, while mem2

Table 2.2 Low cost solutions vs. orig. cost = 5

B_{system}	C_p^{glob}		Bandwidths
Cost	m1	m2	p
4.100	1.7000	1.2000	12000
4.200	1.7000	1.2000	13000
4.200	1.8000	1.2000	12000
4.200	1.9000	1.1000	12000
4.300	1.7000	1.2000	14000
4.300	1.8000	1.2000	13000
4.300	1.9000	1.1000	13000
4.300	1.9000	1.2000	12000
4.300	2	1	13000
4.300	2	1.1000	12000

ranges up to 20% higher than the original. From a design flexibility point of view, note that for each component, this approach provides the designer with component BW choices in a 15% to 20% range. Table 2.2 is the type of tool output that brings designers' decision-making into the codesign process.

Variations of the performance goal for a single node can be satisfied by linearly scaling system cost. However, when minimum cost is sought, or two or more node performance goals change independently, the resulting cost surface can become nonlinear. For example, the most demanding performance goal can require disproportionate BW for that node compared to other nodes. We refer to this as a *nonlinear* cost function of performance.

2.6.2 Performance Sensitivities and Instabilities

We consider performance (Eq. (2.1)), in Sect. 2.6.2.1 with variable HW/architecture and constant SW/code, and in Sect. 2.6.2.2 with variable SW/code and constant HW/architecture. Exploring the solution space at phase transitions in computations reveals a source of *nonlinear performance* behavior caused by saturation state changes (Sect. 2.5.6). For each HW node i, as a program's $C_{i,j}$ changes across phase j transitions or within phases as data sets vary, the Σ row values change, and as a program is moved from one machine to another, the column values also change. C_p^{glob}, as a function of B_{m_1} and B_{m_2}, exhibits three distinct linear regions in Fig. 2.6, where three planes (breaks at dotted lines) form a performance surface viewed from below (higher is better). The heavy line shows the $B_{m_2} = 0.8$ contour. Discrete choices of node BW values and variations in phase weights are other potential sources of performance sensitivity and instability (in cases of extreme sensitivity).

Figure 2.7 shows a C_p^{glob} vs. B_{m_1} slice through the surface of Fig. 2.6 for $B_{m_2} = 0.775$, cutting across three regions. This value illustrates some difficulties of doing

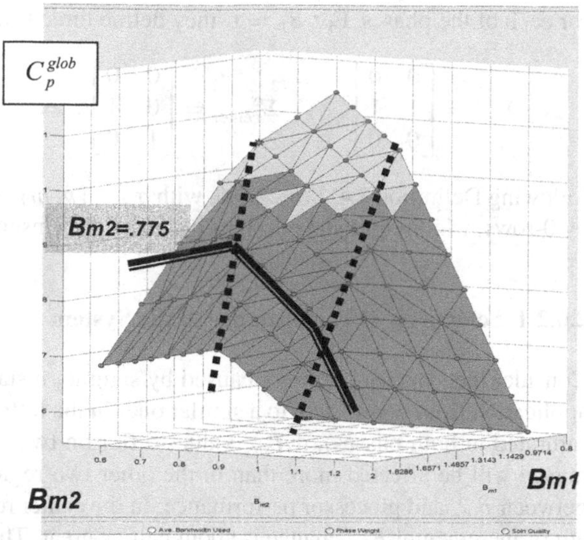

Fig. 2.6 Processor performance vs. memory BW

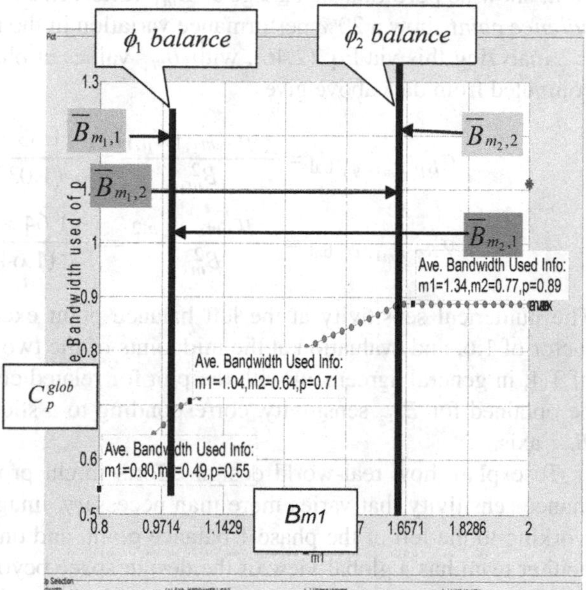

Fig. 2.7 Processor performance vs. B_{m_1}, showing perf regions; $B_{m_2} = 0.775$

reduced-performance and cost redesign of the original system. The * on the right is the original design point, with $C_p^{\mathrm{glob}} = 1.11$. The $\langle m_1, m_2 \rangle$ *balance points* for each phase can be computed using $\mu_{xy} = \alpha_{xy}$ (Definition 2.1). With $B_{m_2} = 0.775$ for phase 1, $\mu_{m_1 m_2} = \alpha_{m_1 m_2}$, so $B_{m_1} = B_{m_2}/\mu_{m_1 m_2,1} = 0.775/0.75 = 1.033$, and for phase 2, $B_{m_1} = 1.64$. These breaks are shown in Fig. 2.7 labeled as balance points for each of the phases. For $B_p = 1$, they define three saturation states,

$$\Sigma^s_{\text{left}} = \begin{bmatrix} 0 & 0 \\ 1 & 1 \\ 0 & 0 \end{bmatrix}, \quad \Sigma^s_{\text{center}} = \begin{bmatrix} 0 & 0 \\ 0 & 1 \\ 1 & 0 \end{bmatrix}, \quad \text{and} \quad \Sigma^s_{\text{right}} = \begin{bmatrix} 0 & 1 \\ 0 & 0 \\ 1 & 1 \end{bmatrix},$$

following Definition 2.4, Sect. 2.5.6, with $\underline{\sigma}_j = [p, m_1, m_2]$. $B^{\text{waste}} > 0$ is indicated by 0-rows, which right-state designs offset by B_{m_1} insensitivity.

2.6.2.1 Sensitivity of Performance to the System

Consider the system sensitivity caused by saturation state transitions in moving an application from one system to a similar one. In the leftmost region of Fig. 2.7 (B_{m_1} saturated in both phases), if B_{m_1} values differ on two similar architectures performance will be affected more than in the other two regions due to a linear tradeoff between B_{m_1} and processor performance. In the center region, the performance benefit of incremental B_{m_1} change is about half as great. The rightmost region is insensitive, as C_p is independent of B_{m_1}. Two machines to the right of the ϕ_2 *balance point* show no performance change as B_{m_1} varies; two machines to the left of the ϕ_1 *balance point* show a 20% performance variation in the region graphed.

Analyzing this via Eq. (2.46), with B_{m_1} values at phase balance points and μ's computed from data above gives

$$\sigma'_{Bp\ Bm1,\ \phi_1\ \text{bal}} = \frac{-\mu_{p\ m_1,1} C_{p,1}}{B^2_{m_1}} = \frac{-1.33 \times 0.7}{(1.033)^2} = -0.87$$

$$\sigma'_{Bp\ Bm1,\ \phi_2\ \text{bal}} = \frac{-\mu_{pm_1,2} C_{p,2}}{B^2_{m_1}} = \frac{-1.64 \times 0.89}{(1.64)^2} = -0.54$$

The numerical sensitivity at the left balance point exceeds that at the right by a factor of 1.6, and evaluation at the midpoints of the two sloping lines yields a ratio of 1.4, in general agreement with the plot for related concepts. Similar results can be obtained for B_{m_2} sensitivity, corresponding to a slice across Fig. 2.6 along the B_{m_2} axis.

To explain how real-world design efforts might produce systems with performance sensitivity that varies more than necessary, imagine two design teams, one working to the left of the phase 1 balance point, and one to the right. Assume that neither team has a global view of the design space beyond what typical simulation studies allow [9]. The first team will be more easily inclined to increase B_{m_1} based on incremental studies, subject to cost constraints. In competition with other teams working in the leftmost region, under fixed cost budgets, team 1 designers could make bigger design errors, by insufficiently incrementing B_{m_1}, than team 2 or other design teams working to the right of the balance point. In general, operating at a balance point is locally optimal, but without global oversight all design teams are likely to err. Adding more-detailed nodes at sensitive points provides zoom-in on hot spot design.

Fig. 2.8 Phase weight sensitivity

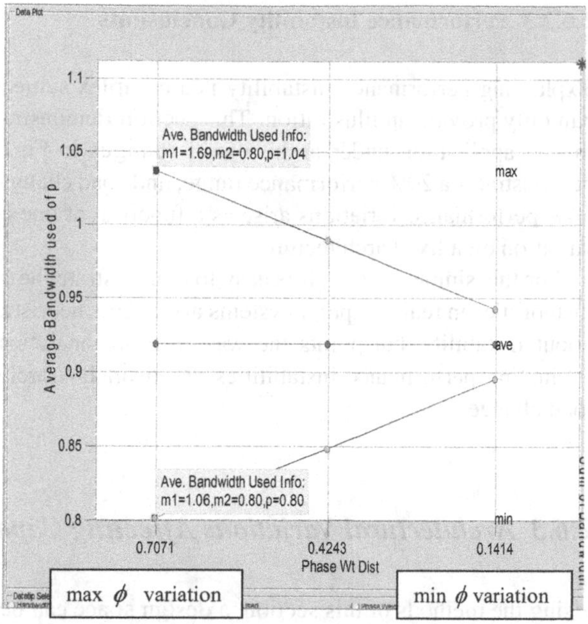

2.6.2.2 Sensitivity of Performance to the SW Load

Future workloads are impossible to know with certainty, but the capacity-based process allows approximating them as changes to current workloads. Varying BW used and codelet weights allows approximation of new paths through existing applications (data set changes), emerging algorithms and applications (codelet variations), and ranges of data rate inputs. Driving the linear analysis with such perturbations also leads to nonlinear performance surfaces. To simulate uncertainty in data-dependent program paths, or application market-importance variation over time, we can vary the phase weights in some range $(UB - LB)$, where $LB \leq \phi_1, \phi_2 \leq UB$, $\sum_j \phi_j = 1$, using a constant increment. The choice of UB and LB depend on specific design constraints. This has the effect of varying $C_{i,j}$ in saturation states.

Figure 2.8 plots the combinatorial magnitude of the distribution of phase weights, ranging here from 0 to 1 in steps of 0.2. The maximum variation of C_p^{glob} in this example is 30%—from 0.80 to 1.04. The details are not shown, but as the phase weight range increases, maximum proc performance variation increases: from 9% (not shown for phase weight range 0.3 to 0.7) to 30% (Fig. 2.8). Comparison at the two balance points (Fig. 2.7) shows greater stability at the ϕ_1 balance point; the opposite of the HW stability conclusion (Sect. 2.6.2.1). The explanation of this may depend on the higher peak and average performance values achieved for higher B_{m_1}. Thus, *load instability* can be manifest on a single system by running two similar applications, or one application with varying data sets.

2.6.2.3 Performance Instability Conclusions

Explaining performance instability is a complex subject, and one simple example can only provide an illustration. This section demonstrated performance variations in one application under architectural changes in Fig. 2.7, where the left region demonstrates a 20% performance range, and load changes in Fig. 2.8, where 10% to 30% performance variations arise as a function of one computation's phase weight variation on a fixed architecture.

For this simple system, it is easy to demonstrate the basic mechanisms by which instabilities in real computer systems arise. This heuristic discussion proves nothing about instability, but points the way to more analytical methods for finding and evaluating performance instabilities based on BW/architectural change as well as load change.

2.6.3 Architectural Variations Affecting Capacity

Using the methods of this section, a design space can be explored for critical architectural changes. For example, given an accurate laptop model for a comprehensive workload, how would a solid-state disk noticeably improve performance. By reducing disk latency appropriately, the shift from hard drive to SSD could be modeled, and those phases (applications) could be discovered for which delivered processor performance increased significantly. Further, designers could examine the potential of small on-chip RAM supplemented by SSD.

The appropriate model could be driven by C and μ values estimated from the original system—on-chip RAM and SSD latencies would be much reduced, while page faults would increase. The tool would show performance improvement per application together with sensitivities to the C and μ parameter estimates. This could quickly provide a crude view of potential architecture vs. market tradeoffs, together with some sensitivities.

2.7 Multirate Nodes

The two types of multirate node have BWs that vary with computational load.

Supernode: A supernode is any connected set of linear nodes. It can be used to denote a subsystem's variable performance behavior in either HW (e.g. memory latency) or architecture (e.g. queues with variable internal latency).

Nonlinear node: A nonlinear node's BW is a nonlinear function of other nodes' capacities. Examples include a parallel processor, cache hierarchy or vector processing unit.

Multirate nodes arise in two ways:

1. A designer chooses a model fidelity that includes supernodes or nonlinear nodes, forcing their analysis.
2. Capacity analysis yields a BW objective exceeding current technology limits, forcing multirate node synthesis.

Analysis: Most RTL-level component performance responses are linear relative to BW or latency. But when architecturally linked and driven by applications imperfectly matched to the architecture, overall performance response can be nonlinear. Multirate node C and μ values generally depend nonlinearly on HW component size metrics and how the computation interacts with the detailed node structure, e.g. loop vectorization or blocking for cache reuse [2].

Synthesis: When linear analysis applied to performance enhancement of a design calls for a node BW that is infeasible using available linear components, a multirate node may be synthesized. Using performance objectives obtained by the capacity-based solver as the multirate node BW requirement, a secondary method can specify its internal structure, e.g. the required number of cores for a multicore component.

2.8 Related Work

Discussion of compute vs. memory or I/O bound programs, the von Neumann architecture bottleneck, and designing systems to match given applications or algorithms, have driven computer design for 50 years [6]. Obtaining, analyzing and interpreting large volumes of performance data present major obstacles that have been addressed in many ways. Deterministic and stochastic models with sampling from the application level (benchmarks) to the trace level (HW performance counters) followed by various discrete event simulators and statistical models are used in specialized or combined ways. Stochastic methods tend to work well for steady-state computations, while discrete event simulation handles all situations but much more slowly. Multiple system types have evolved to cover multiple market needs.

Capacity-based codesign can handle at linear programming speeds, both steady-state and transient system behavior. Codelet coverage is the key need; it can succeed either by reuse of common source- or assembly-codelets. The method explicitly represents the performance of phases and whole computations, so solutions can yield extensive architectural insight (Sect. 2.6). The method's speed of solving codesign problems depends on solving LP problems, doing sensitivity analyses, and exploring design space in various ways. Several statistical methods are emerging [7] that may help in reducing the time and enhancing the insights of design space exploration.

Another codesign issue is application performance enhancement. Potential approaches are given in Sects. 2.5.3 and 2.5.6 to find hotspots by roughening profiles across whole computations. Many papers discuss the two- node case [1] seeks memory-processor *balance* (Definition 2.1) through formulas to analyze loops for compiler transformations. Using two node *capacity* Eq. (2.29b) with variable μ_{xy} and constant C_x, switches the Fig. 2.5 labels (Sect. 2.5.3); [10] explores the

$C_x = C_m = B_m$ case (roofline model). For several algorithms, [8] uses memory access intensity analysis—equivalent to *BW sensitivity* (Eq. (2.47)) with saturated memory—to predict when blocking performance-sensitive loops will be beneficial.

The potential to partition design space and move toward specialized systems for distinct applications exists within this method. This can be done manually by iteratively removing similarly performing phases. Perhaps algebraic analysis may lead to semi-automatic methods of partitioning the computational parameter matrix.

2.9 Conclusions

A number of codesign problems have been posed, together with capacity-based methods of finding BWs of HW system nodes that satisfy given goals, for a given set of computations. System recommendation is a related problem, i.e. for a fixed set of computations, select one of several specific systems as the best in *perf/cost*. Also, codesign can be expressed as solving for SW variables in terms of fixed HW. The ideas presented can be used for many specific problems, but there are some underlying commonalities:

1. Top-down codesign of optimal systems
   - Mixed fidelity modeling allows focus on exactly those parts of the HW system of interest
   - All computations are modeled by weighted combinations of SW repository codelets
2. Simultaneous use of comprehensive load and BW information
   - LP equations are optimized faster than discrete event simulation, combining SW and HW specifics
   - Global sweeps of 3D codesign space show parametric relations among many optimal design points
3. Design of robust, focused-system families under uncertainty
   - Perfect solutions ($B^{\text{waste}} = 0$) for single phases, optimal designs for application classes
   - Pre-Si exploration of design sensitivities; market-segment design partitioning.

Key features of the approach include:

- Rich codelet set relative to benchmark/trace-driven simulation helps prevent application performance regressions
- Capturing system-wide interactions avoids the local optimization traps typical in component-wise design
- Automating the process overcomes design complexities that overwhelm human designers
- Meeting infeasible goals with higher-performance synthesis, only when needed.

The *consequences* are savings of human design and CAE machine time, as well as better system designs.

Acknowledgements David Wong and Ahmed Sameh provided many insights in developing this material; David Wong designed and implemented the CAPE tool.

References

1. Carr, S., Kennedy, K.: Improving the ratio of memory operations to floating-point operations in loops. ACM Trans. Program. Lang. Syst. **16**, 1768–1810 (1994). doi:http://doi.acm.org/10.1145/197320.197366
2. Emer, J., Ahuja, P., Borch, E., Klauser, A., Luk, C.K., Manne, S., Mukherjee, S., Patil, H., Wallace, S., Binkert, N., Espasa, R., Juan, T.: Asim: A performance model framework. Computer **35**, 68–76 (2002). doi:http://doi.ieeecomputersociety.org/10.1109/2.982918
3. Jalby, W., Wong, D., Kuck, D., Acquaviva, J.T., Beyler, J.C.: Measuring computer performance. In this volume
4. Kuck, D.: Computer system capacity fundamentals. Tech. Rep. Technical Note 851, National Bureau of Standards (1974)
5. Kuck, D.: The Structure of Computers and Computations. Wiley, New York (1978)
6. Kuck, D.J., Kumar, B.: A system model for computer performance evaluation. In: Proc. 1976 ACM SIGMETRICS Conf. on Computer Performance Modeling Measurement and Evaluation, pp. 187–199. ACM, New York (1976). doi:http://doi.acm.org/10.1145/800200.806195
7. Lee, B., Brooks, D.: Spatial sampling and regression strategies. IEEE MICRO **27**, 74–93 (2007). doi:http://doi.ieeecomputersociety.org/10.1109/MM.2007.61
8. Liu, L., Li, Z., Sameh, A.: Analyzing memory access intensity in parallel programs on multicore. In: Proc. 22nd Annual Int'l. Conf. Supercomput., ICS '08, pp. 359–367. ACM, New York (2008). doi:http://doi.acm.org/10.1145/1375527.1375579
9. Uhlig, R.A., Mudge, T.N.: Trace-driven memory simulation: A survey. ACM Comput. Surv. **29**, 128–170 (1997)
10. Williams, S., Waterman, A., Patterson, D.: Roofline: an insightful visual performance model for multicore architectures. Commun. ACM **52**, 65–76 (2009). doi:http://doi.acm.org/10.1145/1498765.1498785

Chapter 3
Measuring Computer Performance

William Jalby, David C. Wong, David J. Kuck, Jean-Thomas Acquaviva, and Jean-Christophe Beyler

Abstract Computer performance improvement embraces many issues, but is severely hampered by existing approaches that examine one or a few topics at a time. Each problem solved leads to another saturation point and serious problem. In the most frustrating cases, solving some problems exacerbates others and achieves no net performance gain. This paper discusses how to measure a large computational load globally, using as much architectural detail as needed. Besides the traditional goals of sequential and parallel system performance, these methods are useful for energy optimization.

3.1 Introduction

Overall computer performance time or rate can be measured reasonably well, but breaking out the contributory details of performance, per HW node and computational phase, is much harder. Ideally, performance numbers should result from measurement and modeling procedures that have several *desired properties*:

1. Available for *any HW node*, i.e. any level of HW modeling from RTL to whole system block diagram.

W. Jalby (✉) · J.-T. Acquaviva · J.-C. Beyler
UVSQ/Exascale Computing Research, Versailles, France
e-mail: William.Jalby@uvsq.fr

J.-T. Acquaviva
e-mail: jean-thomas.acquaviva@exascale-computing.eu

J.-C. Beyler
e-mail: jean-christophe.beyler@exascale-computing.eu

D.C. Wong · D.J. Kuck
Intel Corporation, Urbana, USA

D.C. Wong
e-mail: david.c.wong@intel.com

D.J. Kuck
e-mail: david.kuck@intel.com

M.W. Berry et al. (eds.), *High-Performance Scientific Computing*,
DOI 10.1007/978-1-4471-2437-5_3, © Springer-Verlag London Limited 2012

2. Represents *accurate performance* time and rate per node, using unobtrusive measurement techniques.
3. Overall performance should be expressible in terms of node performance.
4. Node performance should be expressible SW terms, e.g. machine and HLL instructions.
5. *Universal modeling* techniques should be used, which are fast and not unique to a given architecture.

Several *issues* have stood in the way of achieving all of these goals.

A. Cycle accurate simulation is very accurate, but prohibitively slow for large applications
B. Using SW methods alone has been thwarted by architectural complexity
C. Performance HW counters offer hope but are notoriously difficult to use in practice (see Sect. 3.3)
D. Performance nonlinearity at a high level (cache, vectors, parallelism) makes the process very complex

This paper proposes an approach that reasonably satisfies the desired properties, but avoids the issues listed above. The methodology rests on several *basic principles* and *new tools* that will be outlined in this paper. While the method has not been demonstrated end-to-end, the key parts have been implemented and tested independently.

The paper discusses the details of obtaining necessary parameters for linear modeling, by using novel ways of measuring program phase performance. This includes simplifying the nonlinearities inherent in some physical nodes that are a function of program parameters, by introducing virtual nodes that combine HW and SW concepts. Section 3.2 surveys problems with using traditional methods, and Sects. 3.3 and 3.4 introduce the computational capacity abstraction and piecewise virtual-node modeling that allows simplifying the problem. Sections 3.5 and 3.6 give more details of the overall process and an example of how the process works. Section 3.7 discusses conclusions, next steps and open questions.

3.2 Traditional Measurement and Modeling

Traditional measurement methods use HW performance registers, and modeling methods include the use of simulation and abstract modeling to understand system performance, energy use, and architectural design. This paper discusses new *measurement methods* based on the virtual-node abstraction, and *modeling methods* that are designed to be much faster than simulation, more accurate than most abstract models, and produce detailed measurements of the modeled system. The issues of performance counters are discussed next.

Over the past 25 years, HW performance counters (HPCs) have become widely used for various purposes. However, the registers themselves as well as their usability still present major problems, as listed below.

1. *HPC fine grain HW focus means low granularity SW information*
   By definition, HPCs are able to measure HW behavior primarily, and SW behavior only indirectly. However, SW performance improvement often requires focused changes to carefully selected instructions in particular parts of a program. HPCs do not have instruction-centric capabilities, instead they give aggregated views of large numbers of instructions.
2. *Linkage of HPCs to performance realities*
   The linkage of the HPC data to overall performance can be blurred by architectural complexities that are missed by the HPC numbers. For example, OOO execution can mask a great deal of the time that may seem to be incurred, from the analysis of only HPC data. Another example is the inability to detect poor spatial locality. For example, a stride-4 load from L2 cache may have equivalent performance to a stride-2 load from L3, so be invisible from an HPC point of view that is stride blind.
3. *HPC register technology per se*
   There are several intrinsic weaknesses of HPC registers, beyond the above. One is that they are neither standardized across manufacturers, nor from system to system or over time within one manufacturer. This makes their use ad hoc by definition. Furthermore, while the sampling methods employed are often adequate, sometimes they seem to deliver very distorted results; quality assurance of the data is not provided by any manufacturer.

This paper offers measurement approaches that are more direct relative to HW and SW, are closely linked to overall performance via capacity (Sect. 3.3), and are not linked to specific architectures.

3.3 Computational Capacity Model

Computational capacity $C_{i,j}$ is the BW actually used in each HW node i, by a specific computational phase j, as defined in [4], and is central to measurement and modeling. Besides being intuitively meaningful in both architecture and SW discussions, it leads to fast and effective analytical techniques, and as shown below, simplifies some measurement activities because it is jointly defined by the HW and SW parameters of a given computation.

3.3.1 Basic Variables and Measurements

Many important characteristics of a computation can be defined in terms of the five independent variables below: three HW, one SW, and one joint HW/SW. The values of the HW variables may be obtained from first principles (perhaps as manufacturers specs) or empirically from a large computational load. The latter method will be discussed in this paper as it may be regarded as more germane, and because in the case of v-nodes (Sect. 3.5) it is the only feasible approach. In general, if linear scaling is used, results obtained either way should be equivalent [4]. The SW and joint variables must be empirically determined.

- *Three HW variables*
  a. HW node BW B measured as [b/s] peak.
  b. Two of: power dissipation W [watts] consumed by B operating at full BW, and idle power W^{idle} [watts] of B or γ [energy/b] consumed when operating.
- *SW variable*
  Each codelet performs O operations during a computation. These can be measured abstractly, e.g. in an algorithm or in terms of a given instruction set.
- *Joint HW/SW variable*
  Capacity C [b/s] used by a computation on each node, is determined by the structure of each program and its linear or nonlinear interactions with the architectural structure of the HW nodes in a given system, which in this paper will be masked during measurement, see Sect. 3.4.1.

From these five independent variables, a running computation yields many dependent variables including overall and per node performance (running time) and energy (operating cost).

An empirically complete set of measurements consists of only three: capacity $C_{i,j}$ and $W_{i,j}$ for node i and phase j, and $O_{i,j}$ or time. The others can be estimated practically from these, using $B_i = \max_i \{C_{i,j}\}$ and several values of $W_{i,j}$ plotted vs. $C_{i,j}$ give approximate values for W_i and W_i^{idle}. Alternatively, measuring only W_i^{max} suffices if $\gamma_{i,j}$ (slope of power model [energy/bit]) is known from first principles.

3.3.2 Single Phase CAPE Simulation

With the measurements described in Sect. 3.3.1, performance is projected for scenarios where the BWs of HW nodes are changed using *CAPE simulation*. Unlike traditional simulation, CAPE simulation does not simulate or emulate individual execution steps. It projects performance by analyzing the utilization of individual components in a macroscopic manner. This section describes how to do CAPE simulation for a single phase workload. Section 3.3.3 shows how to extend this to multiple phase workloads. Consider an m node system shown in Fig. 3.1 with the following measurements:

- Original bandwidths: $B_1, B_2, \dots, B_m$
- Original capacities: $C_1, C_2, \dots, C_m$
- Original execution time: T

Using a new set of bandwidth inputs $B'_1, B'_2, \dots, B'_m$, CAPE can compute new capacities $C'_1, C'_2, \dots, C'_m$ and overall time for the new system to execute the original load.

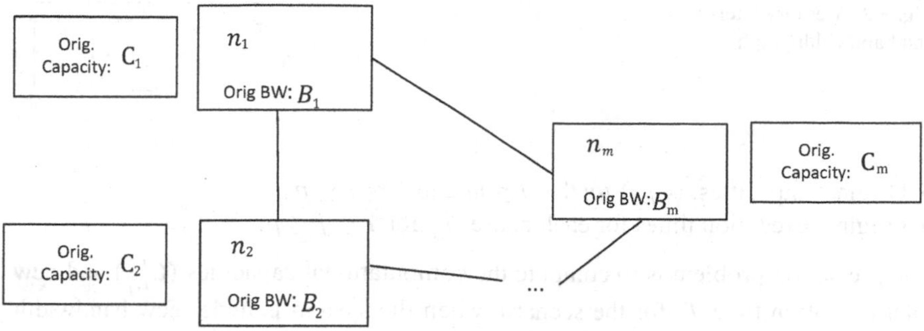

Fig. 3.1 Original single phase m node system

For each node i, the relative saturation, $\sigma_i = C_i / B_i'$ tells us how far node i is from saturation in the new scenario. Multiplying $1/\sigma_i$ by the capacity of n_i will saturate the node. As shown in [6], using $s = \min(1/\sigma_i)$ as the scaling factor for all the capacity values ($C_i' = sC_i$), the resulting system will have the following properties:

- For any node i, $C_i \leq B_i'$
- There will be at least one saturated node, namely node k where $1 \leq k \leq m$ with relative saturation $1/s$, i.e. $C_k / B_k' = 1/s$. For that node, the new capacity is $C_k' = sC_k = B_k'$.

With the above established properties, we have a simple algorithm to compute the capacities given new bandwidth settings $(B_1', B_2', \ldots, B_m')$:

Algorithm:
COMPUTE-ONE-PHASE-CAPACITY$(B_1', B_2', \ldots, B_m', C_1, C_2, \ldots, C_m)$

$s \leftarrow \infty$;
for $i \leftarrow 1$ **to** m **do**
  do $s \leftarrow \min(s, \frac{C_i}{B_i'})$;
for $i \leftarrow 1$ **to** m **do**
  do $C_i' \leftarrow sC_i$
$T' \leftarrow \frac{T}{s}$;

3.3.3 Multiple Phase CAPE Simulation

To generalize the single phase problem to the multi-phase case, the same procedure is applied to each phase. Following are the inputs of the n node p phase problem:

- Original Bandwidths: $(B_{i,j})$ for $1 \leq i \leq m$ and $1 \leq j \leq p$
- New Bandwidths: $(B_{i,j}')$ for $1 \leq i \leq m$ and $1 \leq j \leq p$

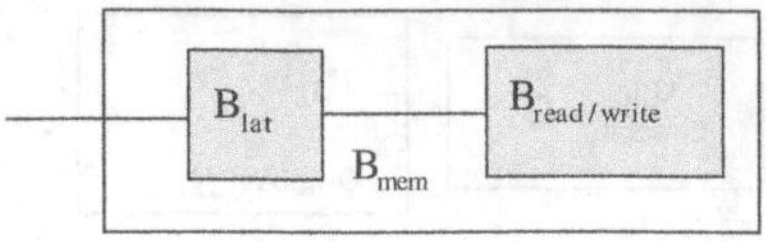

Fig. 3.2 Memory latency and bandwidth graph

- Original capacities: $(C_{i,j})$ for $1 \le i \le m$ and $1 \le j \le p$
- Original execution times for each phase: T_j for $1 \le j \le p$.

The prediction problem is to compute the computational capacities $(C'_{i,j})$ and new total execution time T' for the scenario when the system is under new bandwidth settings.

Following is a generalized version of the algorithm described in the previous section:

Algorithm:
COMPUTE-ALL-PHASE-CAPACITY($B'_{1,1}, \ldots, B'_{m,p}, C_{1,1}, \ldots, C_{m,p}$)

for $j \leftarrow 1$ **to** p **do**
  do $s_j \leftarrow \infty$;
  for $i \leftarrow 1$ **to** m **do**
    do $s_j = \min(s_j, \frac{C_{i,j}}{B'_{i,j}})$

$$\begin{pmatrix} C'_{1,1} & \cdots & C'_{1,p} \\ \vdots & \ddots & \vdots \\ C'_{m,1} & \cdots & C'_{m,p} \end{pmatrix} \leftarrow \begin{pmatrix} C_{1,1} & \cdots & C_{1,p} \\ \vdots & \ddots & \vdots \\ C_{m,1} & \cdots & C_{m,p} \end{pmatrix} \begin{pmatrix} s_1 & 0 & 0 \ldots 0 & 0 \\ 0 & s_2 & 0 \ldots 0 & 0 \\ \vdots & & \ddots & \vdots \\ 0 & 0 & 0 \ldots 0 & s_p \end{pmatrix};$$

$T' \leftarrow (T_1 \ldots T_p) \cdot (1/s_1 \ldots 1/s_p)^t$

3.4 Dealing with Nonlinearity

3.4.1 Nonlinear Multirate Performance Model

The total bandwidth provided in complex computer system nodes can be fully exploited only under ideal operating conditions within the node. Two types of multirate node are described here.

Variable latency arises due to the load on a bus or network, memory access conflicts, or disk rotation time. Fig. 3.2 is a memory example, which following Eq. (II.2) [4] and using capacity terms, leads to Eq. (3.1).

Since BW and latency (λ) are reciprocals, Eq. (3.1) can be rewritten as Eq. (3.2):

$$\boxed{C_{\text{mem}} = \frac{C_{\text{lat}} C_{r/w}}{C_{\text{lat}} + C_{r/w}}} \tag{3.1}$$

using $\lambda_{\text{lat}} = 1/C_{\text{lat}}$. This memory supernode has nonlinear capacity C_{mem} depending on memory read/write capacity, but is also subject to the variable nature of the node's latency ($C_{\text{mem}} = C_{r/w}$, if $\lambda_{\text{lat}} = 0$):

$$C_{\text{mem}} = \frac{C_{r/w}}{1 + C_{r/w}/C_{\text{lat}}} = \frac{C_{r/w}}{1 + \lambda_{\text{lat}} C_{r/w}} \tag{3.2}$$

Two other examples of complex nodes that may be modeled directly as nonlinear nodes follow. For best performance, caches require high hit ratios and parallel processors require high SW parallelization, which depend on how well program structures and compiler algorithms match a given architecture. To model a node that can operate at various speeds, we use a model whose bandwidth B varies, and write Eq. (3.3). The computation time for a *saturated node* is $T^u = O/C$,

$$B_{\min} \le C \le B_{\max} \tag{3.3}$$

(Eq. (I.5) [4]) and the *speedup* in fully exploiting a node, over its slowest running time is Eq. (3.4). Using $\rho = B_{\min}/B_{\max} \le 1$, and $\theta_{\max} \le 1$

$$1 \le S = T_{\max}/T^u = \frac{O/B_{\min}}{O/C} = \frac{C}{B_{\min}} \le \frac{B_{\max}}{B_{\min}} \tag{3.4}$$

as the fraction of total time $T_{\max}$ run at bandwidth $B_{\max}$ leads to $S \le \frac{T_{\max}}{\theta_{\max}\rho T_{\max} + (1-\theta_{\max})T_{\max}} = \frac{1}{1+\theta_{\max}(\rho-1)}$, and replacing $\theta_{\max}$ by θ_{av} (a weighted average across all p values in a computation) yields the equality of Eq. (3.5):

$$S = \frac{1}{1 + \theta_{av}(\rho - 1)} \tag{3.5}$$

For parallel processors, $\rho = 1/P$, for a P-processor parallel system and $\theta_{av} =$ fraction of serial code execution time run fully parallel on P processors, with the P-processor speedup S_p of Eq. (3.6).

$$\boxed{P \ge S_p = \frac{1}{1 + \theta_{av}(1/P - 1)} \ge 1, \quad 0 \le \theta_{av} \le 1.} \tag{3.6}$$

For cache hierarchies, to look at slowdown, we "invert" the interpretation of Eq. (3.5) using the BW ratio of two cache levels $\rho = B_{Li}/B_{Li+1} > 1$, or cache/memory $\rho = B_c/B_m > 1$, and $\theta_{av} = m =$ cache miss ratio, $0 \le m \le 1$, so the *cache slowdown factor* σ_c^u is given by Eq. (3.7).

$$\boxed{\sigma_c^u = 1 + m_i(\rho - 1) \ge 1, \quad 0 \le m_i \le 1, \ \rho > 1} \tag{3.7}$$

Equations (3.2), (3.6), and (3.7) are basic performance formulas for parallelism (Amdahl's Law) and memory hierarchy, respectively. Equation (3.7) can be expanded to cover multiple levels and applied using multi-level cache measurement data. Parallelism is usually regarded as a speedup for P processors, but by measuring the amount of 2, 3, 4, etc. degree parallelism for $< P$ processors, could be useful in computing θ_{av}. For parallel computing, using T_1 and T_P for uniprocessor and P-processor times, respectively, and B_1 and B_P similarly for BW, we can write Eq. (3.8):

$$T_p = T_1/S_p, \quad \text{and} \quad B_p = B_1 S_P \tag{3.8}$$

3.4.2 Piecewise-Linear Approximation

Equations (3.6) and (3.7) define nonlinear performance curves that are functions of *SW parameters*. If a discrete set of SW parameters is chosen for analysis, each of these nonlinear performance functions can be approximated by a discrete set of performance points, leading to a piecewise-linear plot. If one knew the values of applicable parameters for each program phase encountered, then the equations of Sect. 3.4.1 could be used directly to get capacity values. Determining these values online or offline would be a complex and open-ended process that is practically infeasible.

Alternatively, if many program phases are analyzed *offline*, automatically, and decomposed into codelets [4], each of which has linear performance characteristics, then linear analysis may be used *online* for each codelet. Furthermore, the overall nonlinear performance function is dealt with by using multiple codelets to piecewise-linearize the original node's nonlinear performance function. The off line choice of codelets is a closed process, with occasional exceptions for new codelets that must be entered into the repository.

This defines a two-step process for linearizing the nonlinear performance curves of Sect. 3.4.1.

1. Decompose any phase into a discrete set of codelets that provide sufficient coverage to define a reasonable piecewise-linear approximation of the original nonlinear performance curve.
2. Guarantee that each resulting codelet is itself a linear function of the B and C values for the virtual node used by the codelet.

To implement this process following the procedure of Sect. 3.5, assume that the codelets chosen above are macro instantiations and the SW parameters are those that produce macros and SV-nodes. Each point chosen must be linear, so the decomposition of a given nonlinear node is continued until the *linearity test* shows that the decomposition is complete.

Note that "minor nonlinearities" are acceptable in the sense that the overall piecewise-linear curve gives a good approximation of the true nonlinear curve. For example, when measuring cache level Li and assuming all hits, the error included by allowing a few misses is acceptable. (In the worst case, this is leads to ignoring the nonlinear curve defined in Sect. 3.4.1 and connecting its ends with a straight line.) The effect will be to shift the slopes of $\mu_{L,x}$ values per phase [4] and thereby skew design results in proportion to the degree of "minor nonlinearity" acceptable. A quality test that can be made per codelet is to test for steady-state behavior by comparing overall codelet capacity values with sampled capacity values throughout the phase running time.

In general, this leads to a set of non-uniformly spaced SW parameter points that each behave linearly, and collectively yield a piecewise-linear approximation of the performance curve, which improves with the number of points chosen. Choosing intuitively important codelets (macros) to cover the code segments expected in real programs will lead to effective coverage of each nonlinear curve and results that are easy to interpret. The final result is a set of linear SV-nodes that provide piecewise linear coverage of the important parts of codesign space.

The general shape of nonlinear performance curves may be smooth, as above, or may have *periodic irregularities*, as in vector performance that is punctuated by sawteeth corresponding to vector register size [2]. Such phenomena may be modeled by choosing v-nodes at the sawteeth. Most difficult are *sporadic punctuations*, caused by irregular events that may have serious performance effects (e.g. in saturated v-nodes). The following is a list of well-known problems that may be corrected to prevent skewing of the modeling results, or if uncorrected should be regarded as potential sources of error in the modeling. Each is followed by short description of methods for detecting and mitigating the problems.

1. Data alignment: depending upon the starting address of a block of 128 bits, different SSE instructions (with different performance) have to be used. Such problems can be easily detected by an automated analysis of assembly code and can be solved by array padding or array reshaping.
2. Cache bank conflicts: two memory references can hit the same bank during the same cycle, forcing one of them to be delayed while the other one is serviced. In general (arbitrary memory reference patterns), such problems are hard to detect. Fortunately for most HPC codes, array access are very regular and cache conflicts can be easily detected by memory tracing techniques. In general, techniques such as array padding and/or reshaping will alleviate the problem.
3. 4K Load Store aliasing: a load and a store instruction accessing two different memory locations having the same identical low order 12 bits. As for cache bank conflicts, regular array access very frequent in HPC codes allows to use memory tracing for detecting such cases. Instruction reordering (moving far apart the load and store colliding) will reduce performance impact.
4. Associativity conflicts: although modern caches use a high degree of associativity, cache conflicts can still occur due to associativity. There, the problem can be fairly difficult to detect because in general it will involve the mapping from virtual pages to physical pages. Furthermore the phenomenon can be hard to reproduce because it will depend upon the initial state of memory fragmentation. In general techniques such as array restructuring (for minimizing cache foot print) will reduce performance impact.
5. TLB misses: when walking through too many pages at the same time, the TLB can thrash, generating costly misses. As for cache bank conflicts, memory tracing techniques will allow quick identification of the issue. Very often, a high

degree of TLB misses occurs due to access along the "wrong" dimensions in multidimensional arrays and the problem can be resolved by array restructuring.
6. False sharing; this phenomenon occurs when two cores are writing repetitively to the same cache. Again memory tracing will allow quick identification of the issue and array restructuring to avoid cache line split between different cores will resolve most of the cases.

It should be noted that most of the problems listed above are due to very specific "local" and "periodic" properties of the memory access stream: in general, simply changing array sizes will make the problem disappear or at least be drastically reduced. As such, they should be considered as secondary performance effects or more generally as perturbations to the general performance model developed in this paper. Furthermore, their detection and resolution can be decoupled from our general performance analysis.

3.5 Model Synthesis

The two major steps in codesign modeling are model synthesis (to be described in this section) and computational capacity simulation (described in Sects. 3.3.2 and 3.3.3), as shown in Fig. 3.3. Model synthesis is performed by four SW tools—macro generation, microbenchmarking, Maqao (static assembly-code analysis) [1] and DECAN (dynamic binary-code decremental analysis) [3]. This produces, for each v-node chosen, B, C and W power model information. The CAPE tool is then initialized by a designer with simulation study goals, and is used to simulate a wide range of systems. Finally, the CAPE post-processor can be used to filter and interpret the results in terms of either phy-nodes or v-nodes, for use by designers to estimate the performance impact of various design decisions.

3.5.1 Overview

The modeling framework will be described in a series of steps that lead to a block diagram of the codesign process. The concept is to proceed from instruction sets to SW macros to single rate v-nodes (SV-nodes), through computational capacity simulation, to analysis in terms of general, multirate v-nodes. Figure 3.3 shows the CAPE tool inputs $B_{svi,j}$, $C_{svi,j}$, and T_j. CAPE has three sections:

1. Set original node B, C and T values
2. Define codesign experiment
3. Post-process results for designers.

Figure 3.4 shows how the model synthesis tools work together to take instruction set, node definition and apps as inputs to produce $B_{svi,j}$, $C_{svi,j}$, and T_j. The tool decomposes applications into *codelets*, *macros* and *SV-nodes* (not shown in the diagram) with the following definitions:

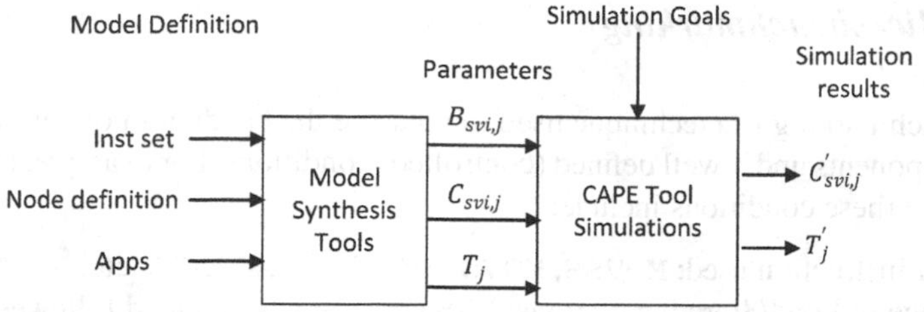

Fig. 3.3 Overall process

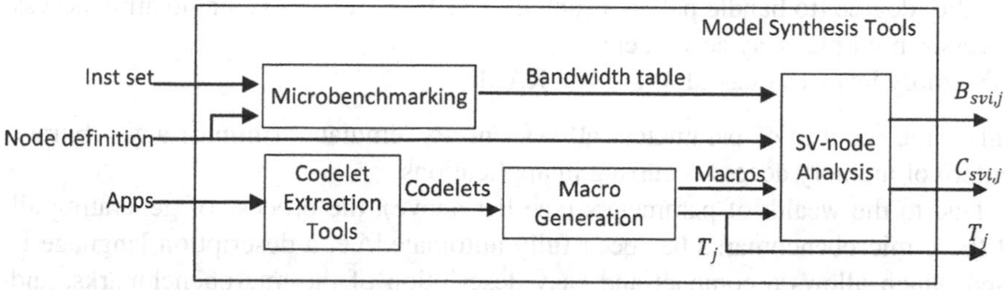

Fig. 3.4 Details of model synthesis tools

- *Codelets* are application fragments with significant execution times.
- *Macros* are decomposed from codelets. The execution of a macro contains uniform/steady streams of operations. Using the algorithm described in Sect. 3.5.4, a generated macro does not overlap with another.
- *SV-nodes* stands for *single-rate virtual nodes*. They are decomposed from macros. Each SV-node contains instructions using the same physical node in a similar access pattern. An SV-node is *single-rate* because given a fixed access pattern, the rate to execute the SV-node instructions is a constant, resulting in a linear capacity model.

The following sections cover individual model synthesis tools:

- Section 3.5.2 describes how *microbenchmarking* generates a bandwidth table used by SV-node analysis. In addition to instruction set knowledge, the definition of SV-nodes in terms of instruction sequences is known a priori. Initially, the definition can be generic, which is then refined as each codesign exercise proceeds; a memory node can be defined as memory instructions operating on several levels of cache. The node definition refinement is guided by the need of designers and developers, and may be implemented in several ways.
- Section 3.5.3 describes application decomposition using the *codelet extraction tools*.
- Section 3.5.4 describes how a *macro generation* algorithm uses execution time to decompose a codelet further into macros. The algorithm also determines the execution time of each macro as an output.
- Section 3.5.5 describes how *SV-node analysis* uses the bandwidth table, the macro and its execution time to determine the *B* and *C* values.

3.5.2 Microbenchmarking

Microbenchmarking is a technique used to measure the bandwidth of various hardware components under well defined (controlled) conditions. For example, for memory access, these conditions include:

- Type of instruction used: `MOVSS`, `MOVSD`, `MOVUPS`, `MOVAPS`, etc.
- Sequence of Load/Store instructions: LSS corresponds to a load followed by two stores
- Address stream: start address, stride, unrolling degree, short-term temporal locality degree (to handle patterns such as `A(I) + A(I+1)`, single array access versus multiple array access, etc.
- Memory level accessed: L1, L2, L3, RAM

This large number of parameters allows one to "emulate" (mimic) any arbitrary pattern of memory access occurring in applications.

Due to the wealth of parameters (see list above), the process of generating all of these microbenchmarks has been fully automated i.e. a description language is used which allows a compact and easy description of the microbenchmarks, and from that description, the various microbenchmarks are generated. The launching and running of this large set of microbenchmarks has also been fully automated to take into account different running execution parameters such as number of cores, frequencies, etc.

The microbenchmarks produce a large database which can be accessed using instruction patterns, memory stream access characteristics, etc.

3.5.3 Codelet Extraction Tools

To decompose an application into manageable fragments, each application is decomposed into codelets using Astex [5]. The decomposition is based on syntactic structure and codelet execution frequency for a given data set. Astex keeps track of the codelet run time data, so the output of Astex is both the codelet and the associated input data. As the codelet is being processed down the tool chain, the data will be carried along as well. This paper assumes data propagation and focuses on the codelet processing. Also, this paper assumes each codelet is a loop nest.

3.5.4 Macro Generation

We use the linearity condition to identify mutually exclusive operations:

Definition 3.1 (Linearity Condition/Test) For two streams of instructions x and y, if $t(xy) = t(x) + t(y)$ then x and y are mutually exclusive.

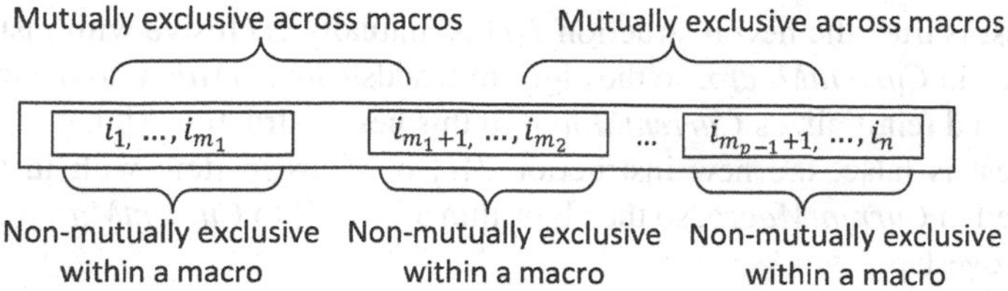

Fig. 3.5 A phase/loop decomposed into macros

This is based on the observation of execution times of each instruction stream. When the condition holds, we observe that executing the two streams of instructions together takes the same amount of time executing them separately. From this, we infer that the execution of the instructions is done mutually exclusively.

The macro generation process visits instructions in a loop sequentially and applies the linearity test to the visited instruction and the instructions before that, to decompose a loop body into distinct instruction streams called macros (see Fig. 3.5).

The MACRO-GENERATION algorithm accepts two inputs, returning a set of macros:

- I is the instruction sequence of the loop to decompose.
- C_{err} is a tolerance value ranging between 0 and 1. When $C_{\text{err}} = 1$, we tolerate no experimental noise. On the other hand, when $C_{\text{err}} = 0$, all instructions will be considered mutually exclusive and the algorithm will return a set of singletons.
- *ExeTime*() is a function used by the algorithm to run the provided instruction sequence and return the measured execution time.

---

1 **Algorithm:** MACRO-GENERATION(I, C_{err})

2 $\mathit{Macros} \leftarrow \emptyset$;

3 $\mathit{CurrentMacro} \leftarrow \{I[1]\}$;

4 **for** $i \leftarrow 2$ **to** $\mathit{length}[I]$ **do**

5     $T_{\text{current}} \leftarrow \mathit{ExeTime}(\mathit{CurrentMacro})$;

6     $T_i \leftarrow \mathit{ExeTime}(I[i])$;

7     $T_{\text{combined}} \leftarrow \mathit{ExeTime}(\mathit{CurrentMacro} \cup I[i])$;

8   **if** $T_{\text{combined}} \times C_{\text{err}} \leq T_{\text{current}} + T_i$ **then**

9         $\mathit{Macros} \leftarrow \mathit{Macros} \cup \{\mathit{CurrentMacro}\}\mathit{CurrentMacro} \leftarrow \{I[i]\}$

10   **else** $\mathit{CurrentMacro} \leftarrow \mathit{CurrentMacro} \cup \{I[i]\}$

11 **if** $\mathit{length}[\mathit{CurrentMacro}] \neq 0$ **then**

12     $\mathit{Macros} \leftarrow \mathit{Macros} \cup \{\mathit{CurrentMacro}\}$

13 **return** *Macros*

The algorithm sequentially visits every instruction $I[i]$ contained in I. It uses *CurrentMacro* to keep track of a set of instructions found to be overlapping. Line 8 is the linearity test using the tolerance value C_{err}:

- If the test is true, the new instruction $I[i]$ is mutually exclusive with instructions contained in *CurrentMacro*, so the algorithm adds *CurrentMacro* to the macro set *Macros* and reinitializes *CurrentMacro* to this new instruction $I[i]$.
- If the test is false, the new instruction $I[i]$ overlaps in time with instructions contained in *CurrentMacro*, so the algorithm adds $I[i]$ to *CurrentMacro* to record they all overlap.

After iterating through all the $I[i]$, the leftover instructions contained in *CurrentMacro* are considered a macro, so the algorithm adds *CurrentMacro* to *Macros*.

The rationale of the algorithm is to group concurrently executing instructions into the same macro. By grouping instructions this way, two instructions from two different macros will be executed mutually exclusively. Since there is no execution time overlap between macros, the total execution time of all macros is just the sum of execution times of individual macros. Using this property, a CAPE simulation of the codelet can be done by adding up the results of CAPE simulations for individual macros. Note that instructions within a macro may operate on distinct physical nodes. A deeper analysis of instructions within a macro is required, as follows.

3.5.5 SV-Node Analysis

Within a macro, the execution of an instruction overlaps with another instruction from the same macro. These instructions may or may not operate on the same physical node(s).[1] To ease reasoning about the execution of the application, it is useful to group together instructions operating on the same physical node to form SV-nodes as described in Sect. 3.5.1. To ensure SV-nodes are executed at a single rate, some of them are parameterized by usage patterns. An example of an SV-node could be "stride one `LOAD`". SV-nodes are useful to application developers and architects. Application developers can simulate various optimizations by changing the usage (bandwidth used) of SV-nodes. Simulated execution can help them to identify the most effective optimization to implement. Architects can simulate the hardware impact by changing the speed (bandwidth) of the SV-nodes.

By construction of the macro, the execution of one SV-node is expected to overlap with the execution of another SV-node. Therefore, changes in SV-node execution time may be (partially) hidden by another SV-node's execution. To model and facilitate CAPE simulation, it is important to quantify this effect and DECAN (described in Sect. 3.5.5.1) is a useful tool for this purpose because it tells us the performance impact of SV-node changes. Section 3.5.5.2 describes how the SV-node saturation is found using DECAN and Sect. 3.5.5.3 describes how to refine the SV-node saturation.

---

[1] In fact, every instruction must use the processor to dispatch the instruction. In some preliminary experiments, we have found that even for `LOAD` instructions, for some access patterns, the instruction can show processor bound behavior.

3.5.5.1 DECAN

DECAN is a tool that allows one to patch binaries. Through patching, specific groups of instructions are modified. Then by comparing the original execution time (corresponding to the unmodified binary) with the execution time of the modified binary, the performance impact of the modified group of instructions can be accurately assessed. For now, the target instructions (to be modified) are SSE instructions (FP as well as data access instructions). Several types of modification are proposed:

- "Iso nopping": the target instruction is replaced by a `NOP` of exactly the same size: the loop body size is not altered. For the loop modified by "iso nopping", the pipeline front end behavior is close to the original loop.
- "Simple nopping": the target instruction is replaced by a simple 1 byte `NOP`. Loop size is altered but the number of instructions for the modified loop is identical to the original one.
- "Suppression": the target instruction is simply deleted. Loop size and pipeline front end behavior are deeply altered.

It should be noted that such modifications do not preserve original code semantics: numerical values produced by the execution of the modified binary are a priori different from the ones produced by the original code. Worse, some patching can result in codes which cannot be executed: for example nopping the load which provides the value for the denominator of a division can result in a division by zero. In practice such cases are rather infrequent and systematic workarounds can be easily developed. Losing the exact code semantics is not an issue since our primary focus is on performance behavior.

Furthermore, due to X86 instruction set characteristics, such modifications as listed above could introduce spurious dependencies between instruction. To prevent the occurrence of such cases (for example when nopping loads), an additional `PXOR` instruction is inserted, which zeroes register content and breaks any dependencies. The rest of this paper refers to these code modifications as *nopping* without distinguishing the details. In practice, the default modification used was "Suppression" of the target instruction.

3.5.5.2 Finding SV-Node Saturation Time and Bandwidth by Nopping All but One SV-Node

To model the execution of SV-nodes inside the macro, we determine the duration when an SV-node is saturated. This is achieved by nopping instructions *not* corresponding to the SV-node and measuring the execution times after this code change, as shown in Fig. 3.6. DECAN is used to perform this analysis.

By nopping all but the instructions of node *svi*, the macro is converted to the microbenchmarking kernel of node *svi* plus extra nops originating from other SV-nodes. The new execution time approximates the saturation duration of the SV-node when it is executed in the original macro.

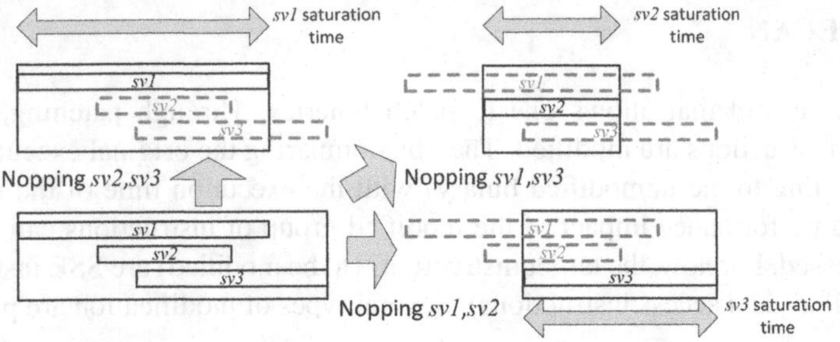

Fig. 3.6 Nopping all but one SV-node reveals overall SV-node saturation duration

Definition 3.2 (Execution time of nopped macro) Given a macro m and a set of SV-nodes S, τ_S^m denotes the execution time of the macro by nopping out instructions corresponding to the SV-nodes in S found in m. For brevity, when the meaning of the macro is clear, we simplify the notation as τ_S. Following are two special cases:

- The execution time obtained by nopping all but the instructions of an SV-node svi is $\tau_{S \setminus \{svi\}}$.
- The execution time obtained by nopping the instructions of an SV-node svi is τ_{svi}.

Suppose the node svi is executed R times during the macro execution, the *pre-adjusted bandwidth value*, $\hat{B_{svi}}$, can be computed using the equation $\hat{B_{svi}} \triangleq \frac{wR}{\tau_{S \setminus \{svi\}}}$ where w is 1 for processor SV-nodes, and is the number of bytes accessed, for memory SV-nodes. It is the rate when the SV-node is fully saturated. The bandwidth value used for CAPE simulation, B_{svi}, is obtained by looking up the bandwidth table generated by microbenchmarking using $\hat{B_{svi}}$. This table look up is performed by matching the characteristics of the SV-node with the microbenchmark parameters (instruction type, load/store sequence, stride, etc.). The capacity for svi, C_{svi}, can be computed using the equation $C_{svi} \triangleq \frac{\tau_{S \setminus \{svi\}}}{T} B_{svi}$ where T is the total execution time of the macro. This value corresponds to the $C_{svi,j}$ value shown in Fig. 3.4.

3.5.5.3 Refining SV-Node Saturation Time by Nopping Single SV-Node

To refine the saturation time of SV-nodes found in Sect. 3.5.5.2, we determine how individual SV-node speed scaling contributes to the execution time of the macro. This is achieved by nopping instructions corresponding to the SV-node and measuring the execution times before and after nopping as shown in Fig. 3.7.

In Fig. 3.7, there are observable execution time changes when we nop $sv1$ and $sv3$, but not for $sv2$. These timing differences give an estimate of the best case we could get by improving the speed of SV-nodes. By nopping SV-nodes one by one, we can break down the execution time of a macro as shown in Fig. 3.8.

It is expected that a portion of the macro execution time cannot be accounted for by nopping a single SV-node. With this abstraction, given m SV-nodes $sv1, \ldots, svm$, we can create an $m+1$ phase model with the following B, C and T values:

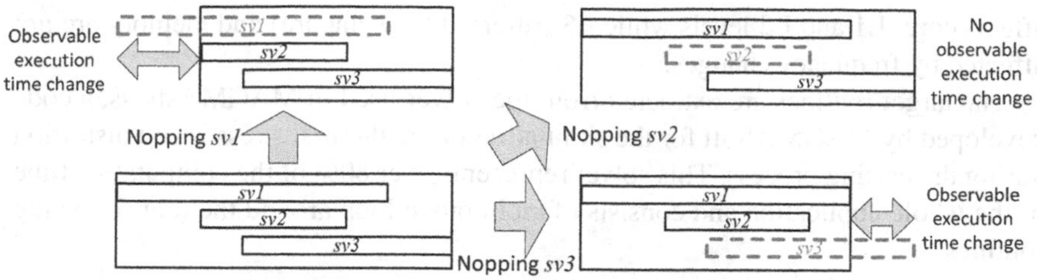

Fig. 3.7 Single SV-node nopping reveals execution time impact

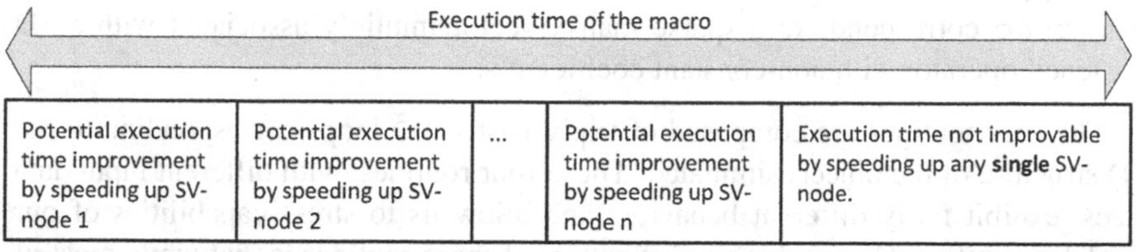

Fig. 3.8 Execution time decomposed by simple SV-node nopping

- For each SV-node svi, the bandwidth value B_{svi} is determined as described in Sect. 3.5.5.2. This bandwidth value is assumed for all the phases, so $B_{svi,j} = B_{svi}$
- The capacity and phase time values are determined as follows:
  - In Phase j where $1 \leq j \leq m$, svj is the only node having non-zero capacity:
    - The phase time, $T_j = \tau_{svj}$.
    - The capacity value, $C_{svi,j} = \begin{cases} B_{svi} & \text{if } i = j \\ 0 & \text{otherwise.} \end{cases}$
  - In Phase $m + 1$, multiple nodes have non-zero capacity.
    - The phase time, $T_{m+1} = T - \sum_{i=1}^{m} \tau_{svi}$ where T is the macro execution time.
    - The capacity value, $C_{svi,m+1} = \frac{\tau_{S\setminus\{svi\}} - \tau_{svi}}{T_{m+1}} B_{svi}$.

3.6 Experimental Results

The following experimental results demonstrate the validity of the v-node approach, using a real code from MAGMAsoft, a leading supplier of metal casting SW (more details about the code used are given in the following Sect. 3.6.1).

3.6.1 Experimental Setting: Hardware, Software and Methodology

The target machine is a dual-socket (2 × 6 cores) Nehalem architecture. Caches L1 and L2 are private to each while L3 is shared by six cores. Frequency scaling only affects core, L1 and L2 levels while L3 (referred to as uncore) and memory are not affected by frequency changes.

The target routines are extracted from the solver used in MAGMAstress, a code developed by MAGMASoft for the simulation of the thermal stresses and distortion during the casting process. This solver represents over 80% of the computation time of the whole application and consists of an iterative loop around the four following routines:

- `scalprod` corresponds to scalar products computation.
- `scalpnorm` corresponds to vector norms computation.
- `saxpy2` corresponds to linear combination of vectors computation.
- `matvec` corresponds to a sparse matrix-vector multiply associated with a 3D stencil operator with non-constant coefficients.

The four routines are composed of triply nested `DO` loops corresponding to the 3D structure of the objects simulated. These four routines, with different input data sizes, exhibit fairly different behaviors and allow us to stress capabilities of our modeling system. One of the key difficulties of these routines is that while most of the operands are single-precision floating-point, many computations are performed in double-precision, inducing a fairly complex code organization for converting back and forth between the two formats. The compiler used was IFC 12.0 with the `-O3` flags: it achieved perfect vectorization and fully used all of the power of the SSE instruction set: in particular throughout the four routines, most of the variants (scalar/packed, single/double precision were generated by the compiler.

Measurements were performed on a standalone system, every experiment was run 30 times, and standard statistical metrics were computed. Overall, across these 30 runs, performance numbers were stable, less than 3% performance variation.

3.6.2 Model

Runs were made on a 12-core system with a medium-size data set, such that each routine reached a steady state behavior, called a phase in the sequel. Our model simply used two v-nodes: core operations (arithmetic instructions using registers) and data access. Such a coarse model was satisfactory because for each routine (depending upon frequency setting), one of the v-nodes was saturated. The performance was measured at two core frequencies, 1.6 and 2.7 GHz, the memory frequency remaining constant. This may be regarded as two systems, one with a slow, and the other with a fast core. The data from each measurement set were used to build a high and low frequency model. Then the CAPE tool was used to evaluate the model and obtain the performance of a system at the other frequency: i.e. 1.6 GHz measurements were used to predict/simulate performance at 2.6 GHz (high frequency model) and vice versa (low frequency model). This leads to eight times being computed, one for each of four phases, at each of two frequencies. The CAPE errors in the eight simulated system times were all within 5% of the correct value. However, there are cases with 10% errors in simulated capacity due to high measured variation in μ values; this is unacceptable for detailed design studies of all nodes in a system.

Better results may be obtained by using a more refined model relying on a larger number of v-nodes. For the 4 routines under study, around 20 v-nodes would provide a good coverage with an increased accuracy. Such a refined model should provide less variation in μ from steady state values. Further experiments will report on more details of the simulated capacity and μ results for each node using the lower level v-nodes, which are expected to approach steady state μ values.

3.6.3 Model Validation Results

Using the framework presented in Sects. 3.6.1 and 3.6.2, a 2-node model is built for low and high frequency. The two nodes in the model are Mistream (memory) and Fpistream (floating point). In this study, we create the model as described in Sect. 3.5.5.2 without the refinement described in Sect. 3.5.5.3.

In practice, the computed capacity of a saturated node may be slightly off from the theoretical bandwidth of that node found by microbenchmarking, so an adjustment is done to scale the node capacities and execution times to make the values consistent with our modeling assumptions described in Sect. 3.3.2. Suppose *SN* is a saturated node in the workload and *UN*is the unsaturated node:

$$C_{SN}^{\text{adjusted}} = \frac{B_{SN}}{C_{SN}} C_{SN} = B_{SN} \qquad C_{UN}^{\text{adjusted}} = \frac{B_{SN}}{C_{SN}} C_{UN} \qquad T^{\text{adjusted}} = \frac{B_{SN}}{C_{SN}} T$$

The scaling factor, B_{SN}/C_{SN}, adjusts the capacity value of a saturated node to make it equal to the bandwidth value. The same factor is applied to the capacity values of other nodes to make the μ value unchanged. The original execution time T is multiplied by the reciprocal of the scaling factor, C_{SN}/B_{SN}. This ensures the scaling of capacities cancels out the scaling of time, making the product of capacity and time invariant.

The adjusted capacity values are entered into the CAPE tool, which then evaluates the model and projects the performance of a system at the other frequency. Following are the results of the CAPE simulations.

The rows in each table correspond to one CAPE simulation. For each loop, there are two simulations—"low to high" and "high to low". The "low to high" simulation creates a model using adjusted capacities and time collected from low CPU frequency setting runs (shown in the columns marked as "Measured"). CAPE projects the performance at the high CPU frequency (shown in the columns marked as "Simulated"). The variation between the simulated values from the collected ones are shown in the columns marked as "Error".

The error in time projection is generally small (maximum error: 4.17%). Since we adjust capacity for saturated nodes, their capacities are always equal to the BW of that node, so the errors in saturated node capacity are all zero. For `scalprod`, `scalpnorm` and high to low simulation of `saxpy2`, Fpistream is saturated, so the

Table 3.1 CAPE simulation of four loops from Magmasoft (part 1)

Loop	Simulation mode	Time [Cycle/Iteration]		
		Measured	Simulated	Error
scalprod	high to low	3.35	3.22	4.00%
	low to high	2.01	2.09	4.17%
scalpnorm	high to low	9.51	9.14	3.93%
	low to high	5.71	5.94	4.09%
saxpy2	high to low	4.61	4.43	3.99%
	low to high	3.64	3.51	3.64%
matvec	high to low	28.72	29.71	3.45%
	low to high	28.14	27.20	3.34%

Table 3.2 CAPE simulation of four loops from Magmasoft (part 2)

Loop	Simulation mode	Capacities					
		Mistream [Byte/Cycle]			Fpistream [FP/Cycle]		
		Measured	Simulated	Error	Measured	Simulated	Error
scalprod	high to low	2.73	3.01	10.26%	2.50	2.50	0.00%
	low to high	4.82	4.37	9.37%	4.00	4.00	0.00%
scalpnorm	high to low	0.74	0.83	10.61%	2.50	2.50	0.00%
	low to high	1.32	1.18	12.16%	4.00	4.00	0.00%
saxpy2	high to low	2.80	3.03	8.21%	2.50	2.50	0.00%
	low to high	3.68	3.68	0.00%	3.04	3.29	8.22%
matvec	high to low	3.58	3.58	0.00%	0.48	0.45	6.25%
	low to high	3.78	3.78	0.00%	0.47	0.51	8.51%

errors for Fpistream capacity are all zero. On the other hand, `matvec` and low to high simulation of `saxpy2` have Mistream saturated, so the error for Mistream capacity is zero. The non-zero error columns in Tables 3.1 and 3.2 tell us the accuracy of CAPE simulation w.r.t. capacity projections.

`saxpy2` is an interesting case because for low CPU frequency it is Fpistream saturated while for high CPU frequency, it is Mistream saturated. CAPE simulation properly simulates this behavior. The errors are related to μ variations as shown in Table 3.3.

The μ variations for `scalpnorm`, `saxpy2` and `matvec` across two CPU frequencies are small ($< 3\%$). Unlike these loops, `scalprod` has a much bigger μ variation ($> 20\%$). The correlation between big μ variation and big capacity error is expected because CAPE holds constant the μ values of nodes when simulating for another frequency.

The μ variations for `scalprod` and `scalpnorm` are greater than 10% and the errors in simulated capacities are of that order. For `saxpy2` and `matvec`, the μ variations are less than 10% and so are the simulated capacities errors.

Table 3.3 μ values of loops and their variations

Loop	$\mu_{\text{Fpistream,Mistream}}$ value at low frequency	$\mu_{\text{Fpistream,Mistream}}$ value at high frequency	Variation
`scalprod`	0.916	0.830	10.39%
`scalpnorm`	3.378	3.030	11.49%
`saxpy2`	0.893	0.826	8.08%
`matvec`	0.134	0.124	7.83%

3.7 Conclusions

This paper describes how to measure a large computational load globally and use the measured data to synthesize the computational capacity model. To demonstrate the validity of our approach, we measured the performance of a real code from Magamasoft and created models to describe the workload under high and low CPU frequency. The CAPE tool was used to evaluate the models and obtain the performance of a system at the other frequency. The preliminary results show that the error in time projection is small while the errors in simulated node capacities are higher. We expect to achieve lower performance projection errors by refining our v-node definition to reduce the μ variation. In addition to this, we are extending the CAPE tool for energy optimization.

References

1. Djoudi, L., Barthou, D., Carribault, P., Lemuet, C., Thomas Acquaviva, J., Jalby, W.: MAQAO: Modular assembler quality analyzer and optimizer for Itanium 2. In: Workshop on EPIC, San Jose, USA (2005)
2. Hockney, R.W., Jesshope, C.R.: Parallel Computers. Hilger, Bristol (1981)
3. Koliai, S., Zuckerman, S., Oseret, E., Ivascot, M., Moseley, T., Quang, D., Jalby, W.: A balanced approach to application performance tuning. In: LCPC, pp. 111–125 (2009)
4. Kuck, D.J.: Computational capacity-based codesign of computer systems. In this volume
5. Petit, E., Bodin, F., Papaure, G., Dru, F.: Poster reception—ASTEX: a hot path based thread extractor for distributed memory system on a chip. In: SC, p. 141 (2006)
6. Wong, D., Kuck, D.: Solving for simulation problems and its use in correctness enforcement for optimization problems (2010). Internal document

Chapter 4
A Compilation Framework for the Automatic Restructuring of Pointer-Linked Data Structures

Harmen L.A. van der Spek, C.W. Mattias Holm, and Harry A.G. Wijshoff

Abstract Memory access patterns are a key factor in the performance of data-intensive applications. Unfortunately, changing the layout in memory of pointer-linked data structures is not a trivial task; certainly not if this is to be done by compiler techniques. The presence of type-unsafe constructs complicates this problem even further. In this chapter, we describe a new, generic restructuring framework for the optimization of data layout of pointer-linked data structures. Our techniques are based on two compiler techniques, pool allocation and structure splitting. By determining a type-safe subset of the data structures of the application, addressing can be done in a logical way (by pool, object identifier and field) instead of traditional pointers. This enables tracing and restructuring per data structure. Further, we describe and evaluate our restructuring methodology, which involves compile-time analysis, run-time rewriting of memory regions and updating referring pointers on both the heap and the stack. Our experiments show that restructuring of pointer-linked data structures can significantly improve performance, while the overhead incurred by the tracing and rewriting is worth paying for.

4.1 Introduction

Predictability in memory reference sequences is a key requirement for obtaining high performance on applications using pointer-linked data structures. This contradicts the dynamic nature of such data structures, as pointer-linked data structures are often used to represent data that dynamically change over time. Also, different traversal orders of data structures cause radical differences in behavior.

Thus, having control on data layout is essential for getting high performance. For example, architectures like the IBM Cell and GPU architectures each have their

H.L.A. van der Spek · C.W.M. Holm (✉) · H.A.G. Wijshoff
LIACS, Leiden University, Niels Bohrweg 1, 2333 CA, Leiden, The Netherlands
e-mail: holm@liacs.nl

H.L.A. van der Spek
e-mail: hvdspek@liacs.nl

H.A.G. Wijshoff
e-mail: harryw@liacs.nl

M.W. Berry et al. (eds.), *High-Performance Scientific Computing*,
DOI 10.1007/978-1-4471-2437-5_4, © Springer-Verlag London Limited 2012

own characteristics and if algorithms using pointer-structures are to be executed on such architectures, the programmer must mold the data structure in a suitable form. For each new architecture, this means rewriting code manually. Another common pattern in code using pointer-linked data structures is the use of custom memory allocators. Drawbacks of this approach are that such allocators must be implemented for various problem domains and that the allocators depend on the knowledge of the programmer, not on the actual behavior of the program. Our restructuring framework is a first step in the direction to liberate the programmer from having to deal with domain specific memory allocation and rewriting of data structures.

In this chapter, we present a compiler transformation chain that determines a type-safe subset of the application and enables *run-time* restructuring of type-safe pointer-linked data structures. This transformation chain consists of type-safety analysis after which disjoint data structures can be allocated from separate memory pools. At runtime, accesses to the memory pools are traced temporarily, in order to gather actual memory access patterns. From these access patterns, a permutation is generated which enables the memory pool to be reordered. Note that these traces are not fed back into a compiler, but are rather used to restructure data layout at *run-time* without any modification of the original application. Pointers in the heap and on the stack are rewritten if the target they are pointing to has been relocated. After restructuring, the program resumes execution using a new data layout.

Restructuring of linked data structures cannot be performed unless a type-safe subset of an application is determined. This information is provided by Lattner and Adve's Data Structure Analysis (DSA), a conservative whole-program analysis reporting on the usage of data structures in applications [10, 11]. The analysis results of DSA can be used to segment disjoint data structures into different memory regions, the memory pools. Often, memory pools turn out to be type-homogeneous, i.e. they store only data of a specific (structured) type.

For type-homogeneous pools, we have implemented *structure splitting*, similar to MPADS [3], the memory-pooling-assisted data splitting framework by Curial et al. This changes the physical layout of the structures, but logically they are still addressed in the same way (any data access can be characterized by a *pool, objectid and field* triplet). Structure splitting is not a strict requirement for restructuring, but it simplifies the implementation and results in higher performance after restructuring.

In order to restructure, a permutation vector must be supplied. This permutation vector is obtained by tracing memory pool accesses. Tracing does have a significant impact on performance, so in our framework tracing can be disabled after a memory pool has been restructured. The application itself does not need to be aware of this process at all. While in principle such a trace can also be used as feedback to the memory allocator the next time the application is executed, we have not done so at this moment. It is important to note that tracing and restructuring all happen within *a single run* of an application.

In order to illustrate the need for restructuring, it is interesting to have a look at what could potentially be achieved by controlling data layout. For this, we used SPARK00 [17, 19], a benchmark set in which the initial data layout can be explicitly controlled. Figure 4.1 shows the potential speedups on an Intel Core 2 system (which

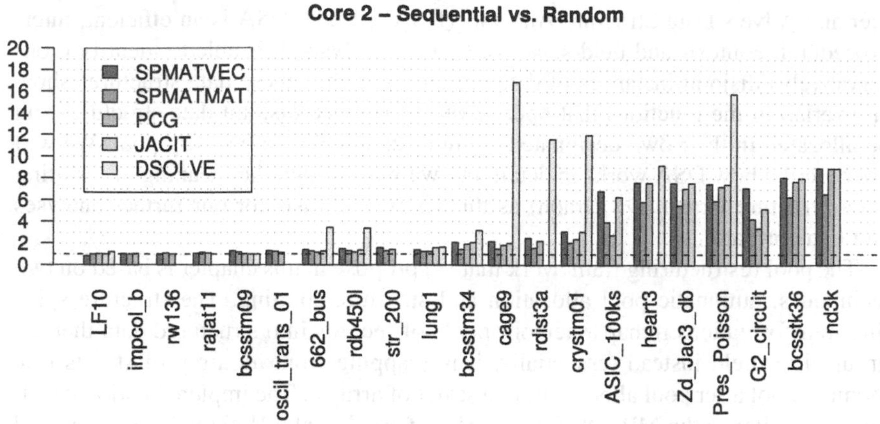

Fig. 4.1 Speedup on SPMATVEC when using data layout with sequential memory access vs. layout with random memory access on the Intel Core 2 architecture

is also used in the other experiments, together with its successor, the Core i7) if the data layout is such that the pointer traversals result in a sequential traversal of the main memory, compared to a layout that results in random memory references. This figure illustrates the potential for performance improvements if data layout could be optimized. Our framework intends to exploit this potential for performance improvements.

Section 4.2 starts with an explanation of work on Data Structure Analysis that our restructuring framework depends on. Section 4.3 describes the compile-time parts of our framework, while Sect. 4.4 treats the run-time components. Section 4.5 contains the experimental evaluation of our framework. Considerable speedups are shown on the SPARK00 benchmarks. The challenge of SPARK00 lies in closing the performance gap between pointer traversals resulting in random access behavior and traversals resulting in perfectly sequential access behavior. As such, it illustrates the potential, but it does not guarantee that such speedups will be obtained for any application. The section also goes into the overhead of the tracing mechanism. It is shown that the performance gains do compensate for this overhead within relatively few consecutive uses of the restructured data structure. Different mechanisms for stack management and their implications are discussed and evaluated. Further, the improved address calculations, compared to the address calculations in Curial's work [3], used for addressing split memory pools are discussed. Related work is discussed in Sect. 4.6. Future work and conclusions are given in Sect. 4.7. Part of this work has appeared in a previous paper [18].

4.2 Preliminaries

The restructuring framework presented in this chapter relies on the fact that a type-safe subset of the program has been identified. This is achieved by applying Lattner and Adve's Data Structure Analysis (DSA) [9–11]. DSA is an efficient, interprocedural, context- and field-sensitive pointer analysis. It is able to identify (conservatively) disjoint instances of data structures even if these data structures show an overlap in the functions that operate on them. Such disjoint data structures can be allocated in their own designated memory area, called a memory pool. We will not describe how DSA works in detail, but we will explain the meaning of resulting Data Structure Graph (DS Graph) as this forms the basis for our further analyses and transformations.

The pool restructuring framework that we propose in this chapter is based on two techniques: automatic pool allocation and structure splitting. The structure splitting transformations remaps memory pools of records into structured data that are grouped by field instead (essentially, it is mapping from an array of structs (the memory pool after pool allocation) to a struct of arrays). The implementation developed is similar to the MPADS framework of Curial et al. [3], though we optimized the address calculations for commonly occurring structure layouts (Sect. 4.5).

In this section, both DSA and structure splitting, which our analysis passes and transformations depend on, are explained in further detail.

Data Structure Analysis Data Structure Analysis (DSA) provides information on the way data structures are actually used in a program. DSA determines which data structures can be proved disjoint in memory. Such a data structure can be a linked list, a tree, a graph or any other pointer-linked data structure, DSA does however not determine the shape.

The result of DSA is the Data Structure Graph (DS Graph). Within this graph, the nodes represent memory objects. A node is described as follows [9]:

> Each DS graph node represents a (potentially unbounded) set of dynamic memory objects and distinct nodes represent disjoint sets of objects, i.e., the graph is a finite, static partitioning of the memory objects. Because we use a unification-based approach, all dynamic objects which may be pointed to by a single static pointer variable or field (in some context) are represented as a single node in the graph.

Construction of the DS graph occurs in three phases. The *Local Analysis Phase* during which the DS graphs are constructed for all functions, taking only local information into account. The *bottom-up phase*, combines the information on the local functions with results from their callees, by propagating this information bottom-up. This phase is context sensitive. The *top-down phase*, is not used by our restructuring framework. We use the result from the bottom-up phase.

Figure 4.2 shows a part of the main function of SPMATVEC, one of the benchmarks used in the evaluation of our method (see Sect. 4.5). Figure 4.3 shows the associated DSGraph and the two stack variables *%tmp* and *%Matrix*. Each of these variables has its own storage space on the stack and is therefore represented by separate nodes. The *MatrixFrame* structure they are both pointing to is one node, indicating that the analysis cannot prove that they are pointing to disjoint structures. The *MatrixFrame* structure contains three pointers to arrays that point to the start of a row, the start of a column and the diagonal elements. The *MatrixElement* structure is the structure containing the matrix data. It has two self references, which are the two pointers used to traverse the matrix row- and column-wise.

```
int main( int argc, char **argv )
{
    ...
    MatrixPtr tmp = ReadMatrixPtrRow( matrixFile );
    MatrixPtr Matrix = MatrixToFormat( tmp, format );
    ...
    for( i = 0; i < iterations; i++ )
        MatrixMultiplyVec( Matrix, right, result );
    ...
}
```

Fig. 4.2 Code excerpt of main function of SPMATVEC

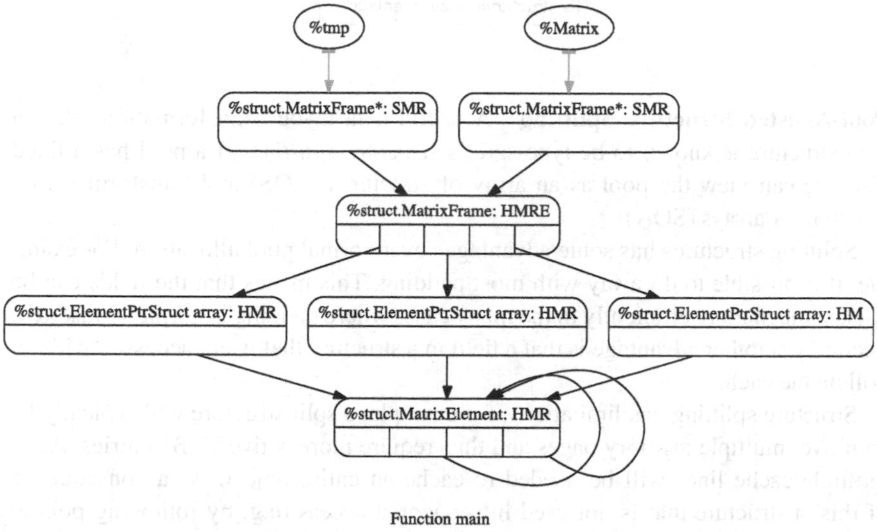

Fig. 4.3 DSGraph for main function of SPMATVEC benchmark

Each function has its own bottom-up DS graph. Nodes that do not correspond to a formal argument depict data structures that are instantiated within this function. These nodes incorporate all information on how they are used in all callees (including whether the usage is type-safe or not).

Automatic Pool Allocation On top of DSA, Lattner et al. implemented *automatic pool allocation* [10, 11]. Pool allocation is a transformation that replaces calls to memory allocation functions by custom pool-based memory allocators. Each provably disjoint data structure may then be placed in separate type homogeneous pools. Pool allocated structures allow for precise control on data layout, as it is known that all allocated elements within a particular region have the same type. We use this property to modify the way structures are laid out in main memory.

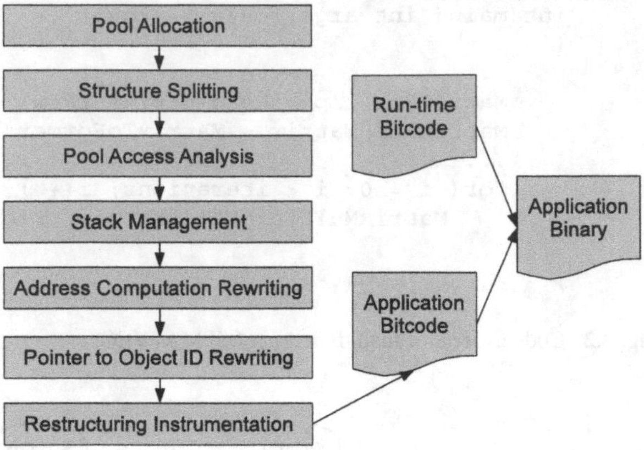

Fig. 4.4 Overview of the pool restructuring compilation chain. GEPI refers to the LLVM GetElementPtrInst instruction

Pool-Assisted Structure Splitting A useful data layout transformation when a data structure is known to be type-safe is *structure splitting*. If a pool has a fixed size, we can view the pool as an array of structures (AOS) and transform it to a structure of arrays (SOA).

Splitting structures has some advantages over normal pool allocation. For example, it is possible to do away with most padding. This means that the fields can be packed much more efficiently in the many cases where padding is normally inserted. Secondly, another advantage is that a field in a structure that is not accessed will not pollute the cache.

Structure splitting has limitations, for example, a split structure will typically be split over multiple memory pages and thus require more active TLB[1] entries. Also, multiple cache lines will be needed to cache an entire object. As a consequence of this, a structure that is not used in sequential access (e.g. by following pointer chains), is not likely to yield any performance benefits when split.

The implementation of our structure splitting transformation is similar to the DSA-based implementation of Curial et al. [3], who implemented structure splitting in the IBM XL compiler.

4.3 Compile-Time Analysis and Transformation

At compile time, a whole program transformation is applied in order to rewrite pools to use a split structure layout that supports run-time restructuring. Figure 4.4 shows an overview of the entire compilation chain for our framework. In this section, the analyses and compiler passes that rewrite the code are discussed.

[1] Translation Look-aside Buffer.

4.3.1 Structure Splitting

Our analysis and transformation chain starts at the point where DSA has been performed on a whole program and pool allocatable data structures have been determined. We then start at the *main* function and traverse all reachable functions, cloning each function that needs to be rewritten to support the data layout of split structures. Note that cloning is only done along execution paths that are known to have type-safe data structures that can be split safely. Functions are cloned because there might also be calling contexts in which splitting cannot be applied, and these cases must also be dealt with correctly (see Lattner and Adve's work [10, 11]). We also use the their technique for the identification of the memory pools. It is not possible to split pools that are not type homogeneous as addressing of object fields would become ambiguous and fields of different types and length would introduce aliasing of field values. This information is however available from the DSA and pool allocation passes.

Rewriting of other instructions, such as address calculations are deferred to a later stage, because they are nothing more than a simple rewrite of the address calculation instruction (*GetElementPtrInst*) in LLVM.

Some information is gathered in the structure splitting analysis pass to be used in subsequent passes. All loads and stores to pool data are identified as well as all loads and stores that store a pointer into a pool. These loads are needed to support the use of *object identifiers* instead of pointers (see Sect. 4.3.5). The structure splitting pass ensures that all the address calculation expressions (*GetElementPtrInst* in LLVM) whose result points to data in split pools are identified. These expressions must be rewritten before the final code generation. The rewriting is deferred as later passes may need to do additional analysis on the *get element pointer* instructions.

4.3.2 Pool Access Analysis

Pool access analysis is a pass in which all pool accesses (loads and stores) are analyzed. The result of the analysis is that instead of using a specific pointer, a triplet (*pool, object, field*) is used to represent location read from or written to. *Pool* is the pool descriptor used at runtime, *object* the pointer to the object the data belong to and *field* is the field number that is accessed. This is analogous to data access in a database (table, row and column).

For each load and store from a split pool we determine the underlying base object by following the chain of instructions attached to the load/store instruction. The previous analysis will already have associated a pool descriptor with the base object, and the accessed field can be determined by inspecting the get element pointer instruction used by the load/store.

Note that for each access to a pool, it must be possible to determine which field is accessed. This property cannot always be proved if the address of one of the fields is taken, and therefore we do not allow that *any address* of a field is written to

Table 4.1 The three stack management options and their individual pros and cons

Method	Advantages	Disadvantages
Pointer Tracking	Simple	Slow
	Portable	Interferes with IR
Shadow Stack	Fast	Interferes with IR
	Portable	
Stack Map	Fast	Backend modifications
	No IR Interference	Stack walking not portable

any memory location using the LLVM *StoreInst*. For example, the following C-code snippet will never be restructured:

```
obj->ptr = &p1->y;
    ...
*obj->ptr = val;
```

This might be a bit over-conservative, and in a future version, we might define this more precisely. Lattner and Adve's pointer compression applies the same restriction on field accesses [12].

4.3.3 Stack Management

The primary requirement for structure splitting to work (in terms of code modifications) is the remapping of address calculation expressions so that data are read and written to the relocated location in the split pool. However, if reordering of the pool contents is to be accomplished this is not sufficient. Other pools may for example contain references to the reordered pool (which means that those references need to be updated). However, these on-heap pointers are not the only references to pool objects that the system needs to deal with. The other type of references that need to be managed are pointers that are stored on the stack and that point into the pool. This problem is similar to what garbage collectors have to do, and in their terminology, the on-stack pointers are known as roots. Tracking the on-heap pointers can be done by adding additional meta data to the pool descriptor, these meta data are derived from the DSA (that keeps track of connectivity information between pools).

Three different alternatives to accurate stack management were explored and evaluated. These approaches include *explicit pointer tracking*, *shadow stacks* and *stack maps*. However, only the first method was fully implemented for reasons that will become clear later on. The three different investigated methods for stack management are summarized in Table 4.1.

Explicit Pointer Tracking One approach to the stack root issue, is to ensure that all pointers are explicitly tracked at the LLVM level. We call this technique pointer tracking. When a pool descriptor is allocated, a special segment of data is acquired that will be used to track all stack local pointers pointing into the pool, whenever a pointer is allocated on the stack, the location of this pointer is inserted in the per-pool stack tracking block. A frame marker is in this case also needed to enable the removal of all the pointer tracking entries associated with a returning function. In LLVM this means that any pointer that is an SSA register must explicitly stored on the stack. The following LLVM function illustrates this a bit further:

```
void @func(pooldesc *pool0) {
entry:
bb0:
  %x = load {i32, i32}** %heapObjectAddr
  call void @foo %x
  ret
}
```

The function listed above is then transformed into the following:

```
void @func(pooldesc *pool0) {
entry:
  %xptr = alloca {i32, i32}**
  call void @split_st_reg_stack_obj %pool0, %xptr
  call void @split_st_push_frame %pool0
bb0:
  %x = load {i32, i32}* %heapObjectAddr
  store %x, %xptr
  %x_foo_arg = load {i32, i32}* %xptr
  call void @foo %x_foo_arg
  call void @split_st_pop_frame %pool0
  ret
}
```

In the transformed function the pointer %x is explicitly backed by a stack variable and this variable is then registered with the run-time function *split_st_reg_stack_obj*. After the pointer registrations, a call to the run-time function *split_st_push_frame* is executed; this function will close the stack frame for the current function in order to speed up the pop operation of the stack. These run-time functions are very short (a few instructions) and will be inlined and thus not induce any function-calling overhead. Figure 4.5 shows how the pointer tracking block is constructed at run-time.

In order to reduce this overhead, an approach where stack tracking is disabled in certain functions has been chosen. The pseudo code in the following example illustrates why this is useful:

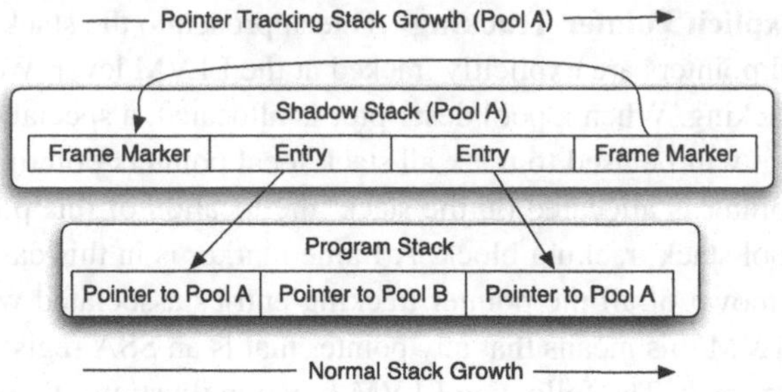

Fig. 4.5 Pointer tracking layout

```
Pool pool;
Matrix *mtx = readMatrix(pool);

doMatrixOperation(pool, mtx);

PoolRestructure(pool, mtx);

for (int i = 1 ; i < N ; i ++) {
 doMatrixOperation(pool, mtx);
}
```

Here the critical code is the *doMatrixOperation*, but if this operation does not call the *PoolRestructure* function, then this function does not need to track the pointers.

The most important point with explicit pointer tracking is that it is very easy to get running. In terms of implementation effort it was small compared to the methods described further on in Sects. 4.3.3 and 4.3.3. Though, while the method (as shown in experiments later on) is not a good choice for a fielded deployment, it takes very little code to implement both *passes* and the run-time support for the explicit pointer tracking.

The pointer tracking method was selected for implementation due to its implementation simplicity, but not before two other stack managing methods had been evaluated.

Shadow Stacks The second approach that we investigated for tracking pointers on the stack, was the utilization of a shadow stack. This technique is based on the garbage-collection method described by Henderson [7]. To implement shadow stacks the compiler creates a per-function data structure where pointers that are stored on the stack will be stored as a group, such that each pointer can be addressed relative to the base of this data structure. When a function is called, such a structure is allocated on the stack and this structure is then registered with the run-time. This pre-registration cuts down on the additional registration overhead compared to the pointer tracking, by only inferring one registered pointer per function call.

Stack Maps The third alternative is the construction of stack maps (for example described by Agesen in [1]). Stack maps are structures that are generated statically for each function; these structures describe the stack frames of the corresponding functions. The maps are computed during the code generation phase and contain information about, for example, frame-pointer offsets of the pointers allocated by the function. The main advantage of delaying this to the code generation phase is that the transformation will not interact in any way with earlier optimizations. The main drawback is that the stack walking will become platform dependent and this may not necessarily suit every compiler.

4.3.4 In-Pool Addressing Expression Rewriting

For non-split structures, the derived pointers to fields in the structures can easily be computed by adding a constant offset to the base pointer of the structure. For split structures however, this is not possible anymore. In a split structure the field addresses no longer have constant offsets from the base pointer of the structure.

It is obvious that calculating addresses for the fields in the structures must be very efficient. This fact was already stressed by Curial [3], but he did not optimize the address calculation expressions and their selection rules to the same extent as we did. If this calculation is inefficient, it potentially nullifies much of the performance improvement gained from the more cache-efficient split structure representation. In general, the offset for field *n* can be represented by the following equation:

$$offset_n = k_n + sizeof_n \frac{p\&(sizeof_{pool} - 1)}{sizeof_0} - p\&(sizeof_{pool} - 1) \tag{4.1}$$

where k_n is the constant offset to field array n from the pool base. The sub-expression $p\&(sizeof_{pool} - 1)$ calculates the object pointer p's offset from the pool base and the expression $\frac{p\&(sizeof_{pool}-1)}{sizeof_0}$ calculates the object index in the pool.[2]

When the accessed field is the first field of the structure then $offset_0 = 0$ and if the size of the accessed field is the same as the first field of the structure then $offset_n = k_n$.

We have observed that in many common cases the size difference between the accessed field and the first field is a power of two. Taking this observation into account, we introduce two additional expressions. When the size of the accessed field is greater than the first field of the structure we have

$$offset_n = k_n + (sizeof_n - sizeof_0) \frac{p\&(sizeof_{pool} - 1)}{sizeof_0} \tag{4.2}$$

and when the size of the first field is greater than the accessed field use the following expression:

$$offset_n = k_n - (sizeof_0 - sizeof_n) \frac{p\&(sizeof_{pool} - 1)}{sizeof_0} \tag{4.3}$$

---

[2]Note that in this context, & is the C-operator for a *bitwise AND*.

Equation (4.1) can be viewed as adding the pool base to the offset from the address of the nth field of the first object. Equations (4.2) and (4.3) take into account the linear drift of field n due to the size differences between fields 0 and n, with respect to the object's pool index and the constant offset k_n.

It is assumed that further passes of the compiler will apply strength reduction on all multiply and divides involving a power of two constant. Fog [4] gives the cost for various instructions for a 45 nm Intel Core 2 CPU. These numbers have been used to estimate the cost in cycles for the various equations calculating the offsets. Assuming that the expressions have been simplified as much as possible through, for example, constant folding and evaluation, we see that when neither $sizeof_0$ nor $sizeof_n$ are powers of two, Eq. (4.1) will take 26 cycles. If $sizeof_0$ is a power of two the same equation will take six cycles (as the very costly divide will be reduced to a shift) and if both sizes are powers of two it will take four cycles. Equation (4.3) will take three cycles, and Eq. (4.2) will take three cycles in the normal case (or two cycles if $sizeof_n - sizeof_0 = sizeof_0$).

The address calculations as defined by Eq. (4.1) and the elimination of calculations if accessing the first field are already used in MPADS [3], but our additional Eqs. (4.2) and (4.3), have some important properties. They allow the calculation of the field offsets to be reduced to two or three instructions instead of four, as the code generator will merge the divide and the multiplication operation into a single shift operation and that the third term in Eq. (4.1) has been eliminated. Note that for Eq. (4.2) when $size_n - size_0 = size_0$, LLVM will automatically eliminate the multiply and the divide instruction, giving even more savings.

The most notable equation cost (26 cycles) come from the existence of a divide instruction in the expression. This will, for example, happen when the first field of a structure is an array of three 32-bit values (arrays are not split since they are already sequential) and the next element is a 32 or 64 bit value. In those cases up to 23 cycles may be saved on the address calculation because the divide instruction has been eliminated through strength reduction introduced by Eq. (4.3).

Overall it can be said that a compiler that splits structures should also reorder the fields in a structure so that address calculations are made as simple as possible. For example, if a structure contains three fields of lengths 1, 2 and 4 bytes, then the field ordering should place the 2-byte element first under the condition that the access frequency of the fields is the same. Though, at this moment our implementation does not do this and this field reordering remains on the future work list.

4.3.5 Converting Between Pointers and Object Identifiers

Instead of storing pointers in split memory pools, object identifiers are used. Object identifiers can be used in type-homogeneous pools to uniquely identify an object within a pool. Together with a field number, each data element can be addressed. Object identifiers are indices within a pool and thus a more compact representation than pointers or byte offsets in the pool, as used in Latttner and Adve's *static pointer compression* [12]. Their *dynamic pointer compression* transformation also uses object identifiers. In that case, it provides a representation independent of the size of fields, whereas byte offsets would need to be rewritten if field sizes change.

Our motivation to use object identifiers is different. While our framework would also benefit from pointer compression (currently object identifiers are stored as 64-bit unsigned integers), we use object identifiers because they can be used as indices in permutation vectors and because they provide position independence for data structures.

Section 4.3.2 described how all loads and stores to memory pools can be represented as a (*pool, object, field*) triplet. In the case that *field* is a field that is pointing to pool-allocated data (whether this defines a recursive data structure or a link to another data structure does not matter), the pointer value that will be stored into the memory pool needs to be converted to an object identifier before it is stored. When such a pointer value is loaded from a memory pool, it must be converted from an object identifier to a pointer. Loads and stores to the stack are unaffected and thus will contain real pointers. As no pointers to fields, but only pointers to objects will be stored to the memory pools, we only need conversion functions for object pointers.

For stores the value to store is rewritten as follows:

```
uintptr_t ptr_to_objid(split_pooldesc_t *pool, void *obj)
{
  uintptr_t objIdx;
  if( obj == 0 ) return 0;  // Special case: NULL pointer
  else {
    uintptr_t poolBase = (uintptr_t)pool->data;
    uintptr_t objOffset = poolBase - (uintptr_t)obj;
    objIdx = objOffset / sizeof_field(0);
  }
  return objIdx;
}
```

And for loads the loaded value is rewritten as follows:

```
void *objid_to_ptr(split_pooldesc_t *pool, uint64_t *objIdx)
{
  if( objIdx == 0 ) return 0; // Special case: NULL pointer
  else {
    uintptr_t poolBase = (uintptr_t)pool->data;
    uintptr_t objOffset = objIdx * sizeof_field(0);
    uintptr_t obj = poolBase + objOffset;
    return (void *)obj;
  }
}
```

Note that the actual implementation uses LLVM bit code and uses a bitmask instead of an if-statement to handle the NULL pointer.

Compared to the description of object indexing used in the pointer compression transformation by Lattner and Adve [12], our implementation differs in some ways. In their work, object indices are not only present in the heap, but are also used on the stack. Pointer comparisons and assignments do not need the object identifier to be expanded to a full pointer in their framework. In our framework, only loads and stores of pointers (only to pool objects) to split pools need rewriting, and the rest of the code will run unchanged; this substantially simplifies the restructuring step.

4.3.6 Restructuring Instrumentation

Pool tracing and restructuring of data structures requires instrumentation of the code with calls to the tracing runtime. During pool access analysis, all loads and stores to pools have been identified and are represented using the triplet (pool, object and field). All these instructions can be instrumented such that a per pool and field trace of object identifiers is recorded. Currently, we only trace load instructions.

Tracing is only done for one execution of a function as continuous tracing would be very costly. After this first tracing, the data are restructured and tracing is disabled. This is accomplished by generating two versions of the function, one with and one without tracing. Selecting the proper function is done through a global function pointer that is set to the non-traced version after a trace has been obtained.

4.4 Run-Time Support

Extracting a type-safe subset of the program and replacing its memory allocation by a split-pool-based implementation requires run-time support, similar to the runtime provided for regular pool allocation. The split-pool runtime provides *create* and *destroy* split pool functions and memory allocation and deallocation functions. In addition, some common operations implemented in the standard C library are also provided, such as a split *memcpy* function (which needs to copy data from multiple regions due to the split layout).

Tracing and Permutation Vector Generation In order to restructure a memory pool a permutation must be supplied to the restructuring runtime. The pool access analysis pass (Sect. 4.3.2) provides the compile-time information (pool, object and field) about all memory references and these memory references can all be traced. Traces are generated per pool, per field. For each pool/field combination, this results in a trace of object identifiers. From any of these traces, a permutation vector can be derived which can be used to permute a pool. The permutation vector is computed by scanning the trace sequentially and appending the object identifiers encountered to the vector, avoiding duplicates:

```
perm[0] = 0;
permLen = 1;
for (i = 0; i < maxTraceEntry; i++)
   if ( !perm[trace[i]] ) {
     perm[trace[i]] = permLen;
     permLen++;
   }
```

Element 0 is reserved to represent the NULL pointer and is thus never permuted.

Since tracing is expensive, it should be avoided if it is not necessary. For the evaluation of our restructuring method we choose to trace the first execution of a specified function (compiler option specifies which function). After the trace, the pools are restructured and tracing disabled. In a future implementation, this will be dynamic and tracing could be triggered if a decrease in performance is detected (for example by using hardware counters).

Pool Reordering One of the more important parts of our system is the pool-rewriting support. Rewriting in this context means that a pool is reordered in memory, so that it is placed in a hopefully more optimal way with respect to memory access sequences. This is done at runtime, and the rewriting is based on passing in a permutation vector generated during runtime as described in Sect. 4.4. We have implemented a copying rewriting system that uses permutation vectors that specify the new memory order of the pool. Although this vector could be generated automatically in some cases (e.g. for single linked lists) this is not done at this point in time.

The pools rewriting algorithm has three distinct phases:

1. *Pool rewrite*, where the actual pool-objects are being reordered
2. *Referring pool rewrite*, where pointers in other pools that refer to the rewritten pool are updated to the new locations
3. *Stack update*, where the on-stack references to objects in the rewritten pool are updated

The basic algorithm for the interior pool update is as follows:

```
newData = mmap(pool.size);
foreach field in pool
  foreach element in field
    if field contains recursive pointers
      newData[field][permVec[element]]
        = permVec[pool.data[field][element]];
    else
      newData[field][permVec[element]]
        = pool.data[field][element];
munmap(pool.data);
pool.data = newData;
```

In this case each field in the split pool is copied into the new address space, and relocated according to the permutation specified in the permutation vector. If the value in the field is itself a pointer to another object in the pool, that pointer is remapped to its new value. For the second phase where all the referring pools are updated, the rewrite is even simpler:

```
foreach referrer in pool.referrers
  foreach entry in referrer.field
    referrer.field[entry] = permVec[referrer.field[entry]];
```

Here, each pool that refers to the rewritten pool will have the field containing those pointers updated with the new locations.

The algorithm detailed here assumes that each pool descriptor has information available regarding the pool connectivity (i.e. which fields in other pools that point out objects in the rewritten pool). This information can be derived from the DSA discussed earlier. This connectivity information is therefore registered as soon as the pool is created.

Stack Rewriting As already discussed in Sect. 4.3.3, the program stack is managed through explicit pointer tracking. When a pool descriptor is allocated, a special segment of data is acquired that is used to track all pointers on the stack pointing into the pool. Whenever a pointer is allocated on the stack, the location of this pointer is inserted in the per-pool stack-tracking block together with some meta information such as whether the pointer is a base or derived pointer and in the latter case the field.

When a pool is rewritten, the current stack will be traversed and all base and derived pointers to locations within the pool are rewritten to reflect the new location of the object.

4.5 Experiments

The challenge of a restructuring compiler is to generate code that will automatically restructure data, either at compile time or runtime, in order to achieve performance that matches the performance when an optimal layout would be used. In the introduction the potential of restructuring was shown. In the experiments here, we ideally want to obtain similar performance gains, but by automatic restructuring of data layout of the pointer-linked data structures.

We use the benchmark set SPARK00 which contains pointer benchmarks whose layout can be controlled precisely [16, 17, 19]. The pointer-based benchmarks used are: SPMATVEC (sparse matrix times vector), SPMATMAT (sparse matrix times matrix), DSOLVE (direct solver using forward and backward substitution), PCG (preconditioned conjugate gradient) and JACIT (Jacobi iteration).

These benchmarks implement matrices using orthogonal linked lists (elements are linked row-wise and column-wise). All of them traverse the matrix row-wise, except DSOLVE, which traverses the lower triangle row-wise and the upper triangle column-wise.

For all benchmarks, one iteration of the kernel is traced, after which the data layout is restructured and tracing disabled. This all happens at run-time, without any hand-modification the application itself.

The experiments have been run on two platforms. The first is the Intel Xeon E5420 2.5 GHz processor with 32 GB of main memory, running Debian 4.0. The other system is an Intel Core i7 920 2.67GHz based system with 6 GB of main memory, running Ubuntu 9.04.

Pool Reordering As shown in the introduction, being able to switch to an alternative data layout can be very beneficial. We applied our restructuring transformations to the SPARK00 benchmarks and show that in ideal cases, speedups exceeding 20 times are possible by regularizing memory reference streams in combination with structure splitting. Of course, the run-time introduces a considerable amount of overhead and is a constant component in our benchmarks. We will consider this overhead separately in Sect. 4.5 to allow a better comparison between the different data sets.

We first determine the maximal improvement possible, by using an initial layout that causes random memory access. Figure 4.6a and 4.6b show the results of restructuring on the pointer-based SPARK00 benchmarks (except DSOLVE, which is treated separately), if the initial data layout causes random memory access, on the Intel Core 2 and Core i7, respectively. The data set size increases from left to right. As shown in previous work [17, 19], optimizing data layout of smaller data sets is not expected to improve performance that much and this fact is reflected in the results. As expected, restructuring had no significant effect for data sets fitting into L1 cache and will therefore not be included in the figures. For sets fitting in the L2 and L3 cache levels, speedups of 1–6× are observed. The Core i7 has a 8 MB L3 cache, whereas the Core 2 only has two cache levels. This explains the difference in behavior for the matrix *Sandia/ASIC_100ks*, which shows higher speedups for the Core 2 for most benchmarks. However, it turns out that the Core i7 runs almost 3× faster when no optimizations are applied on SPMATVEC for this data set. Therefore, restructuring is certainly effective on this data set, but the greatest benefit is obtained when using data sets that do not fit in the caches.

An interesting case is DSOLVE, in which the lower triangle of the matrix is traversed row-wise, but the upper triangle is traversed column-wise. As the available data layouts of the matrices are row-wise sequential (CSR), column-wise sequential (CSC) or random (RND), none of these orders matches the traversal order used by DSOLVE. Figure 4.7a and 4.7b show the results for DSOLVE using the different memory layouts on the Core 2 and Core i7, respectively. The matrices are ordered differently than in the other figures, as DSOLVE uses LU-factorized matrices as its input, which have different sizes depending on the number of fill-ins generated during factorization. The matrices have been ordered from small to large (in the case of DSOLVE, this is the size after LU-factorization).

For the *lung1* data set, a decrease in performance is observed, but for the larger data sets, restructuring becomes beneficial again. Speedups of over 6× are observed for the Core i7, using CSC (column-wise traversal would yield a sequential memory access pattern) as initial data layout. In principle, the RND (initial traversal yields a random memory reference sequence) data set could achieve much higher speedups if after restructuring the best layout has been chosen. Currently, this is not the case for DSOLVE and we attribute this to the very simple permutation vector generation algorithm that we use (see Sect. 4.4).

Tracing- and Restructuring Overhead Our framework uses tracing to generate a permutation vector that is used to rewrite the memory pool. Currently, the trace to be used is specified as a compiler option, but this could potentially be extended to a system that autonomously selects an appropriate trace.

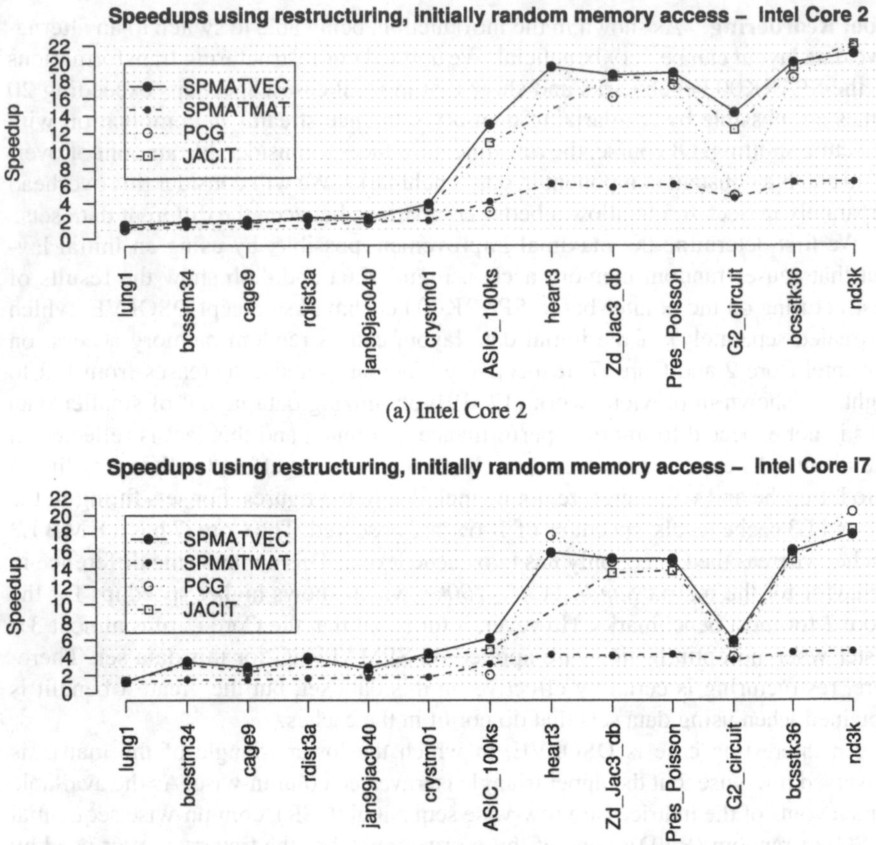

(a) Intel Core 2

(b) Intel Core i7

Fig. 4.6 Speedups obtained using restructuring on the SPARK00 benchmarks. The initial data layout is random

In the benchmarks, we choose to only trace the first iteration of the execution of the kernel due to the time and memory overhead coming from the tracing. In order to minimize the overhead, the trace will only contain object identifiers, as described in Sect. 4.4. For instance, if a linked list contains a floating point field and this list is summed using a list traversal, then if both the pointer field and the floating-point field are traced there is an overhead of two trace entries per node visited. The structure operated on is 32 bytes in size and tracing the above-mentioned traversal would add 16 bytes per node extra storage requirements (when using 64-bit object identifiers). Subsequently, the memory pool is restructured using the information of the trace which relates to the field that contains the floating point values of the linked list nodes.

The overhead of the tracing and restructuring has been estimated by running a single iteration of each kernel with and without tracing and restructuring enabled, using a data layout causing random memory access. Figure 4.8 shows the interpo-

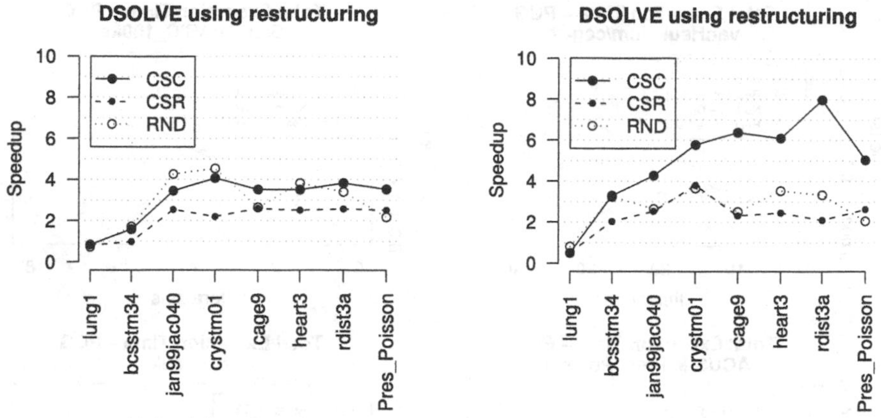

(a) Intel Core 2 (b) Intel Core i7

Fig. 4.7 Speedups obtained using restructuring on DSOLVE for all different initial layouts. Input data sets are ordered by size (after LU-factorization)

lated execution times of the benchmark PCG, both with and without restructuring for the Core 2 and Core i7 architectures. The initial data layout produces random memory access behavior of the application, which is eliminated after the restructuring. Four different matrices have been used which are representative in terms of performance characteristics (see Fig. 4.6a and 4.6b). The break-even points for all matrices are included in Table 4.2.

The figures show that tracing does come with an additional cost, but for most (larger) data sets the break-even point is reached within only a few iterations. For instance, for all data sets shown in Fig. 4.8, the break-even point is reached within four iterations, except for *cage9*, which is the smallest data set depicted. Interestingly, on the Core i7, the break-even point is reached even quicker, making restructuring more attractive on this architecture.

Although we have shown that the additional costs of tracing are manageable, it should be noted that we only showed this on computational kernels. In general, it is not recommended to trace a full application code. Therefore, as we have noted earlier in this chapter, tracing should only be turned on selectively on some functions.

Run-Time Stack Overhead In order to quantify the overhead from the stack management that is needed when restructuring a pool, a few custom programs have been written. The interesting overhead in this case will be a measurement of per-function and per-pointer overhead.

An experiment was carried out where a function is called that declares (and links together) a certain number of pointers that point into a pool. This was repeated for a multiple number of pointers and for both a version of the program built without the semi-managed stack and one version that was built with the semi-managed stack enabled. The function was in turn executed a couple of million times.

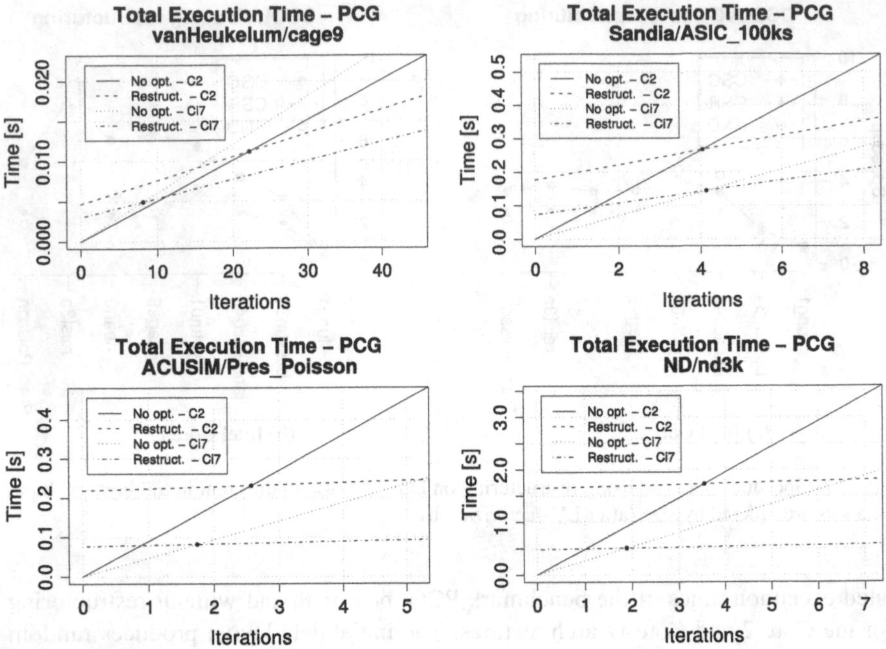

Fig. 4.8 Execution times with and without restructuring. The break-even points are marked with a *dot*

The following code demonstrates how this experiment was conducted:

```
listelem_t*
nextElem(list_t *list)
{
  if (list->current)
  list->current = list->current->next;

#pragma MAKE_POINTERS

  return list->current;
}
```

where the MAKE_POINTERS pragma was replaced by

```
listelem_t *a0 = list->current;
listelem_t *a1 = a0;
...
listelem_t *aN-1 = aN;
```

The execution time for the loop calling the *nextElem* function was measured and the difference between the managed version and unmanaged version should thus represent the overhead introduced for that number of pointers.

Figure 4.9 shows the execution time on a 2.5 GHz *Intel Core 2 Duo*, of 4 million calls to the function above. The data evaluate to a base cost of five cycles per

Table 4.2 Number of iterations for the break-even points when tracing and restructuring is enabled, when using an initial random data layout. The matrices are ordered by increasing size. The lower part of the table contains the larger data sets, which do not fit in the caches. DSOLVE performs worse using *lung1* therefore a break-even point is not applicable. The missing entries for JACIT are due to zero elements on the diagonal. For DSOLVE the missing entries are due to matrices that take too long to factorize

Matrix	*spmatvec*		*spmatmat*		*pcg*		*jacit*		*dsolve*	
	C2	Ci7	C2	Ci7	C2	Ci7	C2	Ci7	C2	Ci7
lung1	42.1	51.8	113.9	58.3	388.5	31.5	98.8	55.4	N/A	N/A
bcsstm34	24.3	6.2	53.6	29.5	22.7	5.6	27.2	6.7	19.8	4.0
cage9	21.0	8.1	44.3	26.1	22.1	8.1	28.6	10.3	2.9	2.0
rdist3a	17.9	5.6	39.5	21.1	17.7	5.2	–	–	3.2	2.1
jan99jac040	16.0	8.0	16.3	15.3	17.8	8.2	–	–	1.1	1.3
crystm01	8.3	4.9	17.1	17.0	9.1	4.9	10.8	5.8	2.2	1.8
ASIC_100ks	2.3	3.9	4.4	5.0	4.0	4.1	2.4	4.4	–	–
heart3	2.4	1.7	4.6	4.8	2.4	1.5	–	–	3.1	2.2
Zd_Jac3_db	2.5	1.6	4.6	4.8	2.6	1.7	2.6	1.9	–	–
Pres_Poisson	2.6	1.7	4.7	5.0	2.6	1.7	2.7	2.0	3.8	3.0
G2_circuit	2.6	4.7	4.6	4.7	5.0	5.1	2.6	5.7	–	–
bcsstk36	3.0	1.7	5.1	5.0	3.1	1.8	3.1	2.0	–	–
nd3k	3.5	1.9	5.4	5.2	3.5	1.9	3.6	2.1	–	–

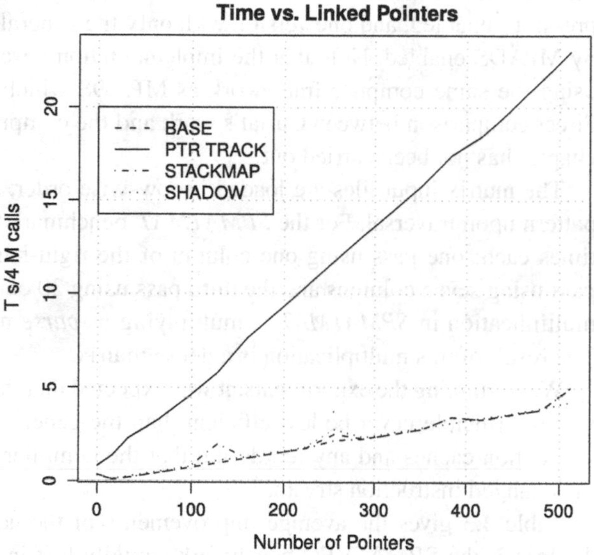

Fig. 4.9 Execution time of a function with different stack management approaches

pointer being linked, for the pointer tracking alternative the cost is around 27 cycles per pointer being registered and linked. This gives the penalty of explicit pointer-tracking to 22 cycles per pointer being tracked. This overhead is obviously quite substantial, but the compilation chain employed a simple optimization where pointer tracking was only enabled when needed. For example, descendant functions from the one that calls the restructuring run-time will not need pointer tracking.

Since the shadow stack and stack map strategies where not implemented, these strategies have not been evaluated using compiler generated code, a hand-written implementation of these strategies has been used to estimate the overhead of these techniques.

Shadow stacks work by pooling all the pointers associated with a pool in a function into a single per-function data structure, it is possible to eliminate all per-pointer overhead associated with registering each pointer. In this case, only the address of the record containing all the pointers would need to be registered (in this case on a shadow stack). This has its own problems, as it prevents certain optimizations such as the elimination of unused pointers (though the pointer tracking suffers from the same issue).

The stack map approach offers none of the run-time overhead (except during the stack walks when program counter entries on the stack are translated into function ids), but does on the other hand require modifications in the compiler's backend.

Address Calculations The address calculation expressions used are an improved variant of those introduced by Curial et al. [3]. These improvements have been verified experimentally by running two versions of the pointer-based applications from the SPARK00 benchmark suite, one with the new optimized address calculation expressions enabled, and one version with only the general addressing equations used by MPADS enabled. Note that the implementation described in this chapter is not using the same compiler framework as MPADS which is based on XLC. Thus a direct comparison between Curial's work and the compiler chain introduced in this chapter has not been carried out.

The matrix input files are loaded in row-wise order, leading to a regular access pattern upon traversal. For the *SPMATMAT* benchmark the matrices are used three times each: one pass using one column of the right-hand side matrix, the second pass using seven columns and the third pass using 30 columns. Note that the matrix multiplication in *SPMATMAT* is multiplying a *sparse* matrix with a *dense* matrix. The result of this multiplication is a dense matrix.

When running the experiments, it was expected that the new field offset equations will in principle never be less efficient than the generic ones, excluding effects on instruction caches and any reordering that the compiler may or may not do due to the changed instruction stream.

Table 4.3 gives the average improvements of the addressing optimizations. In Table 4.3, the SPMATVEC benchmark actually lost in performance, this was due to instruction cache conflicts in the new code. From the SPMATVEC example we can also see that the improvements go down when the memory usage goes up as the programs get more bounded to memory latency.

Table 4.3 Performance gain averages in percent for pool allocation and the improved field offset equations. Note that SPMATVEC has a negative improvement due to instruction cache conflicts. For SPMATMAT, different figures are given in parentheses for 1, 7 and 30 columns in the right hand matrix

Bench Name	*Address Calc Improvements*
DSOLVE	4.87%
JACIT	4.59%
PCG	1.99%
SPMATMAT	3.81% (6.22%/4.16%/1.05%)
SPMATVEC	−6.11%

4.6 Related Work

Optimization of data access in order to improve performance of data-intensive applications has been applied extensively, either by automatic transformations or by hand-tuning applications for efficient access. In some cases, memory access patterns can be determined symbolically at compile-time and in such cases, the traditional transformations such as loop unrolling, loop fusion or fission and loop tiling can be applied. For applications using pointer-linked data structures, such techniques can in general not be applied.

The traditional methods mentioned above change the order of instruction execution such that data are accessed in a different way, without affecting the result. One might as well change the underlying data layout, without affecting the computations. This is exactly what has been done on pointer-linked data structures in this chapter.

In order to be able to automatically control the layout within type-unsafe languages such as C, a type-safe subset must be determined. The Data Structure Analysis (DSA) developed by Lattner and Adve does exactly that [10, 11]. It determines how data structures are used within an application. This has been discussed in Sect. 4.2.

DSA should not be confused with shape analysis. Shape analysis concerns the shape (e.g. tree, DAG or cyclic graph) of pointer-linked data structures. Ghiya and Hendren proposed a pointer analysis that classifies heap directed pointers as a tree, a DAG or a cyclic graph [5]. Hwang and Saltz realized that it is of more importance how data structures are actually traversed instead of knowing the exact layout of a data structure. They integrated this idea in what they call *traversal-pattern-sensitive* shape analysis [8]. Integrating such an approach in our compiler could help in reducing the overhead introduced by the pool access tracing by traversing data structures autonomously in the run-time.

Type-safety is essential for data restructuring techniques. Two other transformations that use information provided by the DSA are structure splitting and pointer compression. Curial et al. implemented structure splitting in the IBM XL compiler, based on the analysis information provided by the DSA [3]. Hagog and Tice have implemented a similar method in GCC [6]. The GCC-based implementation does not seem to provide the same information as DSA. Strictly taken, structure splitting is not necessary for dynamic remapping of pointer structures, but it simplifies tasks like restructuring and relocation considerably. Moreover, splitting simply has performance benefits because data from unused fields will not pollute the cache.

Data layout optimization can also be provided by libraries. Bender and Hu proposed an *adaptive packed-memory array*, which is a sparse array that allows for efficient insertion and deletion of elements while preserving locality [2]. Rubin et al. take a similar approach by grouping adjacent linked-list nodes such that they are colocated in the same cache line. They call this approach *virtual cache lines* (VCL) [13]. In their abstract, they state that they believe that compilers will be able to generate VCL-based code. We believe our pool restructuring achieves this automatic remapping on cache lines (albeit in a different way). In addition, cache usage is very efficient after restructuring a memory pool because our implementation employs full structure splitting,

Rus et al. implemented their Hybrid Analysis that integrates static and run-time analysis of memory references [14]. Eventually, such an approach might be useful in conjunction with our restructuring framework to describe access patterns of pointer traversals. Saltz et al. describe the run-time parallelization and scheduling of loops, which is an inspector/executor approach [15]. Our tracing mechanism is similar to this approach, as it inspects and then restructures. The future challenge will be to extend the system such that it inspects, restructures and parallelizes.

4.7 Conclusions

In this chapter, we presented and evaluated our restructuring compiler transformation chain for pointer-linked data structures in type-unsafe languages. Our transformation chain relies on run-time restructuring using run-time trace information, and we have shown that the potential gains of restructuring access to pointer-based data structures can be substantial.

Curial et al. mention that relying on traces for analysis is not acceptable for commercial compilers [3]. For static analysis, we agree. For dynamic analysis, relying on tracing is not necessarily undesirable and we have shown that the overhead incurred by the tracing and restructuring of pointer-linked data structures is usually compensated for within a reasonable amount of time when data structures are used repetitively.

The restructuring framework described in this chapter opens up more optimization opportunities that we have not explored yet. For example, after data restructuring extra information on the data layout is available and could be exploited in order to apply techniques such as vectorization on code using pointer-linked data structures. This is a subject of future research.

Data structures that are stored on the heap contain object identifiers instead of full pointers. This makes the representation position independent, which provides new means to distribute data structures over disjoint memory spaces. Translation to full pointers would then be dependent on the memory pool location and the architecture.

This position independence using object identifiers has been mentioned before by Lattner and Adve in the context of pointer compression [12]. However, with the pool restructuring presented in this chapter, a more detailed segmentation of the pools can be made and restructuring could be extended to a distributed pool restructuring framework.

The implementation presented in this chapter uses some run-time support functions to remap access to the proper locations for split pools. The use of object identifiers implies a translation step upon each load and store to the heap. These run-time functions are efficiently inlined by the LLVM compiler and have a negligible effect when applications are bounded by the memory system. The run-time support could in principle be implemented in hardware and this would reduce the run-time overhead considerably. We envision an implementation in which pools and their layout are exposed to the processor, such that address calculations can be performed transparently. Memory pools could then be treated similarly to virtual memory in which the processors also take care of address calculations.

We believe the restructuring transformations for pointer-linked data structures that have been described in this chapter do not only enable data layout remapping, but also provide the basis for new techniques to enable parallelizing transformations on such data structures.

References

1. Agesen, O., Detlefs, D., Moss, J.E.: Garbage collection and local variable type-precision and liveness in Java virtual machines. ACM SIGPLAN Not. **33**(5), 269–279 (1998)
2. Bender, M.A., Hu, H.: An adaptive packed-memory array. In: PODS '06: Proceedings of the Twenty-Fifth ACM SIGMOD-SIGACT-SIGART Symposium on Principles of Database Systems, pp. 20–29 (2006)
3. Curial, S., Zhao, P., Amaral, J.N., Gao, Y., Cui, S., Silvera, R., Archambault, R.: MPADS: memory-pooling-assisted data splitting. In: ISMM '08: Proceedings of the 7th International Symposium on Memory Management, pp. 101–110 (2008)
4. Fog, A.: The microarchitecture of Intel and AMD CPU's an optimization guide for assembly programmers and compiler makers (2008). http://www.agner.org/optimize/microarchitecture.pdf
5. Ghiya, R., Hendren, L.J.: Is it a tree, a DAG, or a cyclic graph? A shape analysis for heap-directed pointers in C. In: POPL '96: Proceedings of the 23rd ACM SIGPLAN-SIGACT Symposium on Principles of Programming Languages, pp. 1–15 (1996)
6. Hagog, M., Tice, C.: Cache aware data layout reorganization optimization in GCC. In: Proceedings of the GCC Developers' Summit, pp. 69–92 (2005)
7. Henderson, F.: Accurate garbage collection in an uncooperative environment. In: Proceedings of the 3rd International Symposium on Memory Management, ISMM '02, pp. 150–156 (2002)
8. Hwang, Y.S., Saltz, J.H.: Identifying DEF/USE information of statements that construct and traverse dynamic recursive data structures. In: LCPC '97: Proceedings of the 10th International Workshop on Languages and Compilers for Parallel Computing, pp. 131–145 (1998)
9. Lattner, C.: Macroscopic data structure analysis and optimization. Ph.D. thesis (2005). http://llvm.cs.uiuc.edu
10. Lattner, C., Adve, V.: Automatic pool allocation for disjoint data structures. ACM SIGPLAN Not. **38**(2), 13–24 (2003) supplement

11. Lattner, C., Adve, V.: Automatic pool allocation: improving performance by controlling data structure layout in the heap. ACM SIGPLAN Not. **40**(6), 129–142 (2005)
12. Lattner, C., Adve, V.: Transparent pointer compression for linked data structures. In: MSP '05: Proceedings of the 2005 Workshop on Memory System Performance, pp. 24–35 (2005)
13. Rubin, S., Bernstein, D., Rodeh, M.: Virtual cache line: A new technique to improve cache exploitation for recursive data structures. In: CC '99: Proceedings of the 8th International Conference on Compiler Construction, Held as Part of the European Joint Conferences on the Theory and Practice of Software, ETAPS'99, pp. 259–273 (1999)
14. Rus, S., Rauchwerger, L., Hoeflinger, J.: Hybrid analysis: static & dynamic memory reference analysis. Int. J. Parallel Program. **31**(4), 251–283 (2003)
15. Saltz, J.H., Mirchandaney, R., Crowley, K.: Run-time parallelization and scheduling of loops. IEEE Trans. Comput. **40**(5), 603–612 (1991)
16. Van der Spek, H.L.A., Bakker, E.M., Wijshoff, H.A.G.: SPARK00 source code (2007). http://www.liacs.nl/~hvdspek/SPARK00/SPARK00.html
17. Van der Spek, H.L.A., Bakker, E.M., Wijshoff, H.A.G.: Characterizing the performance penalties induced by irregular code using pointer structures and indirection arrays on the Intel Core 2 architecture. In: CF '09: Proceedings of the 6th ACM Conference on Computing Frontiers, pp. 221–224 (2009)
18. Van der Spek, H.L.A., Holm, C.W.M., Wijshoff, H.A.G.: Automatic restructuring of linked data structures. In: LCPC '09: Proceedings of the 22nd International Workshop on Languages and Compilers for Parallel Computing (2009)
19. Van der Spek, H.L.A., Wijshoff, H.A.G.: SPARK00: A benchmark package for the compiler evaluation of irregular/sparse codes. Tech. Rep. LIACS 2007-06 (2007)

Chapter 5
Dense Linear Algebra on Accelerated Multicore Hardware

Jack Dongarra, Jakub Kurzak, Piotr Luszczek, and Stanimire Tomov

Abstract Design of systems exceeding 1 Pflop/s and inching towards 1 Eflop/s forced a dramatic shift in hardware design. Various physical and engineering constraints resulted in introduction of massive parallelism and functional hybridization with the use of accelerator units. This paradigm change brings about a serious challenge for application developers as the management of multicore proliferation and heterogeneity rests on software. And it is reasonable to expect that this situation will not change in the foreseeable future. This chapter presents a methodology of dealing with this issue in three common scenarios. In the context of shared-memory multicore installations, we show how high performance and scalability go hand in hand when the well-known linear algebra algorithms are recast in terms of Direct Acyclic Graphs (DAGs) which are then transparently scheduled at runtime inside the Parallel Linear Algebra Software for Multicore Architectures (PLASMA) project. Similarly, Matrix Algebra on GPU and Multicore Architectures (MAGMA) schedules DAG-driven computations on multicore processors and accelerators. Finally, Distributed PLASMA (DPLASMA), takes the approach to distributed-memory machines with the use of automatic dependence analysis and the Direct Acyclic Graph Engine (DAGuE) to deliver high performance at the scale of many thousands of cores.

J. Dongarra (✉) · J. Kurzak · P. Luszczek · S. Tomov
Department of Electrical Engineering and Computer Science, University of Tennessee, Knoxville, USA
e-mail: dongarra@cs.utk.edu

J. Kurzak
e-mail: kurzak@eecs.utk.edu

P. Luszczek
e-mail: luszczek@eecs.utk.edu

S. Tomov
e-mail: tomov@eecs.utk.edu

M.W. Berry et al. (eds.), *High-Performance Scientific Computing*,
DOI 10.1007/978-1-4471-2437-5_5,

5.1 Introduction and Motivation

Among the various factors that drive the momentous changes occurring in the design of microprocessors and high end systems [14], three stand out as especially notable:

1. the number of transistors per chip will continue the current trend, i.e. double roughly every 18 months, while the speed of processor clocks will cease to increase;
2. we are getting closer to the physical limit for the number and bandwidth of pins on the CPUs and
3. there will be a strong drift toward hybrid/heterogeneous systems for petascale (and larger) systems.

While the first two involve fundamental physical limitations that the state-of-art research today is unlikely to prevail over in the near term, the third is an obvious consequence of the first two, combined with the economic necessity of using many thousands of CPUs to scale up to petascale and larger systems.

More transistors and slower clocks means multicore designs and more parallelism required. The fundamental laws of traditional processor design—increasing transistor density, speeding up clock rate, lowering voltage—have now been blocked by a set of physical barriers: excess heat produced, too much power consumed, too much energy leaked, useful signal overcome by noise. Multicore designs are a natural response to this situation. By putting multiple processor cores on a single die, architects can overcome the previous limitations, and continue to increase the number of gates per chip without increasing the power densities. However, since excess heat production means that frequencies cannot be further increased, deep-and-narrow pipeline models will tend to recede as shallow-and-wide pipeline designs become the norm. Moreover, despite obvious similarities, multicore processors are not equivalent to multiple-CPUs or to SMPs. Multiple cores on the same chip can share various caches (including TLB—Translation Look-aside Buffer) and they compete for memory bandwidth. Extracting performance from such configurations of resources means that programmers must exploit increased thread-level parallelism (TLP) and efficient mechanisms for inter-processor communication and synchronization to manage resources effectively. The complexity of parallel processing will no longer be hidden in hardware by a combination of increased instruction level parallelism (ILP) and pipeline techniques, as it was with superscalar designs. It will have to be addressed at an upper level, in software, either directly in the context of the applications or in the programming environment. As portability remains a requirement, clearly the programming environment has to drastically change.

A thicker memory wall means that communication efficiency will be even more essential. The pins that connect the processor to main memory have become a strangle point, with both the rate of pin growth and the bandwidth per pin slowing down, if not flattening out. Thus the processor to memory performance gap, which is already approaching a thousand cycles, is expected to grow, by 50% per year according to some estimates. At the same time, the number of cores on a single chip is expected to continue to double every 18 months, and since limitations on space will keep the cache resources from growing as quickly, cache per core ratio will continue to go down. Problems with memory bandwidth and latency, and cache fragmentation will, therefore, tend to become more severe, and that means that communication costs will present an especially notable problem. To quantify the growing cost of communication, we can note that time per flop, network bandwidth (between parallel processors), and network latency are all improving, but at significantly different rates: 59%/year, 26%/year and 15%/year, respectively [18]. Therefore, it is expected to see a shift in algorithms' properties, from computation-bound, i.e. running close to peak today, toward communication-bound in the near future. The same holds for communication between levels of the memory hierarchy: memory bandwidth is improving 23%/year, and memory latency only 5.5%/year. Many familiar and widely used algorithms and libraries will become obsolete, especially dense linear algebra algorithms which try to fully exploit all these architecture parameters. They will need to be reengineered and rewritten in order to fully exploit the power of the new architectures.

In this context, the PLASMA project [24] has developed several new algorithms for dense linear algebra on shared memory system based on tile algorithms. Here, we present DPLASMA, a follow up project related to PLASMA that operates in the distributed-memory environment. DPLASMA introduces a novel approach to schedule dynamically dense linear algebra algorithms on distributed systems. It, too, is based on tile algorithms, and takes advantage of DAGuE [8], a new generic distributed Direct Acyclic Graph Engine for high performance computing. This engine supports a DAG representation independent of problem-size, overlaps communications with computation, prioritizes tasks, schedules in an architecture-aware manner and manages micro-tasks on distributed architectures featuring heterogeneous many-core nodes. The originality of this engine resides in its capability of translating a sequential nested-loop code into a concise and synthetic format which it can interpret and then execute in a distributed environment. We consider three common dense linear algebra algorithms, namely: Cholesky, LU and QR factorizations, to investigate through the DAGuE framework their data driven expression and execution in a distributed system. We demonstrate through performance results at scale that our DAG-based approach has the potential to bridge the gap between the peak and the achieved performance that is characteristic in the state-of-the-art distributed numerical software on current and emerging architectures. However, the most essential contribution, in our view, is the ease with which new algorithmic variants may be developed and how they can be simply launched on a massively parallel architecture without much consideration to the underlying hardware structure. It is due to the flexibility of the underlying DAG scheduling engine and straightforward expression of parallel data distributions.

5.2 PLASMA

Parallel Linear Algebra Software for Multicore Architectures (PLASMA) is a numerical software library for solving problems in dense linear algebra on systems multicore processors and multi-socket systems of multicore processors. PLASMA offers routines for solving a wide range of problems in dense linear algebra, such as: non-symmetric, symmetric and symmetric positive definite systems of linear equations, least square problems, singular value problems and eigenvalue problems (currently only symmetric eigenvalue problems). PLASMA solves these problems in real and complex arithmetic and in single and double precision. PLASMA is designed to give high efficiency on homogeneous multicore processors and multi-socket systems of multicore processors. As of today, the majority of such systems are on-chip symmetric multiprocessors with classic *super-scalar* processors as their building blocks (x86 and alike) augmented with short-vector SIMD extensions (SSE and alike). PLASMA has been designed to supersede LAPACK, principally by restructuring the software to achieve much greater efficiency on modern computers based on multicore processors.

The interesting part of PLASMA from the multithreading perspective is the variety of scheduling mechanism utilized by PLASMA. In the next two subsections, the different options for programming multicore processors are briefly reiterated and then the main design principles of PLASMA introduced.

5.2.1 PLASMA Design Principles

The main motivation behind the PLASMA project are performance shortcomings of LAPACK and ScaLAPACK on shared memory systems, specifically systems consisting of multiple sockets of multicore processors. The three crucial elements that allow PLASMA to achieve performance greatly exceeding that of LAPACK and ScaLAPACK are: the implementation of *tile algorithms*, the application of *tile data layout* and the use of *dynamic scheduling*. Although some performance benefits can be delivered by each one of these techniques on its own, it is only the combination of all of them that delivers maximum performance and highest hardware utilization.

Tile algorithms are based on the idea of processing the matrix by square tiles of relatively small size, such that a tile fits entirely in one of the cache levels associated with one core. This way a tile can be loaded to the cache and processed completely before being evicted back to the main memory. Of the three types of cache misses, *compulsory*, *capacity* and *conflict*, the use of tile algorithms minimizes the number of capacity misses, since each operation loads the amount of data that do not "overflow" the cache.

Tile layout is based on the idea of storing the matrix by square tiles of relatively small size, such that each tile occupies a continuous memory region. This way a tile can be loaded to the cache memory efficiently and the risk of evicting it from the cache memory before it is completely processed is minimized. Of the three types of cache misses, *compulsory*, *capacity* and *conflict*, the use of tile layout minimizes the number of conflict misses, since a continuous region of memory will completely fill out a set-associative cache memory before an eviction can happen. Also, from the standpoint of multithreaded execution, the probability of *false sharing* is minimized. It can only affect the cache lines containing the beginning and the ending of a tile.

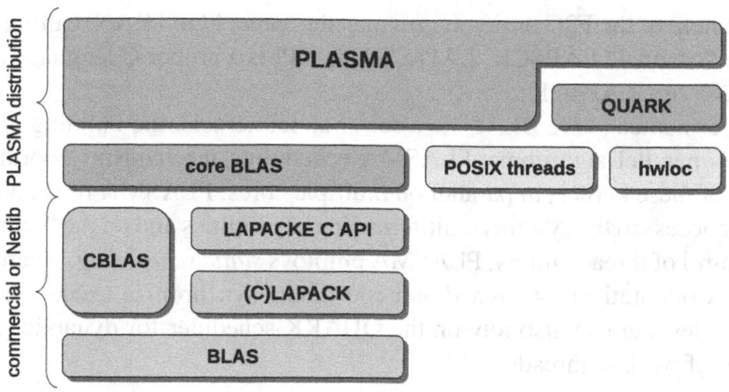

Fig. 5.1 The software stack of PLASMA, version 2.3

Dynamic scheduling is the idea of assigning work to cores based on the availability of data for processing at any given point in time and is also referred to as *data-driven* scheduling. The concept is related closely to the idea of expressing computation through a task graph, often referred to as the DAG (*Direct Acyclic Graph*), and the flexibility exploring the DAG at runtime. Thus, to a large extent, dynamic scheduling is synonymous with *runtime scheduling*. An important concept here is that of the *critical path*, which defines the upper bound on the achievable parallelism, and needs to be pursued at the maximum speed. This is in direct opposition to the *fork-and-join* or *data-parallel* programming models, where artificial synchronization points expose serial sections of the code, where multiple cores are idle, while sequential processing takes place.

5.2.2 PLASMA Software Stack

Starting from the PLASMA, version 2.2, released in July 2010, the library is built on top of standard software components, all of which are either available as open source or are standard OS facilities. Some of them can be replaced by packages provided by hardware vendors for efficiency reasons. Figure 5.1 presents the current structure of PLASMA's software stack. Following is a brief bottom-up description of individual components.

Basic Linear Algebra Subprograms (BLAS) is a, *de facto* standard, set of basic linear algebra operations, such as vector and matrix multiplication. CBLAS is the C language interface to BLAS. Most commercial and academic implementations of BLAS also provide CBLAS. *Linear Algebra PACKage* (LAPACK) is a software library for numerical linear algebra, a direct predecessor of PLASMA, providing routines for solving linear systems of equations, linear least square problems, eigenvalue problems and singular value problems. CLAPACK is a version of LAPACK available from Netlib, created by automatically translating FORTRAN LAPACK to

C with the help of the F2C utility. It provides the same, FORTRAN, calling convention as the "original" LAPACK. LAPACKE C API is a proper C language interface to LAPACK (or CLAPACK).

The layer named "core BLAS" is a set of serial kernels, the building blocks for PLASMA's parallel algorithms. PLASMA scheduling mechanisms coordinate the execution of these kernels in parallel on multiple cores. PLASMA relies on POSIX threads for access to the systems multithreading capabilities and on the hwloc library for the control of thread affinity. PLASMA employs *static scheduling*, where threads have their work statically assigned and coordinate synchronize execution through progress tables, but can also rely on the QUARK scheduler for dynamic (runtime) scheduling of work to threads.

5.2.3 PLASMA Scheduling

By now, multicore processors are ubiquitous in both low-end consumer electronics and high-end servers and supercomputer installations. This let to the emergence of a myriad of multithreading frameworks, both academic and commercial, embracing the idea of task scheduling: Cilk, OpenMP (tasking features), Intel Threading Building Blocks, just to name a few prominent examples. One especially important category are multithreading systems based on dataflow principles, which represent the computation as a *Direct Acyclic Graph* (DAG) and schedule tasks at runtime through resolution of data hazards: *Read after Write* (RaW), *Write after Read* (WaR) and *Write after Write* (WaW). PLASMA's scheduler QUARK is an example of such a system. Two other, very similar, academic projects are also available: StarSs from Barcelona Supercomputer Center and StarPU from INRIA Bordeaux. While all three systems have their strength and weaknesses, QUARK has vital extensions for use in a numerical library.

5.2.4 DAGs for One-Sided Factorizations

The three common one-sided factorizations included in PLASMA are Cholesky, LU, and QR. Their corresponding DAGs are shown in Fig. 5.2. In the figure, each node represents a task. The node label consists of the name of the routine invoked by the task followed by the coordinates of the tiles that the routine operates on. The edges of DAG represent data dependencies between the tasks. To the left of each DAG there are boxes with the step number and the available parallelism at each step. The DAG's are drawn with the minimum number of steps without violating the dependencies between tasks. It turns out that for a 3-by-3 matrix that the minimum height of the DAG is the same for each of the three factorizations. This minimum height of the DAG is the same as the length of the DAGs critical path.

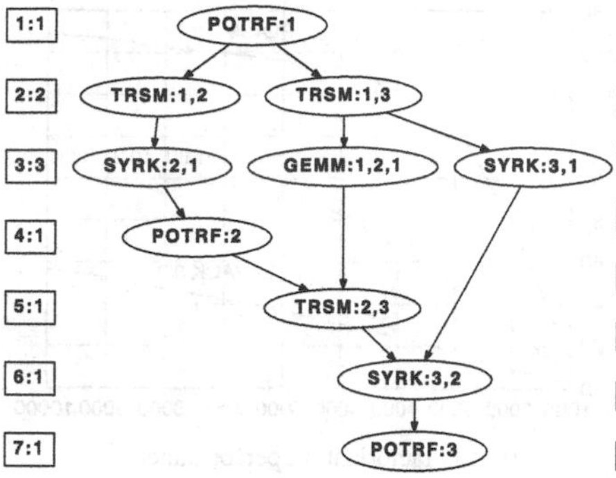

(a) DAG for Cholesky factorization

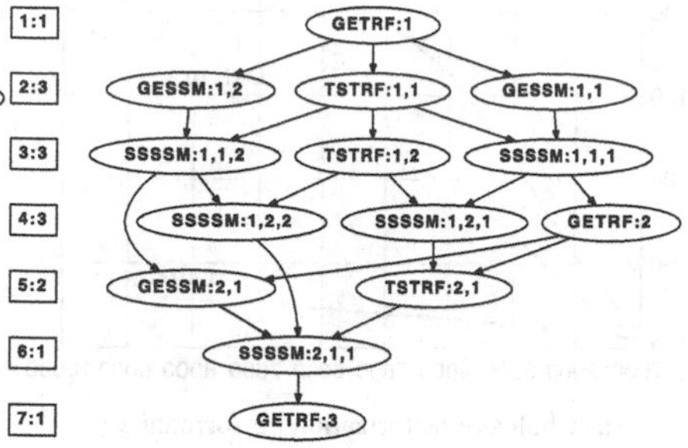

(b) DAG for LU factorization

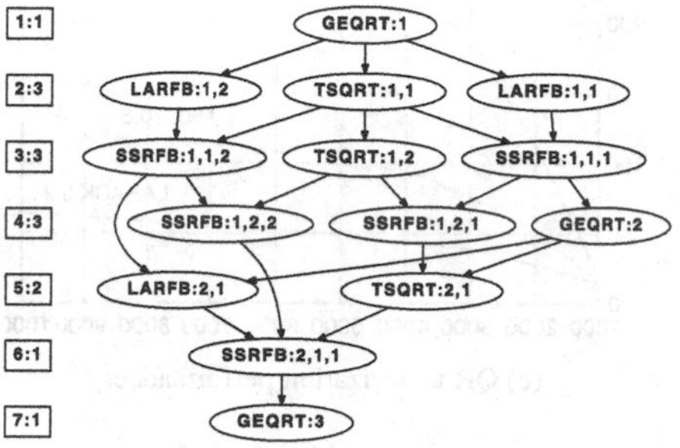

(c) DAG for QR factorization

Fig. 5.2 DAGs of three common one-sided factorizations for a 3-by-3 matrix

5.2.5 Performance for One-Sided Factorizations

To illustrate PLASMA's performance for the three one-sided factorizations we compare it to equivalent implementations[1] from Intel's MKL and the source code distribution of LAPACK from www.netlib.org. Figure 5.3 shows such a comparison on a 48-core machine that features 8 AMD Istanbul processors, each with 6 cores running at 2.8 GHz. PLASMA clearly outperforms the other two libraries by a comfortable performance margin for all the tested matrix sizes. This serves as the validation of its design principles based on dynamic scheduling, tile storage, and tile algorithms.

[1] In terms of numerical accuracy, the incremental pivoting used in PLASMA's LU implementation has a higher upper bound on the backward error than the LU with partial pivoting featured in LAPACK and MKL.

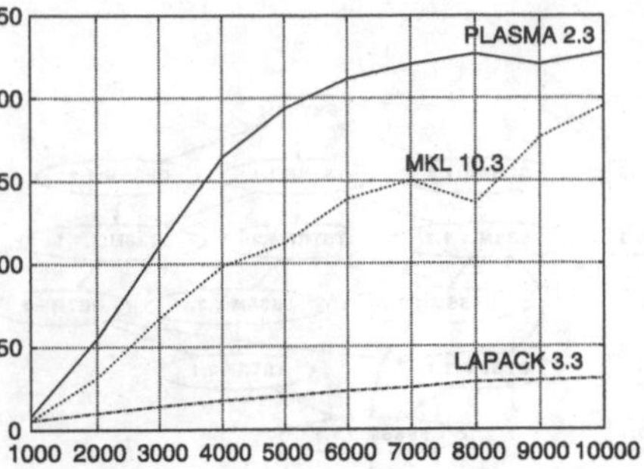

(a) Cholesky factorization performance

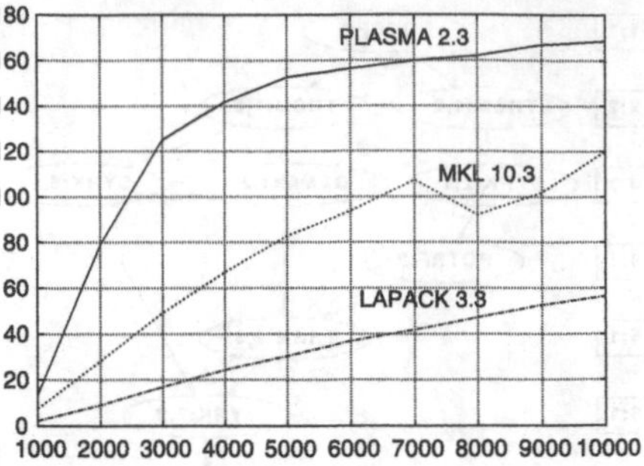

(b) LU factorization performance

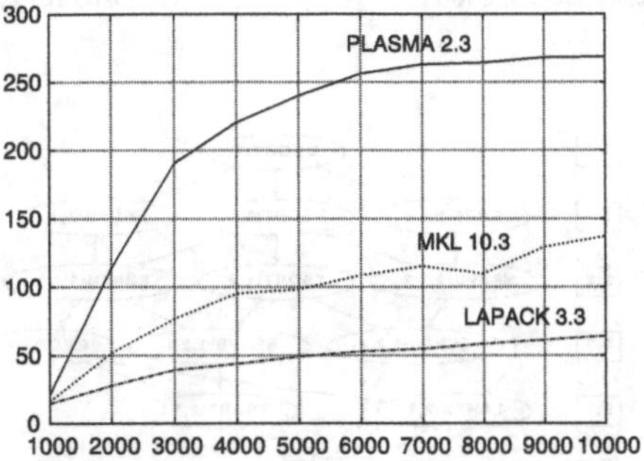

(c) QR factorization performance

Fig. 5.3 Performance for three common one-sided factorizations run on a 48-core machine that features 8 AMD Istanbul processors, each with 6 cores running at 2.8 GHz

5.3 MAGMA

Numerical linear algebra is a fundamental building block in engineering and computational science software applications. It arises in areas such as image and signal processing, data mining, computational biology, materials science, fluid dynamics, and many others. Therefore, highly optimized libraries, implementing the most basic algorithms in numerical linear algebra, are greatly needed. The development and availability of these libraries is crucial for the establishment and use of new architectures for high-performance scientific computing.

In this paper we review and extend some state-of-the-art algorithms and techniques for high performance linear algebra on GPU architectures. The main goal of the development is to map algorithmic requirements to the architectural strengths of the GPUs. For example, GPUs are manycore based architectures and hence, in order to map well, algorithms must be of high parallelism and new data structures must be designed to facilitate parallel (coalescent) memory accesses. The introduction of shared memory and levels of memory hierarchy [19] has enabled memory reuse, giving rise to blocking techniques for increased memory reuse. Furthermore, *hybridization* techniques have been developed are motivated by the fact that GPUs are currently used through CPU applications—queuing computational tasks/requests to be executed on the GPU.

We consider basic dense linear algebra kernels, and higher-level routines such as linear and eigen/singular-value solvers. For dense linear algebra we highlight some Level 2 and 3 BLAS and their use in basic factorizations and solvers.

We show that basic linear algebra kernels achieve a high fraction of their theoretical performance peaks on GPUs. Implementations of most of the dense linear algebra described are available through the MAGMA library [23].

5.3.1 Acceleration Techniques for GPUs

5.3.1.1 Blocking

Blocking is a well known linear algebra optimization technique where the computation is organized to operate on blocks/submatrices of the original matrix. The idea is that blocks are of small enough size to fit into a particular memory hierarchy level so that once loaded, the blocks' data are to be reused in all the arithmetic operations that they are involved in. This idea can be applied for GPUs, using GPUs' shared memory. As it would be demonstrated throughout the paper, the application of blocking is crucial for the performance of numerous kernels and high level algorithms for both sparse and dense linear algebra.

5.3.1.2 Hybridization

Hybridization refers to a methodology that we follow in developing high level linear algebra algorithms:

- Represent the algorithm as a collection of basic kernels/tasks and dependencies among them (see Fig. 5.4):
  - Use parametrized task granularity to facilitate auto-tuning;
  - Use performance models to facilitate the task splitting/mapping.
- Schedule the execution of the basic kernels/tasks over the multicore and the GPU:
  - Schedule small, non-parallelizable task on the CPU and large, parallelizable on the GPU;
  - Define the algorithm's *critical path* and prioritize its execution/scheduling.

5.3.1.3 Data Structures

A great deal of research in GPU computing is on data structures. Data should be organized in a way that facilitates parallel memory accesses (see also Sect. 5.3.1.4 below) so that applications can achieve the inherent for GPUs high memory throughput. This is especially true for sparse matrix-vector kernel, where researchers have

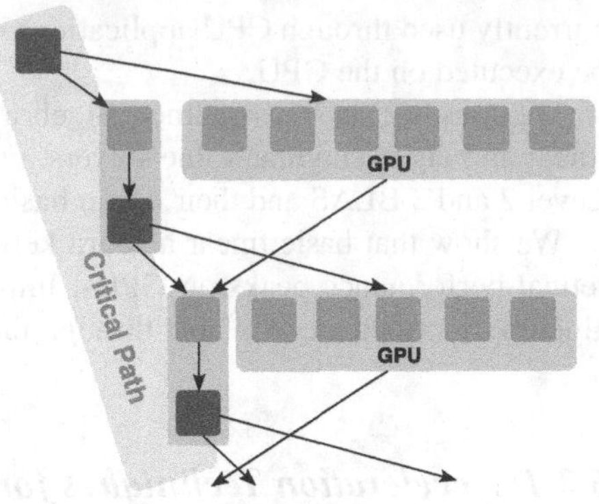

Fig. 5.4 Algorithms as a collection of basic kernels/tasks and dependencies among them (DAGs) for hybrid GPU-based computing

proposed numerous data formats. Even in dense linear algebra, where input/output data structure interfaces are more standardized, there are numerous algorithms where intermediate steps involve rearranging data to facilitate fast memory accesses (see Sect. 5.3.2).

5.3.1.4 Parallel Memory Access

GPU global memory accesses are costly and in general not cached (except in Fermi, the latest generation of GPUs from NVIDIA [19]), making it crucial for the performance to have the right access pattern to get maximum memory bandwidth. There are two access requirements to high throughput [20, 21]. The first is to organize global memory accesses in terms of parallel consecutive memory accesses—16 consecutive elements at a time by the threads of a half-warp (16 threads)—so that memory accesses (to 16 elements at a time) be *coalesced* into a single memory access. This is demonstrated in the kernels' design throughout the paper. Second, the data should be properly aligned. In particular, the data to be accessed by half-warp should be aligned at 16 ∗ `sizeof(element)`, e.g., 64 for single precision elements.

5.3.1.5 Pointer Redirecting

Pointer redirecting is a set of GPU specific optimization techniques that allows to easily remove performance oscillations in cases where the input data are not aligned to directly allow coalescent memory accesses, or when the problem sizes are not divisible by the partitioning sizes required for achieving high performance [17]. For example, applied to the dense matrix–matrix multiplication routines, depending on the hardware configuration and routine parameters, this can lead to two times faster algorithms. Similarly, the dense matrix–vector multiplication can be accelerated more than two times in both single and double precision arithmetic.

5.3.1.6 Padding

Padding, similar to pointer redirecting, is another technique for obtaining high overall performance (no performance oscillations) when input data are not aligned to allow coalescent memory accesses, or when the problem sizes are not divisible by certain blocking sizes. The difference is that in padding the problem dimensions are increased and the extra space is filled up with zeroes (referred to as padding). A drawback of the padding is that it requires extra memory and it may involve extra data copies. Comparing the padding and pointer redirecting approaches for dense matrix–matrix multiplication show that for small matrix sizes the pointer redirecting gives better performance, and for larger matrices the two approaches are almost identical, as it is actually expected. An advantage of using padding is that users may design the input data to achieve highest performance with libraries at hand (e.g., not supporting pointer redirecting, experiencing non-uniform performance).

5.3.1.7 Auto-tuning

Automatic performance tuning (optimization), or auto-tuning in short, is a technique that has been used intensively on CPUs to automatically generate near-optimal numerical libraries. For example, ATLAS [26] and PHiPAC [2] are used to generate highly optimized BLAS.

Auto-tuning can also be applied to tune linear algebra for GPUs. Work in the area of dense linear algebra [15] shows that auto-tuning for GPUs is very practical solution to easily port existing algorithmic solutions on quickly evolving GPU architectures and to substantially speed up even highly hand-tuned kernels.

There are two core components in a complete auto-tuning system:

- *Code generator*: The code generator produces code variants according to a set of pre-defined, parametrized templates/algorithms. The code generator also applies certain state of the art optimization techniques.
- *Heuristic search engine*: The heuristic search engine runs the variants produced by the code generator and finds out the best one using a feedback loop, e.g., the performance results of previously evaluated variants are used as a guidance for the search on currently unevaluated variants.

Autotuning is used to determine best performing kernels, partitioning sizes, and other parameters for the various algorithms described in this paper.

5.3.2 Accelerating Dense Linear Algebra Kernels and Factorizations

Implementations of the BLAS interface are a major building block of dense linear algebra (DLA) libraries, and therefore must be highly optimized. This is true

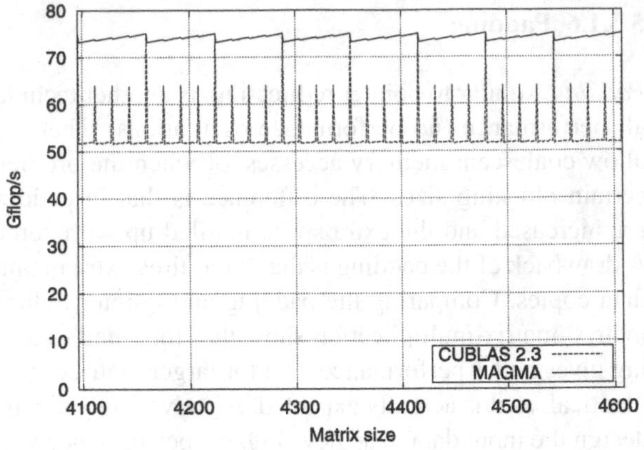

Fig. 5.5 Performance of MAGMA's implementation of matrix–matrix multiplication routine DGEMM on NVIDIA GTX 280

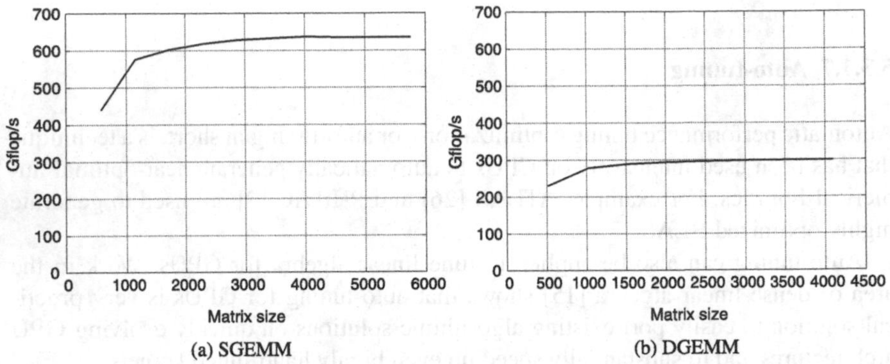

Fig. 5.6 Performance of MAGMA's implementation of matrix–matrix multiplication routine DGEMM on NVIDIA Fermi C2050

for GPU computing as well, especially after the introduction of shared memory in modern GPUs. This is important because it enabled fast Level 3 BLAS implementations, which in turn made possible the development of DLA for GPUs to be based on BLAS for GPUs.

Despite the current success in developing highly optimized BLAS for GPUs [10, 25], the area is still new and presents numerous opportunities for improvements. For example, we address cases of the matrix–matrix and the matrix–vector multiplications, along with discussion on the techniques used to achieve these improvements.

Figure 5.5 shows the performance for GEMM in double precision arithmetic on NVIDIA GTX 280. Note the performance oscillations that CUBLAS experiences for problem sizes not divisible by 32. This performance degradation can be removed using the pointer redirecting technique [17] as the figure clearly illustrates. We see an improvement of 24 Gflops/s in double and 170 Gflops/s in single precision arithmetic. We extended this technique to other Level 3 BLAS kernels to see similar performance improvements. Figure 5.6 show the performance of both single

Table 5.1 Detailed description of NVIDIA Tesla GTX 280 and the server version of NVIDIA Fermi GPU: C2050

Feature	Tesla GTX 280	Fermi C2050
Frequency	602 MHz	1150 MHz
CUDA cores (vertex shaders)	240	448
Streaming Multiprocessors (SM)	30	14
Shared memory per SM	16 kB	16 kB or 48 kB
L1 cache per SM	–	16 kB or 48 kB
L2 cache per SM	–	768 kB
Theoretical peak in single precision	933 Gflop/s	1030 Gflop/s
Theoretical peak in double precision	80 Gflop/s	515 Gflop/s
Address width	32 bits	64 bits

and double precision GEMM routines on NVIDIA Fermi C2050. Comparison of specification of both platforms is summarized in Table 5.1.

Padding can be applied as well but in many cases copying user data may not be feasible. This is the case for algorithms that involve Level 2 BLAS that has to be applied for matrices of continuously decreasing sizes, e.g., in the bidiagonal reduction for the symmetric eigenvalue problem. This motivated us to extend the technique to Level 2 BLAS to see similar improvements.

Conceptually, above the level of basic computational kernels provided by BLAS are the factorization one-sided factorization codes. Most commonly used ones are Cholesky, LU, and QR factorizations. When the aforementioned hybridization techniques are combined with good quality BLAS then a high fraction of the peak performance may indeed be achieved as shown in Fig. 5.7.

5.4 DPLASMA

DPLASMA (Distributed PLASMA) is a framework for developing dense linear algebra algorithms that seamlessly scales to thousands of cores. Its goals can be achieved with a use of a novel generic distributed Direct Acyclic Graph Engine (DAGuE). The engine has been designed for high performance computing and thus it enables scaling of tile algorithms, originating in PLASMA, on large distributed memory systems. The underlying DAGuE framework has many appealing features when considering distributed-memory platforms with heterogeneous multicore nodes: DAG representation that is independent of the problem-size, automatic extraction of the communication from the dependencies, overlapping of communication and computation, task prioritization, and architecture-aware scheduling and management of tasks. The originality of this engine lies in its capacity to translate a sequential code with nested-loops into a concise and synthetic format which can then be interpreted and executed in a distributed environment. We present three common dense linear algebra algorithms from PLASMA (Parallel Linear Algebra for

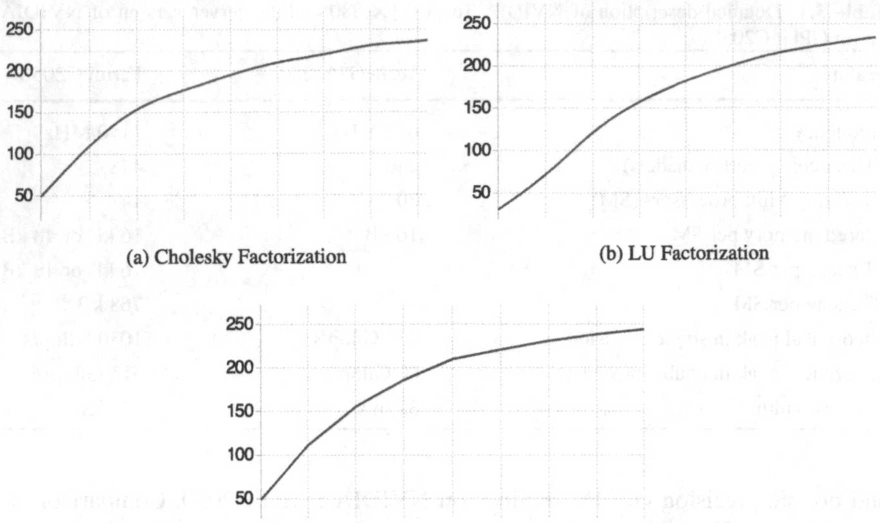

(a) Cholesky Factorization

(b) LU Factorization

(c) QR Factorization

Fig. 5.7 Performance of MAGMA's one sided factorizations on Fermi C2050

Fig. 5.8 Pseudo code of the tile Cholesky factorization (right-looking version)

```
FOR k = 0..TILES-1
  A[k][k] ← DPOTRF(A[k][k])
  FOR m = k+1..TILES-1
    A[m][k] ← DTRSM(A[k][k], A[m][k])
  FOR n = k+1..TILES-1
    A[n][n] ← DSYRK(A[n][k], A[n][n])
    FOR m = n+1..TILES-1
      A[m][n] ← DGEMM(A[m][k], A[n][k], A[m][n])
```

Scalable Multi-core Architectures), namely: Cholesky, LU, and QR factorizations, to investigate their data driven expression and execution in a distributed system. We demonstrate through experimental results on the Cray XT5 Kraken system that our DAG-based approach has the potential to achieve sizable fraction of peak performance which is characteristic of the state-of-the-art distributed numerical software on current and emerging architectures.

5.4.1 Dependence Analysis

We will apply the DAGuE framework to three of the most fundamental one-sided factorizations of numerical linear algebra: Cholesky, LU, and QR factorizations. Figure 5.8 shows the pseudo code of the Cholesky factorization (the right-looking variant). Figure 5.9 shows the pseudo code of the tile QR factorization. Figure 5.10

Fig. 5.9 Pseudo code of the tile QR factorization

```
FOR k = 0..TILES-1
  A[k][k], T[k][k] ← DGEQRT(A[k][k])
  FOR m = k+1..TILES-1
    A[k][k], A[m][k], T[m][k] ← DTSQRT(A[k][k], A[m][k], T[m][k])
  FOR n = k+1..TILES-1
    A[k][n] ← DORMQR(A[k][k], T[k][k], A[k][n])
    FOR m = k+1..TILES-1
      A[k][n], A[m][n] ← DSSMQR(A[m][k], T[m][k], A[k][n], A[m][n])
```

Fig. 5.10 Pseudo code of the tile LU factorization

```
FOR k = 0..TILES-1
  A[k][k], T[k][k] ← DGETRF(A[k][k])
  FOR m = k+1..TILES-1
    A[k][k], A[m][k], T[m][k] ← DTSTRF(A[k][k], A[m][k], T[m][k])
  FOR n = k+1..TILES-1
    A[k][n] ← DGESSM(A[k][k], T[k][k], A[k][n])
    FOR m = k+1..TILES-1
      A[k][n], A[m][n] ← DSSSSM(A[m][k], T[m][k], A[k][n], A[m][n])
```

shows the pseudo code of the tile LU factorization. Each of the figures shows the tile formulation of the respective algorithm: a single tile of the matrix is denoted by double-index notation A[i][j].

The DAGuE framework is generic by design and requires from a specific algorithm to be represented as a DAG of dependencies. This may be readily achieved for the three linear algebra factorizations by recasting the linear algebra meaning of the computational kernels into dependence scheduling nomenclature [1] commonly used in the compiler community. To start with a simple example, in Fig. 5.8, the first (and only) invocation of the DPOTRF computational kernel has a form:

```
A[k][k] <- DPOTRF(A[k][k])
```

From the compiler stand point, this operation reads from A[k][k] (input dependence) and writes to A[k][k] (output dependence). To simplify the dependence analysis we could rewrite the operation as:

```
A[k][k] <- A[k][k] + 1
```

The loss of semantics (the new form is not equivalent to the original) may easily be compensated by preserving a reference to the original code. It is trivial for most of mainstream compiler frameworks to analyze the modified form of the statement: it is both input and output dependence—INOUT for short (following the notation borrowed from Fortran 90's function parameter annotation). It is also possible to have input-only dependencies:

```
A[m][k] <- DTRSM(A[k][k], A[m][k])
```

A[k][k] carries input dependence and the whole statement may be rewritten in simpler (but dependence preserving) form:

```
A[m][k] <- A[k][k] + A[m][k]
```

For output-only dependencies:

```
A[k][k], T[k][k] <- DGEQRT(A[k][k])
```

T[k][k] carry output dependence. And the equivalent form could be

```
T[k][k] <- A[k][k] + 1
A[k][k] <- A[k][k] + 1
```

Finally, it is also possible to have SCRATCH designation for temporary storage that doesn't carry any dependence but is necessary for proper functioning of the algorithm (this is again borrowed from Fortran 2008's SCRATCH designation). The SCRATCH parameters allow for dynamic allocation of memory of size not known before runtime (i.e. at compile time). In addition, the allocated memory is automatically deallocated upon exiting the lexical scope where such allocation occurred.

By rewriting the original statement we can simplify the original code and have it accessible for loop-carried dependence analysis. An alternative approach is to use the dependence designation introduced above (IN, OUT, INOUT, and SCRATCH) inserted into the original code and have the rewriting and dependence analysis done automatically. This is in fact the approach taken by the DAGuE framework as it separates the semantics of the domain specific code from its DAG representation required for efficient scheduling. For example, the DPOTRF function is designated to accept a single argument (a matrix tile) that carries input and output dependence:

```
DPOTRF(A : INOUT)
```

And this is the only change required from the end user in the implementation of DPOTRF() which otherwise should just be a standard sequential function: an LAPACK subroutine in this case.

5.4.2 The DAGuE Framework

This section introduces the DAGuE framework [8], a new runtime environment which schedules tasks dynamically in a distributed environment. The tile QR factorization is used as a test case to explain how the overall execution is performed in parallel.

5.4.2.1 Description

The originality of this framework for distributed environment resides in the fact that its starting point is a sequential nested-loop user-application, similar to the pseudo code from Figs. 5.8, 5.9 and 5.10. The framework then translates it in DAGuE's internal representation called JDF (Job Description Format), which is a concise parameterized representation of the sequential program's DAG. This intermediate representation is eventually used as input to trigger the parallel execution by the DAGuE engine. It includes the input and output dependencies for each task, decorated with additional information about the behavior of the task.

For an $NT \times NT$ tile matrix, there are $\mathcal{O}(NT^3)$ tasks. The memory requirement to store the full DAG quickly increases with NT. In order to have a scalable approach however, DAGuE uses symbolic interpretation to schedule tasks without unrolling the JDF in memory at any given time, and thus spares computation cycles to walk the DAG, and memory to keep a global representation. So, basically this synthetic representation allows the internal dependence management mechanism to efficiently compute the flow of data between tasks without having to unroll the whole DAG, and to discover on the fly the communications required to satisfy these dependencies. Indeed, the knowledge of the IN and OUT dependencies, accessible anywhere in the DAG execution, is sufficient to implement a fully distributed scheduling engine for the underlying DAG. At the same time, the concept of looking variants (i.e., right-looking, left-looking, top-looking) in the context of LAPACK and ScaLAPACK becomes irrelevant with this representation: instead of hard-coding a particular variant of tasks ordering, the execution is now data-driven and dynamically scheduled. The issue of which "looking" variant to choose is avoided because the execution of a task is scheduled when the data are available. On the other hand, it is still possible to insist on a particular traversal order of the DAG which would yield a particular "looking" variant. This kind of extension to DAGuE is supported mostly for educational purposes.

Such representation is expected to be internal to the DAGuE framework though, and not a programming language at user disposal. The framework, as described here, does not automate the computation of the data and task distribution. The user is thus required to manually add such information in the JDF. The process of such automation is beyond the scope of this writing as we are trying to compare against the established practices of distributed linear algebra software which assumes fixed data distribution. The internal representation of the DAG used by DAGuE is called JDF. It is also a language that is used to describe the DAG of tasks in a synthetic and concise way.

From a technical point of view, the main goal of the distributed scheduling engine is to select a local task for which all the IN dependencies are satisfied, i.e. the data are available locally, select one of the local cores where to run the task and execute the body of the task when it is scheduled. Once executed, the scheduling engine releases all the OUT dependencies of this task, thus making more tasks available to be scheduled, locally or remotely. It is noteworthy to mention that the scheduling mechanism is architecture aware, taking into account not only the physical layout of the cores, but also the way different cache levels and memory nodes are shared between the cores. This allows to determine the best local core, i.e. the one that minimizes the number of cache misses and data movements over the memory bus.

The DAGuE engine is obviously responsible for moving data from one node to another when necessary. These data movements are necessary to release dependen-

```
1   DGEQRT(k)  (high_priority)
2     k = 0..NT-1        // Execution space for k
3     : A(k, k)          // Data Distribution
4     // Data flows
5     V <- (k==0) ? A(0,0) : C2 DSSMQR(k-1,k,k)
6       -> (k==NT-1) ? A(k,k) : R DTSQRT(k,k+1)  [U]
7       -> (k!=NT-1) ? V1 DORMQR(k, k+1..NT-1)   [L]
8       -> A(k,k)                                 [L]
9     T -> T DORMQR(k, k+1..NT-1)                 [T]
10      -> T(k,k)                                 [T]
11
12  DTSQRT(k,m)  (high_priority)
13    k = 0..NT-2        // Execution space for k
14    m = k+1..NT-1      // Execution space for m
15    : A(m, k)          // Data Distribution
16    // Data flows
17    V2 <- (k==0) ? A(m,0) : C2 DSSMQR(k-1,k,m)
18        -> V2 DSSMQR(k,k+1..NT-1,m)
19        -> A(m,k)
20    R  <- (m==k+1) ? V DGEQRT(k) :
21                     R DTSQRT(k,m-1)       [U]
22        -> (m==NT-1) ? A(k, k) :
23                     R DTSQRT(k,m+1)       [U]
24    T  -> T DSSMQR(k,k+1..NT-1,m)          [T]
25        -> T(m,k)                          [T]
26  DORMQR(k,n)  (high_priority)
27    k = 0..NT-2        // Execution space for k
28    n = k+1..NT-1      // Execution space for n
29    : A(k, n)          // Data Distribution
30    // Data flows
31    T  <- T DGEQRT(k)                            [T]
32    V1 <- V DGEQRT(k)                            [L]
33    C1 <- (k==0) ? A(k,n) : C2 DSSMQR(k-1,n,k)
34        -> C1 DSSMQR(k,n,k+1)
35
36  DSSMQR(k,n,m)
37    k = 0   .. NT-2    // Execution space for k
38    n = k+1 .. NT-1    // Execution space for n
39    m = k+1 .. NT-1    // Execution space for m
40    : A(m, n)          // Data Distribution
41    // Data flows
42    V2 <- V2 DTSQRT(k,m)
43    T  <- T DTSQRT(k,m)                          [T]
44    C2 <- (k==0) ? A(m,n) : C2 DSSMQR(k-1,n,m)
45        -> (n==k+1 & m==k+1) ? V DGEQRT(k+1)
46        -> (n==k+1 & k<m-1) ? V2 DTSQRT(k+1,m)
47        -> (k<n-1 & m==k+1) ? C1 DORMQR(k+1,n)
48        -> (k<n-1 & k<m-1) ? C2 DSSMQR(k+1,n,m)
49    C1 <- (m==k+1) ? C1 DORMQR(k,n) :
50                     C1 DSSMQR(k,n,m-1)
51        -> (m==NT-1) ? A (k,n) : C1 DSSMQR(k,n,m+1)
```

Fig. 5.11 Concise representation of tile QR factorization

cies of remote tasks. The framework language introduces a type qualifier called *modifier*, expressed as MPI data types in the current version. It tells the communication engine what is the shape of the data to be transferred from a remote location to another. By default, the communication engine uses a default data type for the tiles (the user defines it to fit the tile size of the program). But the framework has also the capability to transfer data of any shape. Indeed, sometimes, only a particular area of the default data type must be conveyed. Again, at this stage, the user has still to manually specify how the transfers must be done using these modifiers. Moreover, the data tracking engine is capable to understand if the different modifiers overlap, and behaves appropriately when tracking the data dependencies. One should note that the DAGuE engine allows modifier settings on both input and output dependencies, so that one can change the shape of the data on the fly during the communication.

Based on this representation the engine can move in the execution space and easily find the tasks that have to be executed on each computing node, solve their dependencies and take scheduling decisions. However, at any moment during the execution the complete DAG is developed. Each node only unroll its own section of the DAG and this only based on the available inputs.

5.4.2.2 A Test Case: QR Factorization

A realistic example of the DAGuE's internal representation for the QR factorization is given in Fig. 5.11. As stated in the previous section, this example has been obtained starting from the sequential pseudo code shown in Fig. 5.9 using the DAGuE's translation tools. The logic to determine the task distribution scheme has been hard-coded and could be eventually provided by auto-tuning techniques. The tile QR consists of four kernel operations: DGEQRT, DSSMQR, DORMQR, and DTSQRT. For each operation, we define a function (lines 1 to 10 for DGEQRT) that consists of

- a definition space (DGEQRT is parameterized by k, the step of the factorization, which takes values between 0 and NT-1);
- how task distribution maps the data distribution (DGEQRT(k) runs on the process that holds the tile A(k, k));
- a set of data flows (lines 5 to 10 for DGEQRT(k)); and
- a body that holds the effective C-code that will eventually be executed by the scheduling engine (the body has been excluded from the picture). It is a simple C code to call the DGEQRT routine of LAPACK on the variables V and T which are instantiated to the corresponding memory locations of the process by the DAGuE framework before the execution of the body.

Dependencies apply on data that are necessary for the execution of the task, or that are produced by the task. For example, the task DGEQRT uses one data item V as input, and produces two data, a modified version of the input V, and T a data item locally produced by the task. Input data, such as V, are indicated using the left arrow. They can come either from input matrix (local to the task, or located on a remote process), or from the output data of another task (executed either locally, or remotely). For example, the V of DGEQRT(k) comes either from the original matrix located in tile A(0, 0) if k==0, or from the output data C2 of task DSSMQR(k-1, k, k) otherwise. Output dependencies, marked with a right arrow, work in the same manner. In particular, DGEQRT produces V which can be sent to DTSQRT and DORMQR depending on the values of k. These dependencies are marked with a modifier (line 6 and 7) at their end: [U] and [L] for DTSQRT and DORMQR, respectively. This tells the DAGuE engine that the functions DTSQRT and DORMQR only require the strict lower part of V and only the upper part of V as inputs, respectively. The whole tile could have been transferred instead, but this would engender two main drawbacks: (1) communicating more data than required and (2) add extra dependencies into the DAG which will eventually serialize the DORMQR and DTSQRT calls. This works in the same manner for output dependencies. For example, in line 8, only the lower part of V is written and stored on the memory in the lower part of the tile pointed by A(k, k). Also, a data item that is sent to memory is final, meaning that no other task will modify its contents until the end of the DAG execution. However, this does not prevent other tasks from using it as a read-only input.

Figure 5.12 depicts the complete unrolled DAG of a 4x4 tiles QR, as resulting from the execution of the previously described DAG on a 2-by-2 processor grid. The color represents the task to be executed (DGEQRT, DORMQR, DTSQRT and DSSMQR), while the border of the circles represents the node where the tasks has been executed. The edges between the tasks represents the data flowing from one tasks to another. A solid edge indicates that the data are coming from a remote resource, while a dashed edge indicates a local output of another task.

5.4.2.3 DPLASMA and the DAGuE Framework

DPLASMA is an extension of the PLASMA idea using the DAGuE framework. It implements a subset of PLASMA's tile algorithms for some of the linear algebra

Fig. 5.12 DAG of QR for a 4x4 tile matrix

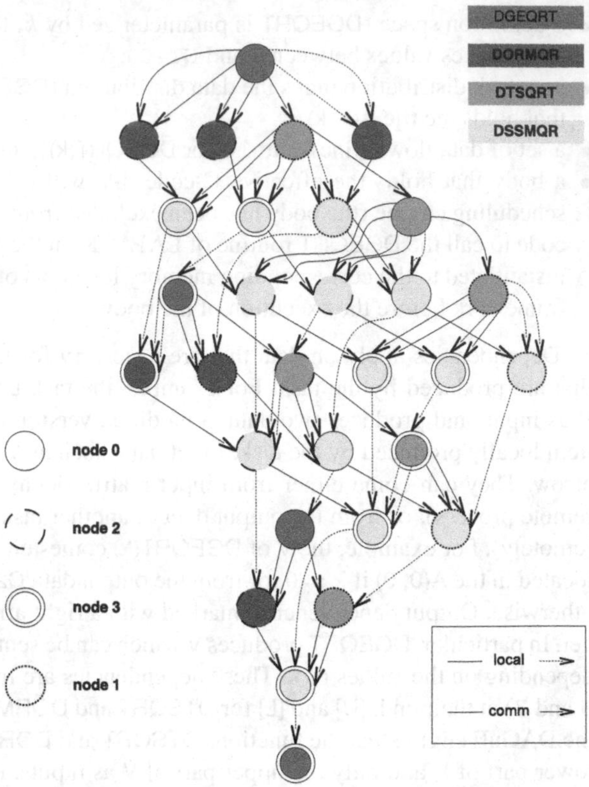

operations of LAPACK inside the DAGuE system. It provides an implementation of these algorithms for a distributed-memory system with multicore nodes. Four operations have been implemented in DPLASMA today: the Cholesky, QR and LU factorizations, as well as the distributed matrix matrix multiply (GEMM). Although DPLASMA is implemented on top of DAGuE, it is usable in any scientific application that uses MPI. Thus, in this context, DPLASMA provides a replacement for ScaLAPACK, as PLASMA replaces LAPACK for shared-memory multicore systems.

5.4.3 Performance of DPLASMA

The performance of the DAGuE runtime have been extensively studied in related publications [5–7, 9]. The goal here is to present the compiler process that is part of the framework of DAGuE. Therefore, we present a summary of these result, to demonstrate that the tool chain achieves its main goals of overall performance, performance portability, and capability to process different non-trivial algorithms.

The experiments we summarize here have been conducted on two different platforms. The Griffon platform is one of the clusters of Grid'5000 [4]. We used 81 dual socket Intel Xeon L5420 quad core processors at 2.5 GHz to gather 648 cores. Each node has 16 GB of memory, and is interconnected to the others by a 20 Gbs Infiniband network. Linux 2.6.24 (Debian Sid) is deployed on these nodes.

The benchmark consists of three popular dense matrix factorization: Cholesky, LU and QR. The Cholesky factorization solves the problem $Ax = b$, where A is symmetric and positive definite. It computes the real lower triangular matrix with positive diagonal elements L such that $A = LL^T$. The QR factorization offers a numerically stable way of solving full rank underdetermined, overdetermined, and regular square linear systems of equation. It computes Q and R such that $A = QR$, Q is a real orthogonal matrix, and R is a real upper triangular matrix. The LU factorization with partial pivoting of a real matrix A has the form $PA = LU$ where L is a real unit lower triangular matrix, U is a real upper triangular matrix, and P is a permutation matrix.

All these three operations are implemented in the ScaLAPACK numerical library [3]. Moreover, the Cholesky factorization has been implemented in a more optimized way in the DSBP software [13], using static scheduling of tasks, and a specific, more efficient, data distribution, and the LU factorization with partial pivoting is also solved by the well known High Performance LINPACK benchmark (HPL) [12], used to measure the performance of high performance computers.

We have re-implemented these operations in DAGuE, using the DAGuE compiler to generate the JDF symbolic representation from the simple sequential algorithms that are given in Figs. 5.8, 5.9, 5.10. Some parameters of the kernels are omitted to increase clarity and reduce the space. Then, we have distributed the initial data following a classical 2D-block cyclic distribution used by ScaLAPACK, and used the DAGuE runtime engine to schedule the operations on the distributed data. The kernels consist of the BLAS operations referenced by the sequential codes, and their implementation was the most efficient available on each of the machine. The same kernel implementation for ScaLAPACK, DAGuE, HPL and DSBP was used on each of the machines.

Figure 5.13 presents the performance measured for DAGuE and ScaLAPACK, and when applicable DSBP and HPL, as function of the problem size. 648 cores participated to the distributed run, and the data were distributed according to a 9×9 2D block-cyclic grid for DAGuE. A similar distribution was used for ScaLAPACK, and the other benchmarks when appropriate, and the block size was tuned to provide the best performance on each setup. As the figures illustrate, on all benchmarks, and for all problem sizes, the DAGuE framework was able to outperform ScaLAPACK, and perform as well as the state of the art, hand-tuned codes for specific problems. The DAGuE solution goes from the sequential code to the parallel run completely automatically, but is still able to outperform DSBP, and competes with the HPL implementation on this machine.

5.5 Summary

The tumultuous changes occurring in the computer hardware space such as flatlining of processor clock speeds after more than 15 years of exponential increases mark

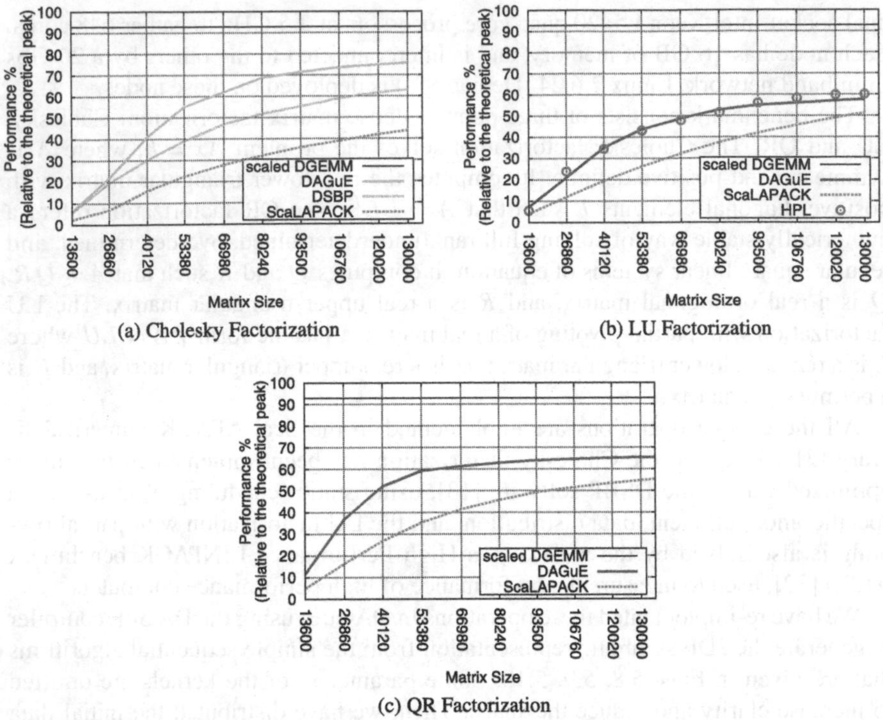

Fig. 5.13 Performance comparison on the Griffon platform with 648 cores

the end of the era of routine and near automatic performance improvements that the research community had previously enjoyed [22]. Three main factors converged to force processor architects to turn to multicore and heterogeneous designs and, consequently, bring an end to the "free ride." First, system builders have encountered intractable physical barriers—too much heat, too much power consumption, and too much leaking voltage—to further increases in clock speeds. Second, physical limits on the number of pins and bandwidth on a single chip mean that the gap between processor performance and memory performance, which was already bad, has gotten increasingly worse. Consequently, the design trade-offs made to address the previous two factors rendered commodity processors, absent any further augmentation, inadequate for the purposes of extreme scale systems for advanced applications. And finally, the exponential growth of transistor count on the heels of the stubbornly alive Moore's law [16] and Dennard's scaling law [11]. This daunting combination of obstacles forced the designers of new multicore and hybrid systems to explore architectures that software built on the old model are unable to effectively exploit without radical modification.

To develop software that will perform well on extreme scale systems with thousands of nodes and millions of cores, the list of major challenges that must now be confronted is formidable:

- dramatic escalation in the costs of intrasystem communication between processors and/or levels of memory hierarchy;
- increased hybridization of processor architectures (mixing CPUs, GPUs, etc.), in varying and unexpected design combinations;
- cooperating processes must be dynamically and unpredictably scheduled for asynchronous execution due to high levels of parallelism and more complex constraints;
- software will not run at scale without much better resilience to faults and far more robustness; and
- new levels of self-adaptivity will be required to enable software to modulate process speed in order to satisfy limited energy budgets.

The software projects presented above meet the aforementioned challenges and allow the users to run their computationally intensive codes at scale and to achieve a significant percentage of peak performance on the contemporary hardware systems that may soon break the barrier of 100 Pflop/s. This is achieved by finding and integrating solutions to problems in two critical areas: novel algorithm design as well as management of parallelism and hybridization.

References

1. Banerjee, U.: Dependence Analysis (Loop Transformation for Restructuring Compilers). Springer, Berlin (1996)
2. Bilmes, J., Asanovic, K., Chin, C.W., Demmel, J.: Optimizing matrix multiply using PHiPAC: a portable, high-performance, ANSI C coding methodology. In: International Conference on Supercomputing, pp. 340–347 (1997). citeseer.ist.psu.edu/article/bilmes97optimizing.html
3. Blackford, L.S., Choi, J., Cleary, A., D'Azevedo, E., Demmel, J., Dhillon, I., Dongarra, J., Hammarling, S., Henry, G., Petitet, A., Stanley, K., Walker, D., Whaley, R.C.: ScaLAPACK Users' Guide. Society for Industrial and Applied Mathematics, Philadelphia (1997)
4. Bolze, R., Cappello, F., Caron, E., Daydé, M., Desprez, F., Jeannot, E., Jégou, Y., Lanteri, S., Leduc, J., Melab, N., Mornet, G., Namyst, R., Primet, P., Quetier, B., Richard, O., Talbi, E.G., Touche, I.: Grid'5000: A large scale and highly reconfigurable experimental grid testbed. Int. J. High Perform. Comput. Appl. **20**(4), 481–494 (2006)
5. Bosilca, G., Bouteiller, A., Danalis, A., Faverge, M., Haidar, A., Herault, T., Kurzak, J., Langou, J., Lemarinier, P., Ltaief, H., Luszczek, P., YarKhan, A., Dongarra, J.: Flexible development of dense linear algebra algorithms on massively parallel architectures with DPLASMA. In: IEEE International Symposium on Parallel and Distributed Processing Workshops and PhD Forum, pp. 1432–1444 (2011)
6. Bosilca, G., Bouteiller, A., Danalis, A., Faverge, M., Haidar, H., Herault, T., Kurzak, J., Langou, J., Lemarinier, P., Ltaief, H., Luszczek, P., YarKhan, A., Dongarra, J.: Distributed-memory task execution and dependence tracking within DAGuE and the DPLASMA project. Tech. Rep. 232, LAPACK Working Note (2010). http://www.netlib.org/lapack/lawnspdf/lawn232.pdf
7. Bosilca, G., Bouteiller, A., Danalis, A., Herault, T., Lemarinier, P., Dongarra, J.: DAGuE: A generic distributed DAG engine for high performance computing. Tech. Rep. 231, LAPACK Working Note (2010). http://www.netlib.org/lapack/lawnspdf/lawn231.pdf
8. Bosilca, G., Bouteiller, A., Danalis, A., Herault, T., Lemarinier, P., Dongarra, J.: DAGuE: A generic distributed DAG engine for high performance computing. In: Proceedings of the 16th International Workshop on High-Level Parallel Programming Models and Supportive Environments (HIPS'11), Anchorage, AL, USA (2011)

9. Bosilca, G., Bouteiller, A., Danalis, A., Herault, T., Lemarinier, P., Dongarra, J.: DAGuE: A generic distributed DAG engine for high performance computing. In: 16th International Workshop on High-Level Parallel Programming Models and Supportive Environments (HIPS-11), Anchorage, AK (2011)
10. CUDA CUBLAS Library. http://developer.download.nvidia.com
11. Dennard, R.H., Gaensslen, F.H., Rideout, V.L., Bassous, E., LeBlanc, A.R.: Design of ion-implanted MOSFET's with very small physical dimensions. IEEE J. Solid-State Circuits **9**(5), 256–268 (1974). doi:10.1109/JSSC.1974.1050511
12. Dongarra, J.J., Luszczek, P., Petitet, A.: The LINPACK benchmark: Past, present and future. Concurr. Comput. **15**(9), 803–820 (2003). doi:10.1002/cpe.728
13. Gustavson, F.G., Karlsson, L., Kågström, B.: Distributed SBP Cholesky factorization algorithms with near-optimal scheduling. ACM Trans. Math. Softw. **36**(2), 1–25 (2009). http://doi.acm.org/10.1145/1499096.1499100
14. Kogge, P., Bergman, K., Borkar, S., Campbell, D., Carlson, W., Dally, W., Denneau, M., Franzon, P., Harrod, W., Hill, K., Hiller, J., Karp, S., Keckler, S., Klein, D., Lucas, R., Richards, M., Scarpelli, A., Scott, S., Snavely, A., Sterling, T., Williams, R.S., Yelick, K.: ExaScale computing study: Technology challenges in achieving exascale systems. Tech. Rep. TR-2008-13, Department of Computer Science and Engineering, University of Notre Dame (2008)
15. Li, Y., Dongarra, J., Tomov, S.: A note on auto-tuning GEMM for GPUs. In: ICCS '09: Proceedings of the 9th International Conference on Computational Science, pp. 884–892. Springer, Berlin (2009). doi:10.1007/978-3-642-01970-8_89
16. Moore, G.E.: Cramming more components onto integrated circuits. Electronics **38**(8) (1965)
17. Nath, R., Tomov, S., Dongarra, J.: Accelerating GPU kernels for dense linear algebra. In: Proc. of High Performance Computing for Computational Science (VECPAR'10), June 22–25, 2010
18. National Research Council Committee on the Potential Impact of High-End Computing on Illustrative Fields of Science and Engineering: The Potential Impact of High-End Capability Computing on Four Illustrative Fields of Science and Engineering. Academies Press, Washington (2008)
19. NVIDIA: NVIDIA's Next Generation CUDA Compute Architecture: Fermi (2009). http://www.nvidia.com/object/fermi_architecture.html
20. NVIDIA: NVIDIA CUDA™ Best Practices Guide Version 3.0. NVIDIA Corporation (2010)
21. NVIDIA: NVIDIA CUDA™ Programming Guide Version 3.0. NVIDIA Corporation (2010)
22. Sutter, H.: The free lunch is over: A fundamental turn toward concurrency in software. Dr. Dobb's Journal **30**(3) (2005). http://www.ddj.com/184405990
23. Tomov, S., Nath, R., Du, P., Dongarra, J.: MAGMA version 0.2 User Guide (11/2009). http://icl.cs.utk.edu/magma
24. University of Tennessee: PLASMA Users' Guide, Parallel Linear Algebra Software for Multicore Architectures, Version 2.2 (2009)
25. Volkov, V., Demmel, J.: Benchmarking GPUs to tune dense linear algebra. In: SC '08: Proceedings of the 2008 ACM/IEEE Conference on Supercomputing, pp. 1–11. IEEE Press, Piscataway (2008). http://doi.acm.org/10.1145/1413370.1413402
26. Whaley, R.C., Petitet, A., Dongarra, J.: Automated empirical optimizations of software and the ATLAS project. Parallel Comput. **27**(1–2), 3–35 (2001)

Chapter 6
The Explicit Spike Algorithm: Iterative Solution of the Reduced System

Carl Christian Kjelgaard Mikkelsen

Abstract The explicit Spike algorithm applies to narrow banded linear systems which are strictly diagonally dominant by rows. The parallel bottleneck is the solution of the so-called reduced system which is block tridiagonal and strictly diagonally dominant by rows. The reduced system can be solved iteratively using the truncated reduced system matrix as a preconditioner. In this paper we derive a tight estimate for the quality of this preconditioner.

6.1 Introduction

A matrix $A = [a_{ij}] \in \mathbb{R}^{n\times n}$ is diagonally dominant by rows if

$$\forall i \ : \sum_{j\neq i} |a_{ij}| \leq |a_{ii}|.$$

If the inequality is sharp, then A is strictly diagonally dominant by rows. If A is nonsingular and diagonally dominant by rows or if A is strictly diagonally dominant by rows, then $a_{ii} \neq 0$, and the dominance factor ϵ given by

$$\epsilon = \max_i \left\{ \frac{\sum_{j\neq i} |a_{ij}|}{|a_{ii}|} \right\}$$

is well defined.

The matrix A has lower bandwidth b_l if $a_{ij} = 0$ for $i > j + b_l$ and upper bandwidth b_u if $a_{ij} = 0$ for $j > i + b_u$. If $b = \max\{b_l, b_u\} \ll n$, then we say that A is narrow banded. Every square banded matrix can be partitioned as a block tridiagonal

C.C.K. Mikkelsen (✉)
Department of Computing Science and HPC2N, Umeå University, Umeå, Sweden
e-mail: spock@cs.umu.se

M.W. Berry et al. (eds.), *High-Performance Scientific Computing*,
DOI 10.1007/978-1-4471-2437-5_6, © Springer-Verlag London Limited 2012

matrix with square diagonal blocks, i.e.

$$A = \begin{bmatrix} A_1 & C_1 & & \\ B_1 & \ddots & \ddots & \\ & \ddots & \ddots & C_{m-1} \\ & & B_m & A_m \end{bmatrix}, \tag{6.1}$$

only the dimension of each diagonal block must be bounded from below by b. In particular, we do not have to choose the same dimension for each diagonal block, even in the exceptional case where b divides n.

The Spike algorithms are designed to solve banded systems on a parallel machine. The central idea was introduced by Sameh and Kuck [9] who considered the tridiagonal case and Chen, Kuck and Sameh [1] who studied the triangular case. Lawrie and Sameh [3] applied the algorithm to the symmetric positive definite case while Dongarra and Sameh [2] considered the diagonally dominant case. Polizzi and Sameh [7, 8] introduced the truncated Spike algorithm for systems which are strictly diagonally dominant by rows. Recently, Manguoglu, Sameh, and Schenk [4] have combined PARDISO with Spike in the PSPIKE package.

The explicit Spike algorithm by Dongarra and Sameh [2] can be used to solve narrow banded linear systems which are strictly diagonally dominant by rows. The algorithm extends naturally to systems which are block tridiagonal. Moreover, the analysis is simplified if we focus on the number of diagonal blocks, rather than the bandwidth of the matrix.

In Sect. 6.2 we state the explicit Spike algorithm for systems which are block tridiagonal and strictly diagonally dominant by rows. The parallel bottleneck is the solution of a reduced system which is block tridiagonal and strictly diagonally dominant by rows. The reduced system can be solved iteratively using the main block diagonal as a preconditioner. We derive a tight estimate for the quality of this preconditioner in Sect. 6.3. This is a special case of a more general theorem by Mikkelsen [5].

6.2 The Explicit Spike Algorithm

In this section we state the explicit Spike algorithm for systems which are block tridiagonal and strictly diagonally dominant by rows. The validity and the basic analysis of the algorithm hinges on the following lemma.

Lemma 6.1 *Let $G = [E, D, F]$ be a matrix such that $[D, E, F]$ is strictly diagonally dominant by rows with dominance factor ϵ. Then G is row equivalent to a unique matrix $K = [U, I, V]$. Moreover, the matrix $[U, V]$ satisfies*

$$\|[U, V]\|_\infty \le \epsilon.$$

$$
[A|f] = \left[\begin{array}{cccc|cccc|cccc|c}
A_1^{(1)} & C_1^{(1)} & & & & & & & & & & & f_1^{(1)} \\
B_2^{(1)} & \ddots & \ddots & & & & & & & & & & \vdots \\
 & \ddots & \ddots & C_{q-1}^{(1)} & & & & & & & & & \vdots \\
 & & B_q^{(1)} & A_q^{(1)} & C_q^{(1)} & & & & & & & & f_q^{(1)} \\
\hline
 & & & B_1^{(2)} & A_1^{(2)} & C_1^{(2)} & & & & & & & f_1^{(2)} \\
 & & & & B_2^{(2)} & \ddots & \ddots & & & & & & \vdots \\
 & & & & & \ddots & \ddots & C_{q-1}^{(2)} & & & & & \vdots \\
 & & & & & & B_q^{(2)} & A_q^{(2)} & C_q^{(2)} & & & & f_q^{(2)} \\
\hline
 & & & & & & & B_1^{(3)} & A_1^{(3)} & C_1^{(3)} & & & f_1^{(3)} \\
 & & & & & & & & B_2^{(3)} & \ddots & \ddots & & \vdots \\
 & & & & & & & & & \ddots & \ddots & C_{q-1}^{(3)} & \vdots \\
 & & & & & & & & & & B_q^{(3)} & A_q^{(3)} & f_q^{(3)}
\end{array}\right]
$$

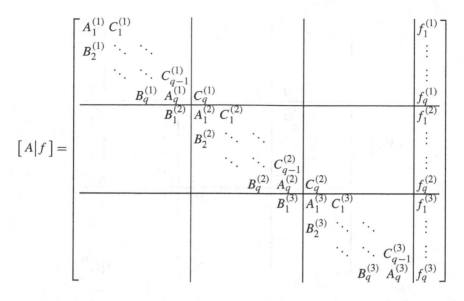

Fig. 6.1 The Spike partitioning for $p = 3$ processors

Proof Mikkelsen and Manguoglu [6] contains an elementary proof. □

Now consider the solution of a block tridiagonal linear system

$$Ax = f$$

on a parallel machine with p processors. Given a small tolerance $\delta > 0$, we shall now seek an approximation y, such that the forward error satisfies

$$\|x - y\|_\infty \leq \delta \|x\|_\infty.$$

We assume that A has $m = pq$ diagonal blocks and we assign q consecutive block rows to each processor. The case of $p = 3$ is illustrated in Fig. 6.1. If A is strictly diagonally dominant by rows, then we can predivide with the main block diagonal in order to obtain an equivalent linear system

$$Sx = g.$$

The case of $p = 3$ is displayed in Fig. 6.2. It is from the narrow columns or spikes protruding from the main diagonal that the original algorithm has derived it name. The matrix S is called the Spike matrix; the vector g is called the modified right hand side. By Lemma 6.1,

$$\|S - I\|_\infty \leq c < 1,$$

so S is strictly diagonally dominant by rows.

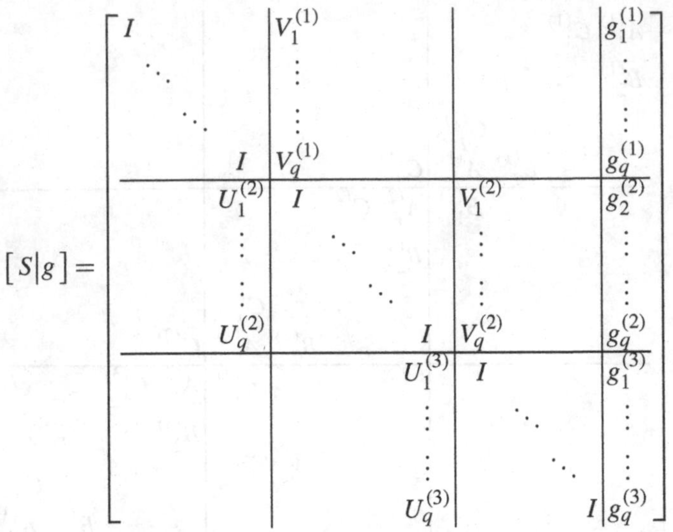

Fig. 6.2 The Spike matrix corresponding to $p = 3$ processors

$$[R|g_r] = \begin{bmatrix} I & V_q^{(1)} & & & & & & g_q^{(1)} \\ U_1^{(2)} & I & & V_1^{(2)} & & & & g_1^{(2)} \\ U_q^{(2)} & & I & V_q^{(2)} & \ddots & & & g_q^{(2)} \\ & & U_1^{(3)} & I & & \ddots & & g_1^{(3)} \\ & & & \ddots & & \ddots & & \vdots \\ & & & & \ddots & & V_1^{(p-1)} & \vdots \\ & & & U_q^{(p-1)} & & I & V_q^{(p-1)} & g_q^{(p-1)} \\ & & & & & U_q^{(p)} & I & g_1^{(p)} \end{bmatrix}$$

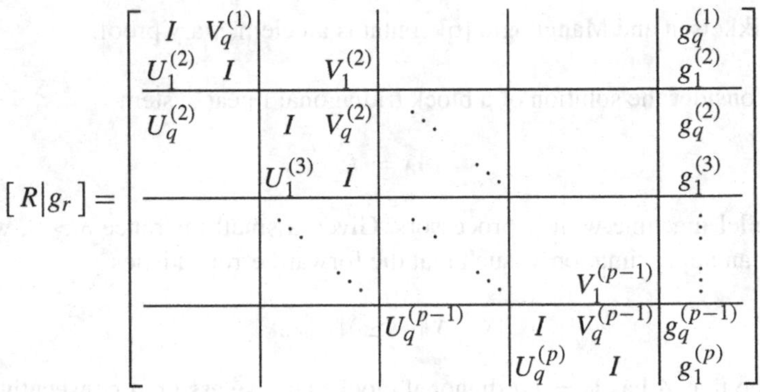

Fig. 6.3 The general structure of the reduced system

The equations within a single block row of each the main partitions lines form a reduced system,

$$R x_r = g_r$$

which can be solved independently. The general structure of the reduced system is given in Fig. 6.3. Once the reduced system has been solved, the solution of the original system can be retrieved by backsubstitution. Specifically, we have

$$x_i^{(j)} = g_i^{(j)} - U_i^{(j)} x_q^{(j-1)} - V_i^{(j)} x_1^{(j+1)}, \quad 1 \le i \le q, \ 1 \le j \le p, \tag{6.2}$$

$$
[T|g_r] = \left[\begin{array}{cc|cc|c|cc|c}
I & V_q^{(1)} & & & & & & g_q^{(1)} \\
U_1^{(2)} & I & & & & & & g_1^{(2)} \\
\hline
 & & I & V_q^{(2)} & & & & g_q^{(2)} \\
 & & U_1^{(3)} & I & & & & g_1^{(3)} \\
\hline
 & & & & \ddots & & & \vdots \\
 & & & & \ddots & & & \vdots \\
\hline
 & & & & & I & V_q^{(p-1)} & g_q^{(p-1)} \\
 & & & & & U_q^{(p)} & I & g_1^{(p)}
\end{array}\right]
$$

Fig. 6.4 The structure of the truncated reduced system

where $U_i^{(1)}$, $V_i^{(p)}$, $x_q^{(0)}$, and $x_1^{(p+1)}$ are undefined and should be taken as zero.

Suppose for the moment that we have somehow solved the reduced system with a small normwise relative forward error, say,

$$\|x_r - y_r\| \leq \delta \|x_r\|_\infty$$

In view of Eq. (6.2) it is natural to partition y_r conformally with x_r, i.e.

$$y_r = \left(x_q^{(1)T}, x_1^{(2)T}, \ldots, x_q^{(p-1)T}, x_1^{(p)T}\right)^T$$

and define a vector $y \in \mathbb{R}^n$ using

$$y_i^{(j)} = g_i^{(j)} - U_i^{(j)} y_q^{(j-1)} - V_i^{(j)} y_1^{(j+1)}, \quad 1 \leq i \leq q,\ 1 \leq j \leq p.$$

Then

$$x_i^{(j)} - y_i^{(j)} = -\left[U_i^{(j)}, V_i^{(j)}\right] \begin{bmatrix} x_q^{(j-1)} - y_q^{(j-1)} \\ x_1^{(j+1)} - y_1^{(j+1)} \end{bmatrix},$$

and it follows immediately that

$$\|x - y\|_\infty \leq \epsilon \|x_r - y_r\|_\infty \leq \epsilon\delta \|x_r\|_\infty \leq \delta \|x\|_\infty.$$

It is clear that we must solve the reduced system accurately in order to achieve a small forward normwise relative error. We now consider the solution of the reduced system.

The reduced system matrix R is block tridiagonal and strictly diagonally dominant by rows. The inequality

$$\|R - I\|_\infty \leq \epsilon < 1$$

is inherited from the Spike matrix S. Frequently, but not universally, the off diagonal blocks are insignificant and can be dropped. This phenomenon is exploited heavily in the truncated Spike algorithm by Polizzi and Sameh [7, 8]. Let T denote the main block diagonal of R, see Fig. 6.4. Mikkelsen and Manguoglu [6] showed that

$$\|T - R\|_\infty \leq \epsilon^q$$

when A is banded and strictly diagonally dominant by rows.

In this paper we consider the significance of the off diagonal blocks *relative* to the main block diagonal. To this end we define an auxiliary matrix $\mathcal{B}$ by

$$\mathcal{B} = T^{-1}(T - R).$$

Now, let x_{tr} be the solution of the truncated reduced system

$$Tx_{\text{tr}} = g_r.$$

Then

$$T(x_r - x_{\text{tr}}) = (R - (T - R))x_r - g_r = (Rx_r - g_r) - (T - R))x_r = -(T - R)x_r$$

from which it immediately follows that, if $x_r \neq 0$, then

$$\frac{\|x_{\text{tr}} - x_r\|_\infty}{\|x_r\|_\infty} \leq \|\mathcal{B}\|_\infty.$$

We have already understood the need to solve the reduced system with a forward normwise relative error of at most δ. If $\|\mathcal{B}\|_\infty \leq \delta$, then we simply drop the off diagonal blocks and approximate x_r with x_{tr}. If $\|\mathcal{B}\|_\infty > \delta$, then we can solve the reduced system iteratively using the main block diagonal as a preconditioner. If we use the stationary iteration

$$Tx_r^{(i)} = (T - R)x_r^{(i-1)} + g_r, \quad i = 1, 2, \ldots,$$

where $x_r^{(0)} = 0$, then

$$\left\|x_r - x_r^{(i)}\right\|_\infty \leq \|\mathcal{B}\|_\infty^i \|x_r\|_\infty$$

and we can stop the iteration whenever

$$\|\mathcal{B}\|_\infty^i \leq \delta.$$

In the next section we establish a tight upper bound on the central parameter $\|\mathcal{B}\|_\infty$. The Spike and the PSPIKE packages both apply BiCG, rather than the stationary iteration. Nevertheless, the size of $\|\mathcal{B}\|_\infty$ remains an interesting question.

6.3 The Main Result

Our purpose is to establish Theorem 6.1.

Theorem 6.1 *The auxiliary matrix $\mathcal{B}$ satisfies*

$$\|\mathcal{B}\|_\infty \leq \epsilon^q,$$

where q is the number of diagonal blocks assigned to each processor and equality is possible.

We shall reduce the problem of proving Theorem 6.1 to a single application of the following theorem.

Theorem 6.2 (Mikkelsen [5]) *Let G_k be a representation of $2k-1$ consecutive block rows of a block tridiagonal matrix A which is strictly diagonally dominant by rows with dominance factor ϵ, i.e.*

$$
G_k = \left[\begin{array}{c|ccc|c|cc|c}
B_{1-k} & A_{1-k} & C_{1-k} & & & & & \\
 & \ddots & \ddots & \ddots & & & & \\
 & & \ddots & \ddots & \ddots & & & \\
 & & & B_{-1} \; A_{-1} & C_{-1} & & & \\
\hline
 & & & B_0 & A_0 & C_0 & & \\
\hline
 & & & & B_1 & A_1 \; C_1 & & \\
 & & & & & \ddots \; \ddots & \ddots & \\
 & & & & & \ddots & \ddots \; \ddots & \\
 & & & & & & B_{k-1} \; A_{k-1} & C_{k-1}
\end{array}\right].
$$

Then G_k is row equivalent to a unique matrix K_k of the form

$$
K_k = \left[\begin{array}{c|ccc|c|ccc|c}
\mathscr{U}_{1-k}^{(k)} & I & & & & & & & \mathscr{V}_{1-k}^{(k)} \\
\vdots & & \ddots & & & & & & \vdots \\
\vdots & & & \ddots & & & & & \vdots \\
\mathscr{U}_{-1}^{(k)} & & & I & & & & & \mathscr{V}_{-1}^{(k)} \\
\hline
\mathscr{U}_{0}^{(k)} & & & & I & & & & \mathscr{V}_{0}^{(k)} \\
\hline
\mathscr{U}_{1}^{(k)} & & & & & I & & & \mathscr{V}_{1}^{(k)} \\
\vdots & & & & & & \ddots & & \vdots \\
\vdots & & & & & & & \ddots & \vdots \\
\mathscr{U}_{k-1}^{(k)} & & & & & & & I & \mathscr{V}_{k-1}^{(k)}
\end{array}\right] \tag{6.3}
$$

where the spikes decay exponentially as we move toward the main block row. Specifically, if we define

$$
Z_i^{(k)} = \begin{bmatrix} \mathscr{U}_{-i}^{(k)} & \mathscr{V}_{-i}^{(k)} \\ \mathscr{U}_{i}^{(k)} & \mathscr{V}_{i}^{(k)} \end{bmatrix}, \quad 0 < i < k,
$$

and

$$
Z_0^{(k)} = \left[\mathscr{U}_0^{(k)}, \mathscr{V}_0^{(k)}\right]
$$

then

$$
\|Z_i^{(k)}\|_\infty \le \epsilon^{k-i}, \quad 0 \le i < k.
$$

Proof The existence and uniqueness of K_k follows immediately from Lemma 6.1. The central inequality can be established using the well ordering principle. The details can be found in a report by Mikkelsen and Kågström [5]. □

We now move to prove the estimate given by Theorem 6.1. It is straightforward to verify that equality is achieved for matrices A given by Eq. (6.1) where

$$B_i = O_k, \quad A_i = I_k, \quad C_i = \epsilon I_k,$$

and O_k is the k by k zero matrix, I_k is the k by k identity matrix and $\epsilon < 1$.

In order to prove the general inequality it suffices to consider the interaction between two neighboring partitions. This follows immediately from the properties of the infinity norm. Let G_k be a compact representation of $2k$ block rows drawn from the original matrix A, i.e.

$$G_k = \left[\begin{array}{c|ccc|ccc|c}
B_{-k} & A_{-k} & C_{-k} & & & & & \\
 & \ddots & \ddots & \ddots & & & & \\
 & & \ddots & \ddots & \ddots & & & \\
 & & B_{-1} & A_{-1} & C_{-1} & & & \\
\hline
 & & & B_1 & A_1 & C_1 & & \\
 & & & & \ddots & \ddots & \ddots & \\
 & & & & & \ddots & \ddots & \ddots \\
 & & & & & B_k & A_k & C_k
\end{array}\right]$$

and let H_k be a compact representation of the corresponding rows of the associated Spike matrix. Then $G_k \sim H_k$ and H_k has the form

$$H_k = \left[\begin{array}{c|ccc|ccc|c}
U_{-k} & I & & & V_{-k} & & & \\
\vdots & & \ddots & & \vdots & & & \\
\vdots & & & \ddots & \vdots & & & \\
U_{-1} & & & I & V_{-1} & & & \\
\hline
 & & & U_1 & I & & & V_1 \\
 & & & \vdots & & \ddots & & \vdots \\
 & & & \vdots & & & \ddots & \vdots \\
 & & & U_k & & & I & V_k
\end{array}\right]. \tag{6.4}$$

Our task is to show that the auxiliary matrix Z_k given by

$$Z_k = \begin{bmatrix} Z_{11} & Z_{12} \\ Z_{21} & Z_{22} \end{bmatrix} = \begin{bmatrix} I & V_{-1} \\ U_1 & I \end{bmatrix}^{-1} \begin{bmatrix} U_{-1} & 0 \\ 0 & V_1 \end{bmatrix} \tag{6.5}$$

satisfies

$$\|Z_k\|_\infty \leq \epsilon^k, \quad k = 1, 2, \ldots, q.$$

We continue to reduce H_k using row operations. We repartition H_k in order to focus our attention on the two central block rows, i.e.

$$G_k \sim \left[\begin{array}{c|ccc|c|ccc|c}
U_{-k} & I & & & V_{-k} & & & & \\
\vdots & & \ddots & & \vdots & & & & \\
\vdots & & & \ddots & \vdots & & & & \\ \hline
U_{-1} & & & I & V_{-1} & & & & \\
 & & & & U_1 & I & & & V_1 \\ \hline
 & & & & \vdots & & \ddots & & \vdots \\
 & & & & \vdots & & & \ddots & \vdots \\
 & & & & U_k & & & I & V_k
\end{array}\right].$$

Then we predivide with the central 2 by 2 block matrix and obtain

$$G_k \sim \left[\begin{array}{c|ccc|c|ccc|c}
U_{-k} & I & & & V_{-k} & & & & \\
\vdots & & \ddots & & \vdots & & & & \\
\vdots & & & \ddots & \vdots & & & & \\ \hline
Z_{11} & & & I & & & & & Z_{12} \\
Z_{21} & & & & & I & & & Z_{22} \\ \hline
 & & & & \vdots & & \ddots & & \vdots \\
 & & & & \vdots & & & \ddots & \vdots \\
 & & & & U_k & & & I & V_k
\end{array}\right]$$

and it is clear that there exists a matrix K_k such that $G_k \sim K_k$ and

$$K_k = \left[\begin{array}{c|cccccccc|c}
\mathcal{U}_{-q} & I & & & & & & & & \mathcal{V}_{-q} \\
\vdots & & \ddots & & & & & & & \vdots \\
\vdots & & & \ddots & & & & & & \vdots \\
\mathcal{U}_{-2} & & & & I & & & & & \mathcal{V}_{-2} \\ \hline
\mathcal{U}_{-1} & & & & & I & & & & \mathcal{V}_{-1} \\
\mathcal{U}_{1} & & & & & & I & & & \mathcal{V}_{1} \\ \hline
\mathcal{U}_{2} & & & & & & & I & & \mathcal{V}_{2} \\
\vdots & & & & & & & & \ddots & \vdots \\
\vdots & & & & & & & & \ddots & \vdots \\
\mathcal{U}_{q} & & & & & & & & I & \mathcal{V}_{q}
\end{array}\right] \tag{6.6}$$

and the matrix Z_k satisfies

$$Z_k = \begin{bmatrix} \mathcal{U}_{-1} & \mathcal{V}_{-1} \\ \mathcal{U}_{1} & \mathcal{V}_{1} \end{bmatrix}. \tag{6.7}$$

At this point we have reduced the problem of proving Theorem 6.1 to a straightforward application of Theorem 6.2.

6.4 Conclusion

The explicit Spike algorithm by Dongarra and Sameh [2] extends naturally to systems which are block tridiagonal and strictly diagonally dominant by rows. Moreover, the analysis of the method is simplified by focusing on the number of diagonal blocks rather than the bandwidth. The parallel bottleneck remains the solution of the reduced system $Rx_r = g_r$ which is strictly diagonally dominant and block tridiagonal. The significance of the off diagonal blocks can be measured using the auxiliary matrix $\mathscr{B}$ given by

$$\mathscr{B} = T^{-1}(T - R) = I - T^{-1}R,$$

where T denotes the main block diagonal of R. If $\|\mathscr{B}\|_\infty$ is sufficiently small, then we can ignore the off diagonal blocks and approximate x_r with the solution of the truncated reduced system $Tx_{\text{tr}} = g_r$. In general, we can solve the reduced system iteratively using the main block diagonal T as a preconditioner and the convergence rate is controlled by the size of $\|\mathscr{B}\|_\infty$. Our main contribution is Theorem 6.1 which establishes a tight upper bound on $\|\mathscr{B}\|_\infty$.

Acknowledgements The work is supported by eSSENCE, a collaborative e-Science programme funded by the Swedish Research Council within the framework of the strategic research areas designated by the Swedish Government. In addition, support has been provided by the Swedish Foundation for Strategic Research under the frame program A3 02:128 and the EU Mål 2 project UMIT.

References

1. Chen, S., Kuck, D., Sameh, A.: Practical parallel band triangular system solvers. ACM Trans. Math. Softw. **4**, 270–277 (1978)
2. Dongarra, J.J., Sameh, A.: On some parallel banded system solvers. Parallel Comput. **1**(3–4), 223–236 (1984)
3. Lawrie, D.H., Sameh, A.H.: The computation and communication complexity of a parallel banded system slover. ACM Trans. Math. Softw. **10**(2), 185–195 (1984)
4. Manguoglu, M., Sameh, A.H., Schenk, O.: PSPIKE: A parallel hybrid sparse linear system solver. In: Proc. 15th Int'l. Euro-Par Conf. on Parallel Proc., Euro-Par '09, pp. 797–808. Springer, Berlin (2009)
5. Mikkelsen, C., Kågström, B.: Analysis of incomplete cyclic reduction for narrow banded and strictly diagonally dominant linear systems. Tech. Rep. UMINF 11.07, Department of Computing Science, Umeå University (2011). Submitted to PPAM-2011
6. Mikkelsen, C.C.K., Manguoglu, M.: Analysis of the truncated Spike algorithm. SIAM J. Matrix Anal. Appl. **30**, 1500–1519 (2008)
7. Polizzi, E., Sameh, A.H.: A parallel hybrid banded system solver: The SPIKE algorithm. Parallel Comput. **32**(2), 177–194 (2006)
8. Polizzi, E., Sameh, A.H.: SPIKE: A parallel environment for solving banded linear systems. Comput. Fluids **36**(1), 113–120 (2007)
9. Sameh, A.H., Kuck, D.J.: On stable parallel linear system solvers. J. Assoc. Comput. Mach. **25**(1), 81–91 (1978)

Chapter 7
The Spike Factorization as Domain Decomposition Method; Equivalent and Variant Approaches

Victor Eijkhout and Robert van de Geijn

Abstract In this paper we present the Spike algorithm of Sameh and Polizzi in the context of domain decomposition methods. We present several variants that differ in their treatment of the separators, showing that one of these is equivalent to the Spike algorithm.

7.1 Introduction

The parallel solution of linear systems has a long history, spanning both direct and iterative methods. While direct methods exist that have great generality, here we consider a subcase of practical importance: that of banded matrices. We note that many PDE problems naturally give rise to banded systems, given a large enough bandwidth.

For any banded matrix, we can impose a block structure such that the matrix is block tridiagonal. This structure gives each processor a contiguous block row of the matrix; we assume that the number of processors is low enough that the part owned by any processor comprises one or more of the blocks that define the block tridiagional structure.

In this paper we present a number of variants on the Spike factorization of Polizzi and Sameh [7], but going back to Sameh and Kuck [8]. Instead of the customary algebraic presentation we view this algorithm as a domain decomposition method, where each processor corresponds to a subdomain, and the problem variables are divided in interior regions and separators. We will make a cost analysis for the case where the algorithm is applied to a finite element type matrix. Note that our analysis is only in terms of flop counting; in practice the merits of the Spike algorithm and other banded solvers are determined to a large extent by memory access patterns and other considerations related to computer architecture.

V. Eijkhout (✉)
Texas Advanced Computing Center, The University of Texas at Austin, Austin, USA
e-mail: eijkhout@tacc.utexas.edu

R. van de Geijn
Computer Science Department, The University of Texas at Austin, Austin, USA
e-mail: rvdg@cs.utexas.edu

M.W. Berry et al. (eds.), *High-Performance Scientific Computing*,
DOI 10.1007/978-1-4471-2437-5_7, © Springer-Verlag London Limited 2012

Fig. 7.1 A one-dimensionally partitioned domain with four subdomains and three separators

Throughout this paper, we will discuss the 1D domain decomposition model problem, pictured in Fig. 7.1. This leads to a matrix of the form:

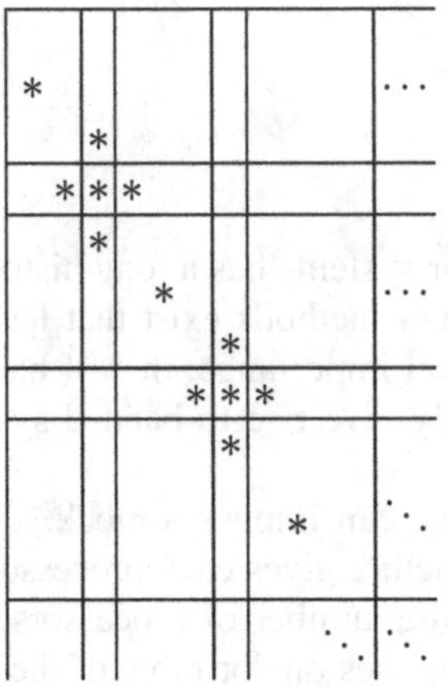

(7.1)

where the large blocks correspond to subdomains and the small ones to separators.

However, this is only for ease of analysis; in practice the only requirement for applicability of our ideas is that the matrix is partitioned with an alternating sequence of separators and subdomain interiors.

We will say that this matrix has block dimension N, where each block comprises a subdomain and a separator. For a model cost analysis, we assume that each subdomain interior consists of m lines of size n each, and that the matrix has a typical sparsity pattern based on some finite difference or finite element scheme. Thus, in the natural ordering, the matrix has dimension $N \times (m+1) \times n \approx Nmn$ and half-bandwith n. The cost of a sequential factorization is then $Nmn \cdot n^2$ muladds and the cost of solving a system with the resulting LU factorization $Nmn \cdot n$ muladds.

7.2 Single Separator Case

We will now describe the factorization of the matrix of Eq. (7.1). The first steps of the factorization are in parallel over the subdomains; we will illustrate the factorization by considering a subdomain with the two surrounding separators.

$$A = \left[\begin{array}{c|c|c} A_{i-1i-1} & A_{i-1i} & \\ \hline A_{ii-1} & A_{ii} & A_{ii+1} \\ \hline & A_{i+1i} & A_{i+1i+1} \end{array}\right], \qquad A_{ii-1} = \begin{pmatrix} * \\ 0 \\ \vdots \\ 0 \end{pmatrix}, \quad A_{ii+1} = \begin{pmatrix} 0 \\ \vdots \\ 0 \\ * \end{pmatrix}.$$

The matrix blocks are of dimensions

$$\begin{cases} A_{ii} & \text{size } mn, \text{ halfbandwidth } n, \\ A_{ii-1}, A_{ii+1} & \text{size } n, \text{ halfbandwidth } O(1). \end{cases}$$

We make an LU factorization of the large subdomain diagonal blocks.

$$L^{-1}A = \left[\begin{array}{c|c|c} 1 & & \\ \hline & L_{ii} & \\ \hline & & 1 \end{array}\right]^{-1}, \qquad A = \left[\begin{array}{c|c|c} A_{i-1i-1} & A_{i-1i} & \\ \hline L_{ii}^{-1}A_{ii-1} & U_{ii} & L_{ii}^{-1}A_{ii+1} \\ \hline & A_{i+1i} & A_{i+1i+1} \end{array}\right].$$

Next the U factor:

$$\left[\begin{array}{c|c|c} 1 & & \\ \hline & U_{ii} & \\ \hline & & 1 \end{array}\right]^{-1} (L^{-1}A) = \left[\begin{array}{c|c|c} A_{i-1i-1} & A_{i-1i} & \\ \hline A_{ii}^{-1}A_{ii-1} & I & A_{ii}^{-1}A_{ii+1} \\ \hline & A_{i+1i} & A_{i+1i+1} \end{array}\right].$$

Next we left-multiply in parallel by a matrix T to eliminate the connection between the subdomain interior and the separators:

$$\left[\begin{array}{c|c|c} I & A_{i-1i} & \\ \hline & I & \\ \hline & A_{i+1i} & I \end{array}\right]^{-1} U^{-1}(L^{-1}A) = \left[\begin{array}{c|c|c} \begin{array}{c} A_{i-1i-1} \\ -A_{i-1i}A_{ii}^{-1}A_{ii-1} \end{array} & \emptyset & -A_{i-1i}A_{ii}^{-1}A_{ii+1} \\ \hline -A_{ii}^{-1}A_{ii-1} & I & -A_{ii}^{-1}A_{ii+1} \\ \hline -A_{i+1i}A_{ii}^{-1}A_{i-1} & \emptyset & \begin{array}{c} A_{i+1i+1} \\ -A_{i+1i}A_{ii}^{-1}A_{ii+1} \end{array} \end{array}\right] \tag{7.2}$$

noting that the cost of forming the various products is limited to $O(n^3)$ since A_{i-1i}, A_{i+1i} are of size $n \times mn$, but have only one nonzero $n \times n$ block.

We now have a factorization $A = LUTS$, where S is called a 'spike' matrix after the dense columns flanking the large identity blocks. Note that the spikes need not be stored explicitly: we can multiply by $A_{ii}^{-1}A_{ii-1}$ by solving a linear system with A_{ii} and multiplying by A_{ii-1}.

The cost analysis of this factorization is as follows:

- The A_{ii} blocks that describe the subdomain interior are of size mn with halfbandwidth n, so factoring them takes mn^2 muladds, ignoring lower order terms: each elimination step adds a row to the n rows below it.

  Solving a system with the matrix A_{ii} takes $2mn^2$ operations: in both the forward and backward solve each element of the factors is touched once, and the factors have size mn and bandwidth n. Note that the factors are dense inside the band, unlike the original matrix.
- The blocks $A_{ii}^{-1}A_{ii\pm 1}$ are of size $mn \times n$ and fully dense, which is larger than storing the coefficient matrix A_{ii} of the interior by a factor $O(n)$. Multiplying by such an explicitly stored matrix takes mn^2 muladds.
- Forming $A_{ii}^{-1}A_{ii\pm 1}$ involves solving n linear systems, for a total cost of $2mn^3$
- Multiplying a vector by an explicitly stored block $A_{ii}^{-1}A_{ii\pm 1}$ takes mn^2 operations; doing this with an implicitly formed block takes $O(n)$ operations for the multiplication by $A_{ii\pm 1}$, and $2mn^2$ operations for the solution with A_{ii}. This is on the same order as the explicit multiplication.
- We note that because of the sparsity pattern of A_{ii+1} only the backward sweep of A_{ii} is needed, halving the cost of applying the spike block $A_{ii}^{-1}A_{ii+1}$.

The preliminary conclusion of this analysis is that the flop count for factoring the subdomain interiors equals (up to lower order terms) that of factoring the matrix sequentially, and performing a system solve on the subdomains has the cost of solving a system sequentially. Clearly, dealing with the spike matrix S is parallel overhead.

Next we factor the matrix S. This is no longer parallel over the subdomains, so we use two subdomains with separators to illustrate the inductive process.

$$S = \left[\begin{array}{c|c|c|c|c}
S_{i-1i-1} & & S_{i-1i} & & \\ \hline
S_{ii-1} & I & S_{ii+1} & & \\ \hline
S_{i+1i-1} & & S_{i+1i+1} & & S_{i+1i+3} \\ \hline
 & & S_{i+2i+1} & I & S_{i+2i+3} \\ \hline
 & & S_{i+3i+1} & & S_{i+3i+3}
\end{array}\right].$$

We sweep the first column with a lower triangular matrix L_{i-1}:

$$\left[\begin{array}{c|c|c|c|c}
I & & & & \\ \hline
S_{ii-1}S_{i-1i-1}^{-1} & I & & & \\ \hline
S_{i+1i-1}S_{i-1i-1}^{-1} & & I & & \\ \hline
 & & & I & \\ \hline
 & & & & I
\end{array}\right]^{-1} \cdot S = \left[\begin{array}{c|c|c|c|c}
S_{i-1i-1} & & S_{i-1i} & & \\ \hline
\emptyset & I & \tilde{S}_{ii+1} & & \\ \hline
\emptyset & & \tilde{S}_{i+1i+1} & & S_{i+1i+3} \\ \hline
 & & S_{i+2i+1} & I & S_{i+2i+3} \\ \hline
 & & S_{i+3i+1} & & S_{i+3i+3}
\end{array}\right]$$

where we note that elements are updated:

$$\tilde{S}_{ii+1} = S_{ii+1} - S_{ii-1} S_{i-1i-1}^{-1} S_{i-1i+1},$$
$$\tilde{S}_{i+1i+1} = S_{i+1i+1} - S_{i+1i-1} S_{i-1i-1}^{-1} S_{i-1i+1}.$$

For this we need to factor the dense blocks S_{i-1i-1}, S_{i+1i+1}.

We are left with a matrix that is (block) upper triangular on the first subdomain, and has the same structure on the next subdomain as what we started out with. This is enough to continue the inductive process. In the end this leaves us with

$$S = \Pi_i L_i \Pi_i U_i.$$

The diagonal blocks $S_{i\pm1,i\pm1}$ are of size $n \times n$ and dense so to solve a system with them they have to be factored. The blocks $S_{i\pm1,i\mp1}$ are of the same size and dense.

Factorization Cost Analysis The operation count for the factorization is as follows.

- The separator block S_{i-1i-1} is dense of size n, so there is a cost of $1/3\,n^3$ muladds in factoring it.
- A further cubic cost of $2n^3$ comes from the update $\tilde{S}_{i+1i+1} = S_{i+1i+1} - S_{i+1i-1} S_{i-1i-1}^{-1} S_{i-1i+1}$.
- The update $\tilde{S}_{ii+1} = S_{ii+1} - S_{ii-1} S_{i-1i-1}^{-1} S_{i-1i+1}$ takes $2mn^3$ muladds.

Taking this together, the factorization of the spike matrix takes the same $2Nmn^3$ operation count as factoring the interiors, which was the same as doing the sequential factorization, making the Spike factorization roughly twice as expensive as the sequential method.

In parallel, the dominant cost of forming $\tilde{S}_{ii+1}$ is not on the path of sequential dependencies, so it can be done in parallel, or by any inactive processors. The remaining cost is then forming the sequence of updated diagonal blocks $\tilde{S}_{i+1i+1}$, which adds a sequential time of $2Nn^3$, which is $O(m)$ lower than the cost of factoring the interiors.

Solution Cost Analysis The solution of a system $Ax = y$ involves the parallel subdomain solves with the A_{ii} blocks, and a sequential solve with the lower and upper factors of S. The important observation is that these carry sequential dependencies only between the separator blocks; the subdomain interiors depend on them, but carry no further dependencies. Hence, their solution can be happen after solving the separators system, or interweaved with it.

In terms of operations counts, the cost is dominated by subdomain solves that occur both in the block diagonal L and U factors, and in multiplying with the S_{ii-1} and S_{ii+1} spike blocks. Note that solving a spike system $Sx = y$ involves solving with the subdomain interiors twice, once in the forward and once in the backward solve, since $S_{ii-1} = -A_{ii}^{1} A_{ii-1}$ and $S_{ii+1} = -A_{ii}^{1} A_{ii+1}$. As argued above, the right spike only needs the backward solve.

Comparison to Domain Decomposition Methods Traditionally, parallel factorizations are a variant of LU applied to a partitioned and permuted domain. This partitioning can be based on a multi-coloring (see [3, 4]), or on a division in subdomains and separators, see for instance [1, 2, 5]. This latter approach leads to a factorization LSU with a similar analysis as we just saw: a parallel solution of the subdomain interiors, and a sequential system that couples the separators. In the traditional case there is a solve on the interior during both the parallel forward and backward sweep, giving a total cost of $2\cdot(|L|+|U|)$, where $|\cdot|$ indicates the cost of applying a matrix. In the Spike factorization there is a parallel interior solve plus an interior solve in multiplying by the spike matrices $A_{ii\pm 1}$, where we note that A_{ii+1} only requires the backward solve.

Yet another banded solver based on single separators can be found in [6].

We can now resume the discussion of parallel solve time that was started in Sect. 7.2. Solving a system with S has a sequential component:

$$T(S)=\sum_{i=1}^{N}\big[T(L_i^{(1)})+T(L_i^{(2)})+T(L_i^{(3)})\big]+T(\tilde S).$$

- Solving $L^{(1)}x=y$ has both a parallel and sequential time $O(1)$ in terms of a line block solve.
- Solving a system $L^{(2)}x=y$ is a parallel operation if S_{i+2i+1} is stored explicitly. However, note that $S_{i+2i+1}=A_{i+2i+2}^{-1}A_{i+2i+1}$, so in practice we will do a subdomain solve, with a solve time $T(L^{(2)})=T(L_{i+1i+1})+T(U_{i+1i+1})$.
- Solving $L^{(3)}x=y$ involves applying S_{i+1i+1}^{-1} which is of the size of a line block, so this operation is parallel.
- Solving $\tilde Sx=y$ is determined by the application of S_{ii+1} which is $A_{ii}^{-1}A_{ii+1}$. As with $L^{(2)}$, this could be stored explicitly, but more likely we will do a subdomain solve with A_{ii}.

7.3 Double Separator Case

One problem with domain decomposition methods using separators is the matter of distributing the separators. Since they are located between two subdomains, their processing does not trivially belong on either. We will now consider a method that allows for simpler work assignment, since it splits each separator in two; see Fig. 7.2.

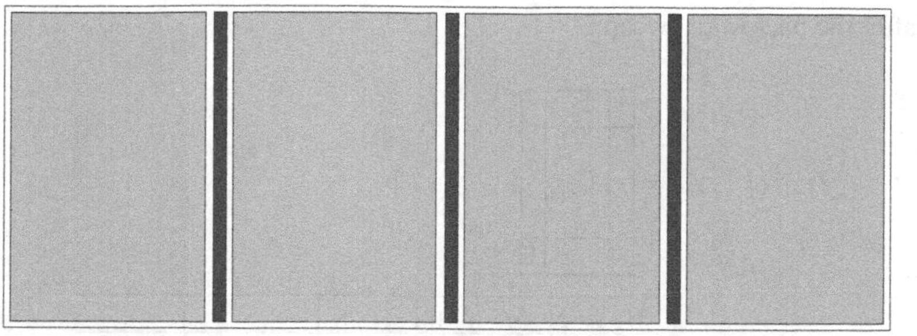

Fig. 7.2 One dimensional partitioning of a domain, with separators divided over the processors

The resulting matrix has the structure

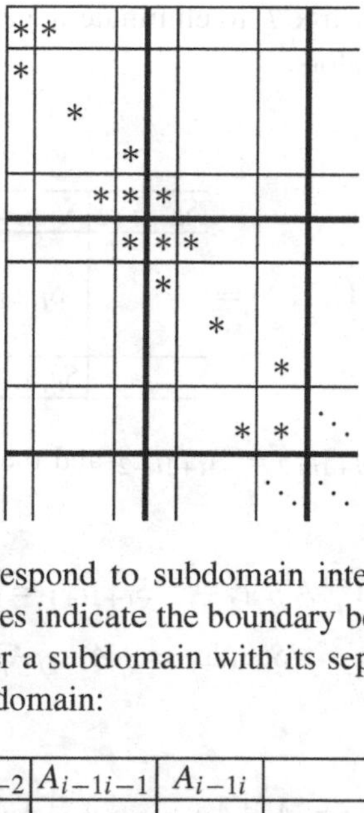

(7.3)

where the large blocks correspond to subdomain interiors and the small ones to separators, and the heavy lines indicate the boundary between processors.

To factor this we consider a subdomain with its separators and the connections to the previous and next subdomain:

$$A = \begin{array}{|c|c|c|c|c|} \hline A_{i-1i-2} & A_{i-1i-1} & A_{i-1i} & & \\ \hline & A_{ii-1} & A_{ii} & A_{ii+1} & \\ \hline & & A_{i+1i} & A_{i+1i+1} & A_{i+1i+2} \\ \hline \end{array}.$$

First we eliminate the subdomain interior. Applying the forward sweep gives

$$L^{-1}A = \begin{array}{|c|c|c|} \hline 1 & & \\ \hline & L_{ii} & \\ \hline & & 1 \\ \hline \end{array}^{-1}, \quad A = \begin{array}{|c|c|c|c|c|} \hline A_{i-1i-2} & A_{i-1i-1} & A_{i-1i} & & \\ \hline & L_{ii}^{-1}A_{ii-1} & U_{ii} & L_{ii}^{-1}A_{ii+1} & \\ \hline & & A_{i+1i} & A_{i+1i+1} & A_{i+1i+2} \\ \hline \end{array}$$

and after the backward sweep:

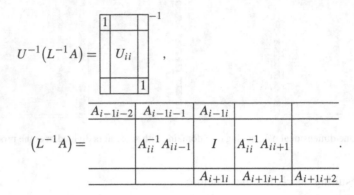

$$U^{-1}(L^{-1}A) = \left[\begin{array}{c|c|c} 1 & & \\ \hline & U_{ii} & \\ \hline & & 1 \end{array}\right]^{-1},$$

$$(L^{-1}A) = \left[\begin{array}{c|c|c|c|c} A_{i-1i-2} & A_{i-1i-1} & A_{i-1i} & & \\ \hline & A_{ii}^{-1}A_{ii-1} & I & A_{ii}^{-1}A_{ii+1} & \\ \hline & & A_{i+1i} & A_{i+1i+1} & A_{i+1i+2} \end{array}\right].$$

Next we left-multiply by a matrix T to eliminate the connection between the subdomain interior and the separators:

$$\begin{aligned} S &= T^{-1}U^{-1}L^{-1}A \\ &= \left[\begin{array}{c|c|c} I & A_{i-1i} & \\ \hline & I & \\ \hline & A_{i+1i} & I \end{array}\right]^{-1} U^{-1}(L^{-1}A) = \left[\begin{array}{c|c|c|c|c} S_{i-1i-2} & S_{i-1i-1} & \emptyset & S_{i-1i+1} & \\ \hline & S_{ii-1} & I & S_{ii+1} & \\ \hline & S_{i+1i-1} & \emptyset & S_{i+1i+1} & S_{i+1i+2} \end{array}\right] \end{aligned} \tag{7.4}$$

where $S_{i-1i-2} = A_{i-1i-2}$, $S_{i+1i+2} = A_{i+1i+2}$ and the following matrices will be explicitly formed:

$$\begin{aligned} S_{i-1i-1} &= A_{i-1i-1} - A_{i-1i}A_{ii}^{-1}A_{ii-1}, \qquad S_{i+1i+1} = A_{i+1i+1} - A_{i+1i}A_{ii}^{-1}A_{ii+1}, \\ S_{i+1i-1} &= -A_{i+1i}A_{ii}^{-1}A_{i-1}, \qquad S_{i-1i+1} = -A_{i-1i}A_{ii}^{-1}A_{ii+1} \end{aligned}$$

the blocks

$$S_{ii-1} = A_{ii}^{-1}A_{ii-1}, \quad A_{ii}^{-1}A_{ii+1}$$

are not explicitly formed, and their application involves a subdomain solve with A_{ii}.

Together we now have a factorization

$$A = LUTS$$

where the first three factors can be processed in parallel, both during the factorization and the system solution. It remains to analyze S. We see that the interior of the subdomain depends on the two separators, but not the other way around: in effect we now have a linear system where only the separators, two per subdomain, are mutually coupled. We can now proceed in two different ways.

7.3.1 Direct Factorization

To make an inductive analysis of the factorization of S we consider, for purposes of illustration, two full subdomains with separators.

$$
\left[\begin{array}{c|c|c||c|c|c}
S_{i-1i-1} & & S_{i-1i} & & & \\
\hline
S_{ii-1} & I & S_{ii+1} & & & \\
\hline
S_{i+1i-1} & & S_{i+1i+1} & S_{i+1i+2} & & \\
\hline\hline
 & & S_{i+2i+1} & S_{i+2i+2} & & S_{i+2i+4} \\
\hline
 & & & S_{i+3i+2} & I & S_{i+3i+4} \\
\hline
 & & & S_{i+4i+3} & & S_{i+4i+4}
\end{array}\right].
$$

We make an LU factorization of this matrix. The first step of forward sweep is applying a matrix $L_i^{(1)}$ to sweep the first column:

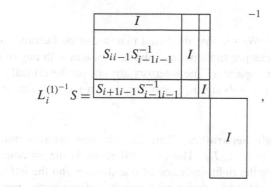

$$
L_i^{(1)^{-1}} S = \left[\begin{array}{c|c|c|c}
I & & & \\
\hline
S_{ii-1}S^{-1}_{i-1i-1} & I & & \\
\hline
S_{i+1i-1}S^{-1}_{i-1i-1} & & I & \\
\hline
 & & & I
\end{array}\right]^{-1},
$$

$$
S = \left[\begin{array}{c|c|c||c|c|c}
S_{i-1i-1} & & S_{i-1i} & & & \\
\hline
\emptyset & I & \tilde{S}_{ii+1} & & & \\
\hline
\emptyset & & \tilde{S}_{i+1i+1} & S_{i+1i+2} & & \\
\hline\hline
 & & S_{i+2i+1} & S_{i+2i+2} & & S_{i+2i+4} \\
\hline
 & & & S_{i+3i+2} & I & S_{i+3i+4} \\
\hline
 & & & S_{i+4i+3} & & S_{i+4i+4}
\end{array}\right]
$$

where

$$
\tilde{S}_{ii+1} = S_{ii+1} - S_{ii-1}S^{-1}_{i-1i-1}S_{i-1i+1},
$$
$$
\tilde{S}_{i+1i+1} = S_{i+1i+1} - S_{i+1i-1}S^{-1}_{i-1i-1}S_{i-1i+1}.
$$

Secondly, we apply a matrix $L_i^{(2)}$ to communicate with the left separator of the next subdomain:

$$\left[\begin{array}{c|c|c||c|c|c}
I & & & & & \\
\hline
 & I & & & & \\
\hline
 & & I & & & \\
\hline\hline
 & & \tilde{S}_{i+2i+1}\tilde{S}_{i+1i+1}^{-1} & I & & \\
\hline
 & & & & I & \\
\hline
 & & & & & I
\end{array}\right]^{-1} L^{(1)^{-1}} S = \left[\begin{array}{c|c|c||c|c|c}
S_{i-1i-1} & & S_{i-1i} & & & \\
\hline
\emptyset & I & \tilde{S}_{ii+1} & & & \\
\hline
\emptyset & & \tilde{S}_{i+1i+1} & S_{i+1i+2} & & \\
\hline\hline
 & & \emptyset & \tilde{S}_{i+2i+2} & & S_{i+2i+4} \\
\hline
 & & & S_{i+3i+2} & I & S_{i+3i+4} \\
\hline
 & & & S_{i+4i+3} & & S_{i+4i+4}
\end{array}\right]$$

where

$$\tilde{S}_{i+2i+2} = S_{i+2i+2} - \tilde{S}_{i+2i+1}\tilde{S}_{i+1i+1}^{-1}S_{i+1i+2}.$$

The second subdomain now has the same connections as the first had, so we can continue inductively.

Parallel Solve Time We see that the spike matrix can be factored as a product of $L^{(1)}, L^{(2)}, U$ matrices per subdomain. Solving a system with any of these takes solving a system on a separator. The interiors are not on the critical path, so the parallel solve consists of solving the separator system sequentially, and then the interiors in parallel.

Comparison to Single Separators This factorization behaves much like the single-separator case in Sect. 7.2. The only difference is the introduction of the $L^{(2)}$ matrix connecting the right separator of one domain and the left separator of the next. This mostly adds one solution with an $n \times n$ dense matrix per subdomain to the sequential time.

7.3.2 Full Elimination of the Subdomain

There is a second way of dealing with the spike matrix, which we show using only a single subdomain:

$$S = \left[\begin{array}{c|c|c|c|c}
\hline
S_{i-1i-2} & S_{i-1i-1} & & S_{i-1i+1} & \\
\hline
 & S_{ii-1} & I & S_{ii+1} & \\
\hline
 & S_{i+1i-1} & & S_{i+1i+1} & S_{i+1i+2} \\
\hline
\end{array}\right].$$

We first sweep the left separator by applying a lower triangular matrix $L^{(s)}$:

$$L^{(s)^{-1}} S = \left[\begin{array}{c|c|c} S_{i-1i-1} & \emptyset & \\ \hline S_{ii-1} & I & \\ \hline S_{i+1i-1} & \emptyset & I \end{array}\right]^{-1}, \quad S = \left[\begin{array}{c|c|c|c|c} S_{i-1i-2} & I & & S_{i-1i+1} & \\ \hline \tilde{S}_{ii-2} & \emptyset & I & S_{ii+1} & \\ \hline \tilde{S}_{i+1i-2} & \emptyset & & S_{i+1i+1} & S_{i+1i+2} \end{array}\right].$$

Next we sweep the second separator with a matrix $U^{(s)}$:

$$U^{(s)^{-1}} L^{(s)^{-1}} S = \left[\begin{array}{c|c|c} I & & S_{i-1i+1} \\ \hline & I & S_{ii+1} \\ \hline & & S_{i+1i+1} \end{array}\right]^{-1}, \quad L^{(s)^{-1}} S = \left[\begin{array}{c|c|c|c|c} S_{i-1i-2} & I & & \emptyset & \tilde{S}_{i-1i+2} \\ \hline \tilde{S}_{ii-2} & \emptyset & I & \emptyset & \tilde{S}_{ii+1} \\ \hline \tilde{S}_{i+1i-2} & \emptyset & & I & S_{i+1i+2} \end{array}\right].$$

The resulting factorization

$$A = LUL^{(s)}U^{(s)}Z$$

gives the matrix Z appearing in the original Spike algorithm, and is fully parallel in the parts just considered: any sequential component is entirely in the Z matrix.

We show the factorization of the Z matrix by considering two subsequent subdomains. The factorization is then

$$\left[\begin{array}{c|c|c|c|c|c} I & & & S_{i-1i+2} & & \\ \hline & I & & S_{ii+2} & & \\ \hline & & I & S_{i+1i+2} & & \\ \hline & & S_{i+2i+1} & I & & \\ \hline & & S_{i+3i+1} & & I & \\ \hline & & S_{i+4i+1} & & & I \end{array}\right] =$$

$$\left[\begin{array}{c|c|c|c|c|c} I & & & & & \\ \hline & I & & & & \\ \hline & & I & & & \\ \hline & & S_{i+2i+1} & I & & \\ \hline & & S_{i+3i+1} & & I & \\ \hline & & S_{i+4i+1} & & & I \end{array}\right] \left[\begin{array}{c|c|c|c|c|c} I & & & & & \\ \hline & I & & & & \\ \hline & & I & & & \\ \hline & & & I - S_{i+2i+1}S_{i+1i+2} & & \\ \hline & & & -S_{i+3i+1}S_{i+1i+2} & I & \\ \hline & & & -S_{i+4i+1}S_{i+1i+2} & & I \end{array}\right] \left[\begin{array}{c|c|c|c|c|c} I & & & S_{i-1i+2} & & \\ \hline & I & & S_{ii+2} & & \\ \hline & & I & S_{i+1i+2} & & \\ \hline & & & I & & \\ \hline & & & & I & \\ \hline & & & & & I \end{array}\right].$$

As before, we observe that only the separators are mutually dependent; the interiors are dependent on the separators but not the other way around.

7.3.3 Original Derivation of the Spike Algorithm

Instead of going through the S matrix, we can also derive Z directly. As before, we eliminate the subdomain interiors, and their connections with the separators; see the derivation of Eq. (7.2). Thus we start with

$$\left(\begin{array}{c|c|c|c|c} A_{i-1i-2} & A_{i-1i-1} & & A_{i-1i+1} & \\ \hline & A_{ii-1} & I & A_{ii+1} & \\ \hline & A_{i+1i-1} & & A_{i+1i+1} & A_{i+1i+2} \end{array}\right).$$

Note that these are no longer the original matrix blocks. We continue factoring by sweeping the column of the first separator: $A^{(1)} = LA^{(2)}$

$$\left(\begin{array}{c|c|c} I & & \\ \hline A_{ii-1}A^{-1}_{i-1i-1} & I & \\ \hline A_{i+1i-1}A^{-1}_{i-1i-1} & & I \end{array}\right) \left(\begin{array}{c|c|c|c|c} A_{i-1i-2} & A_{i-1i-1} & & A_{i-1i+1} & \\ \hline A_{ii-2} & \emptyset & I & A_{ii+1} & \\ \hline A_{i+1i-2} & & & A_{i+1i+1} & A_{i+1i+2} \end{array}\right)$$

where

$$\begin{cases} A_{ii-2} = -A_{ii-1}A^{-1}_{i-1i-1}A_{i-1i-2}, \\ A_{i+1i-2} = -A_{i+1i-1}A^{-1}_{i-1i-1}A_{i-1i-2}, \\ A_{ii+1} \leftarrow A_{ii+1} - A_{ii-1}A^{-1}_{i-1i-1}A_{i-1i+1}, \\ A_{ii+1} \leftarrow A_{ii+1} - A_{ii-1}A^{-1}_{i-1i-1}A_{i-1i+1}. \end{cases}$$

Now we sweep the upper part of the column of the second separator: $A^{(2)} = UA^{(3)}$

$$\left(\begin{array}{c|c|c} I & & A_{i-1i+1}A^{-1}_{i+1i+1} \\ \hline & I & A_{ii+1}A^{-1}_{i+1i+1} \\ \hline & & I \end{array}\right) \left(\begin{array}{c|c|c|c|c} A_{i-1i-2} & A_{i-1i-1} & & & A_{i-1i+2} \\ \hline A_{ii-2} & \emptyset & I & & A_{ii+2} \\ \hline A_{i+1i-2} & & & A_{i+1i+1} & A_{i+1i+2} \end{array}\right)$$

where now blocks in the right spike are newly formed, and ones in the left spike get updated. After normalizing the diagonal blocks on the separators, we now have an identity block for the whole subdomain, and the traditional spikes flanking it, which is the Z matrix of the previous subsection.

7.4 Discussion

We have given three one-sided factorization algorithms, of which one is a computational variant of the Spike algorithm. The factorizations are presented using the interior/separator division of the subdomains that is commonly associated with domain decomposition methods. These methods, applied to sparse finite element type matrices, are seen to have an essentially similar operation count to traditional domain decomposition methods. It should be noted, however, that the actual performance of methods depends on memory access patterns and other matters not considered in this paper.

Acknowledgements This work was sponsored by NSF through awards CCF 0917096 and OCI-0850750. Any opinions, findings and conclusions or recommendations expressed in this material are those of the authors and do not necessarily reflect the views of the National Science Foundation (NSF).

References

1. Bjørstad, P., Widlund, O.: Iterative methods for the solution of elliptic problems on regions partitioned in to substructures. SIAM J. Numer. Anal. **23**, 1097–1120 (1986)
2. Dryja, M.: A capacitance method for elliptic problems on regions partitioned into substructures. Numer. Math. **39**, 51–64 (1982)
3. Jones, M., Plassmann, P.: A parallel graph coloring heuristic. SIAM J. Sci. Stat. Comput. **14**, 654–669 (1993)
4. Luby, M.: A simple parallel algorithm for the maximal independent set problem. SIAM J. Comput. **4**, 1036–1053 (1986)
5. Meurant, G.: Domain decomposition methods for partial differential equations on parallel computers. Int. J. Supercomput. Appl. **2**, 5–12 (1988)
6. Naumov, M., Sameh, A.H.: A tearing-based hybrid parallel banded linear system solver. J. Comput. Appl. Math. **226**(2), 306–318 (2009). http://www.sciencedirect.com/science/article/pii/S037704270800410X. Special Issue: Large scale scientific computations
7. Polizzi, E., Sameh, A.: A parallel hybrid banded system solver: the SPIKE algorithm. Parallel Comput. **32**, 177–194 (2006)
8. Sameh, A.H., Kuck, D.J.: On stable parallel linear system solvers. J. Assoc. Comput. Mach. **25**(1), 81–91 (1978)

Chapter 8
Parallel Solution of Sparse Linear Systems

Murat Manguoglu

Abstract Many simulations in science and engineering give rise to sparse linear systems of equations. It is a well known fact that the cost of the simulation process is almost always governed by the solution of the linear systems especially for large-scale problems. The emergence of extreme-scale parallel platforms, along with the increasing number of processing cores available on a single chip pose significant challenges for algorithm development. Machines with tens of thousands of multicore processors place tremendous constraints on the communication as well as memory access requirements of algorithms. The increase in number of cores in a processing unit without an increase in memory bandwidth aggravates an already significant memory bottleneck. Sparse linear algebra kernels are well-known for their poor processor utilization. This is a result of limited memory reuse, which renders data caching less effective. In view of emerging hardware trends, it is necessary to develop algorithms that strike a more meaningful balance between memory accesses, communication, and computation. Specifically, an algorithm that performs more floating point operations at the expense of reduced memory accesses and communication is likely to yield better performance. We present two alternative variations of **DS** factorization based methods for solution of sparse linear systems on parallel computing platforms. Performance comparisons to traditional **LU** factorization based parallel solvers are also discussed. We show that combining iterative methods with direct solvers and using **DS** factorization, one can achieve better scalability and shorter time to solution.

8.1 Introduction

Many simulations in science and engineering give rise to sparse linear systems of equations. It is a well known fact that the cost of the solution process is almost always governed by the solution of the linear systems especially for large-scale problems. The emergence of extreme-scale parallel platforms, along with the increasing number of processing cores available on a single chip pose significant challenges

M. Manguoglu (✉)
Department of Computer Engineering, Middle East Technical University, 06531 Ankara, Turkey
e-mail: manguoglu@ceng.metu.edu.tr

M.W. Berry et al. (eds.), *High-Performance Scientific Computing*,
DOI 10.1007/978-1-4471-2437-5_8,

for algorithm development. Machines with tens of thousands of multicore processors place tremendous constraints on the communication as well as memory access requirements of algorithms. The increase in number of cores in a processing unit without an increase in memory bandwidth aggravates an already significant memory bottleneck. Sparse linear algebra kernels are well-known for their poor processor utilization. This is a result of limited memory reuse, which renders data caching less effective. In view of emerging hardware trends, it is necessary to develop algorithms that strike a more meaningful balance between memory accesses, communication, and computation. Specifically, an algorithm that performs more floating point operations at the expense of reduced memory accesses and communication is likely to yield better performance.

Significant amount of effort has been devoted to design and implementation of parallel sparse linear systems solvers. Existing parallel sparse direct solvers, such as MUMPS [1–3], Pardiso [39, 40], SuperLU [22], and WSMP [13, 14], are based on **LU** factorization. Therefore, the speed improvements realized by such solvers are often limited due to the inherited limitations of sparse **LU** factorizations and sparse triangular forward-backward sweeps. Iterative solvers, such as preconditioned Krylov subspace methods with sparse approximate inverse or incomplete **LU** factorization based preconditioners, on the other hand, are often more scalable but not as robust as direct solvers.

We present two robust hybrid algorithms based on **DS** factorization for parallel solution of general sparse linear systems. At the cost of increased computation, **DS** factorization for solving the system allows us to minimize the interprocess communications and, hence, enhances concurrency. The remainder of this chapter is organized as follows. In Sect. 8.2, we present banded and sparse variations of **DS** factorization. In Sect. 8.3, we develop two hybrid general sparse linear system solvers that use **DS** factorization to solve the preconditioned system.

8.2 Banded and Sparse Parallel DS Factorizations

A number of banded solvers have been proposed and implemented in software packages such as, LAPACK [4] for uniprocessors, ScaLAPACK [7], and Spike [6, 9, 21, 30, 31, 34, 36, 37] for parallel architectures. The central idea of Spike is to partition the matrix so that each process (or processing element) can work on its own part of the matrix, with the processes communicating only during the solution of the common reduced system. The size of the reduced system is determined by the bandwidth of the matrix and the number of partitions.

Unlike classical sequential **LU** factorization of the coefficient matrix **A**, for solving a banded linear system $\mathbf{A}x = f$, the Spike scheme employs the factorization:

$$\mathbf{A} = \mathbf{DS}, \tag{8.1}$$

where **D** is the block diagonal of **A** for a given number of partitions. The factor **S**, given by $\mathbf{D}^{-1}\mathbf{A}$ (assuming **D** is nonsingular), called the spike matrix, consists of the

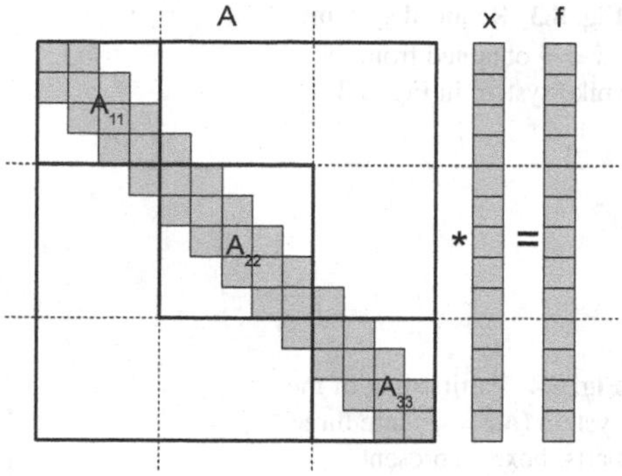

Fig. 8.1 Partitioning of the system ($\mathbf{A}x = f$) into three parts, boxes represent nonzeros

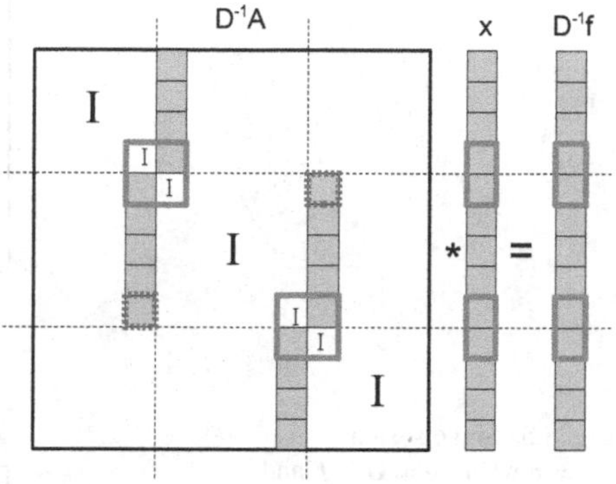

Fig. 8.2 Spike system: $\mathbf{S}x = g$ where $g = \mathbf{D}^{-1} f$ and $\mathbf{S} = \mathbf{D}^{-1}\mathbf{A}$ for the system in Fig. 8.1

block diagonal identity matrix modified by "spikes" to the right and left of each partition. The process of solving $\mathbf{A}x = f$, then, reduces to a sequence of the following steps:

- $g \leftarrow \mathbf{D}^{-1} f$ (modification of the right hand side)
- $\mathbf{S} \leftarrow \mathbf{D}^{-1}\mathbf{A}$ (forming the spike system coefficient matrix)
- $\hat{x} \leftarrow \hat{\mathbf{S}}^{-1}\hat{g}$ (solving a smaller independent reduced system)
- $x \leftarrow \mathbf{S}^{-1} g$ (retrieving the full solution).

All the steps of the solution process can be executed in perfect parallelism with the exception of the solution of the small reduced system. The size of the reduced system will increase as we increase the number of partitions (or processors). Furthermore, each step can be accomplished using one of several available methods, depending on the specific parallel architecture and the linear system at hand. This gives rise to a family of optimized variants of the basic Spike algorithm. For a small banded system we illustrate the partitioning of the system among three processors in Fig. 8.1. Figure 8.2 depicts the spike system $\mathbf{S}x = g$. Finally, Fig. 8.3 shows the smaller reduced system. Dashed boxes in figures show those elements which could be near machine precision if $\mathbf{A}$ is diagonally dominant system. Ignoring those small elements will give rise to "truncated" variation of the algorithm allowing more parallelism in the solution of the reduced system.

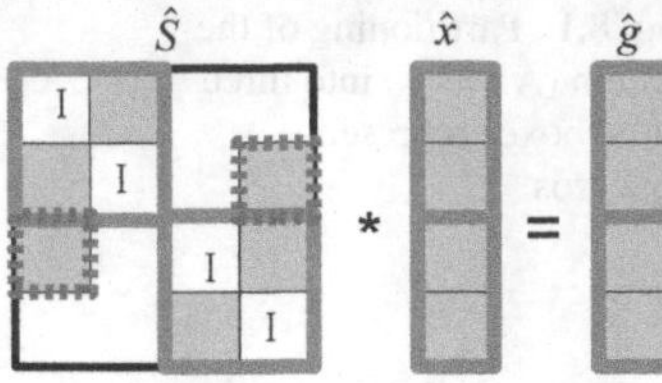

Fig. 8.3 Reduced system: $\hat{\mathbf{S}}\hat{x} = \hat{g}$ obtained from the spike system in Fig. 8.2

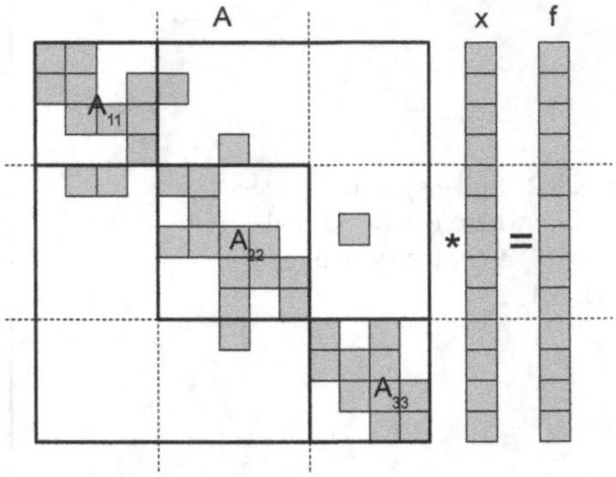

Fig. 8.4 Partitioning of the system ($\mathbf{A}x = f$) into three parts, boxes represent nonzeros

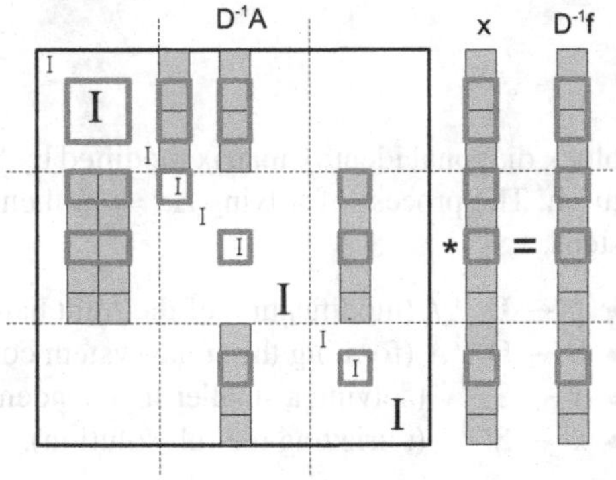

Fig. 8.5 Spike system: $\mathbf{S}x = g$ where $g = \mathbf{D}^{-1} f$ and $\mathbf{S} = \mathbf{D}^{-1}\mathbf{A}$ for the system in Fig. 8.4

For the solution of general sparse linear systems (where **A** is sparse) a new algorithm has been proposed in [24, 28]. Inspired by the banded Spike solver, this algorithm also uses the **DS** factorization and partitioning of the system (see Figs. 8.4 and 8.5). The resulting **S** matrix, however, is not banded and consists of "sparse" spikes. Nevertheless, a smaller reduced system can still be obtained (see Fig. 8.6). Solution stages follow exactly the banded case.

Both banded and sparse **DS** factorization based algorithms can be adapted for efficient and scalable solution of general sparse linear systems as will be discussed in Sect. 8.3.

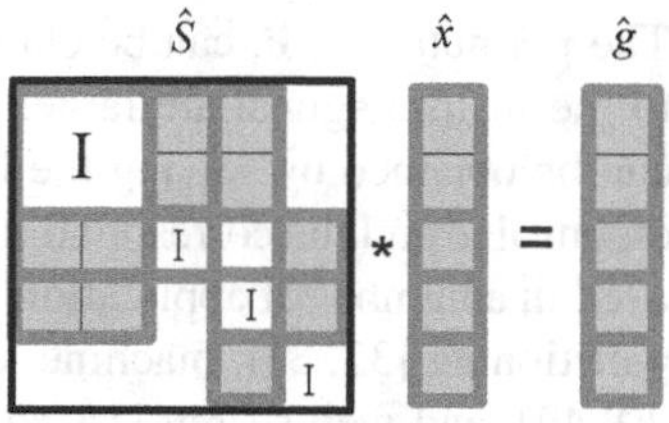

Fig. 8.6 Reduced system: $\hat{\mathbf{S}}\hat{x} = \hat{g}$ obtained from the spike system in Fig. 8.5

8.3 Solution of General Sparse Linear Systems

In this section, we present two hybrid methods for the solution of general sparse linear systems which use an iterative method as the outer layer with preconditioning. In both methods, the solution of preconditioned linear systems are handled by one of the variations of the **DS** factorization based banded or sparse solvers described in Sect. 8.2. The first method relies on reordering systems so that the large entries in the coefficient matrix are moved closer to the main diagonal. After reordering, an effective banded preconditioner, **M**, can be extracted and used for solving the system with an outer iterative layer. **M** could be treated as dense or sparse within the band and hence the diagonal blocks can be handled by a variety of algorithms. The second method, on the other hand, eliminates the need for the reordering with weights by dropping small elements when forming the preconditioner. An outer iterative method is also used for the second method and preconditioned systems are handled by the sparse variation of the **DS** factorization.

8.3.1 Weighted Reordering and Banded Preconditioning

Given a linear system of equations, $\mathbf{A}x = f$, we first apply a nonsymmetric row permutation as follows:

$$\mathbf{QA}x = \mathbf{Q}f. \tag{8.2}$$

Here, **Q** is the row permutation matrix that either maximizes the number of nonzeros on the diagonal of **A**, or the permutation that maximizes the product of the absolute values of the diagonal entries [11]. The first algorithm is known as maximum traversal search, while the second algorithm provides scaling factors so that the absolute values of the diagonal entries are equal to one and all other elements are less than or equal to one. This scaling can be applied as follows:

$$(\mathbf{QD}_2\mathbf{AD}_1)(\mathbf{D}_1^{-1}x) = (\mathbf{QD}_2 f). \tag{8.3}$$

Both algorithms are implemented in subroutine MC64 [10] of the HSL [17] library.

Following the above nonsymmetric reordering and optional scaling, we apply the symmetric permutation **P** as follows:

$$(\mathbf{PQD}_2\mathbf{AD}_1\mathbf{P}^T)(\mathbf{PD}_1^{-1}x) = (\mathbf{PQD}_2 f). \tag{8.4}$$

The permutation, $\mathbf{P}$, can be chosen such that the magnitude of the nonzeros closer to the main diagonal are larger than the ones that are far away. Such a reordering can be obtained by solving the following eigenvalue problem. The second smallest eigenvalue and the corresponding eigenvector of the Laplacian of a graph have been used in a number of application areas including matrix reordering [5, 23, 26], graph partitioning [32, 33], machine learning [29], protein analysis and data mining [16, 20, 42], and web search [15]. The second smallest eigenvalue of the Laplacian of a graph is sometimes called *the algebraic connectivity of the graph*, and the corresponding eigenvector is known as the *Fiedler vector*, due to the work of Fiedler [12].

For a given $n \times n$ sparse symmetric matrix $\mathbf{A}$, or an undirected weighted graph with positive weights, one can form the weighted-Laplacian matrix, $\mathbf{L}_w$, as follows:

$$\mathbf{L}_w(i,j) = \begin{cases} \sum_{\hat{j} \neq i} |\mathbf{A}(i,\hat{j})| & \text{if } i = j, \\ -|\mathbf{A}(i,j)| & \text{if } i \neq j. \end{cases} \tag{8.5}$$

Since the Fiedler vector can be computed independently for disconnected graphs, we assume that the graph is connected. The eigenvalues of $\mathbf{L}_w$ are different than zero except λ_1. The eigenvector x_2 corresponding to smallest nontrivial eigenvalue λ_2 is called the Fiedler vector. Since we assume a connected graph, the trivial eigenvector, x_1, is a vector of all ones. In case the matrix, $\mathbf{A}$, is nonsymmetric one can use $(|\mathbf{A}| + |\mathbf{A}^T|)/2$, instead.

A state of the art multilevel solver [18] called MC73_Fiedler for computing the Fiedler vector is implemented in the Harwell Subroutine Library (HSL) [17]. It uses a series of levels of coarser graphs where the eigenvalue problem corresponding to the coarsest level is solved via the Lanczos method for estimating the Fiedler vector. The results are then prolongated to the finer graphs and Rayleigh Quotient Iterations (RQI) with shift and invert are used for refining the eigenvector. Linear systems encountered in RQI are solved via the SYMMLQ algorithm. We consider MC73_Fiedler as one of the best uniprocessor implementation for determining the Fiedler vector. A new parallel algorithm TraceMin-Fiedler is developed based on the Trace Minimization algorithm (TraceMin) [35, 38], and parallel results comparing it to MC73_Fiedler is presented in [25].

We consider solving the standard symmetric eigenvalue problem

$$\mathbf{L}x = \lambda x \tag{8.6}$$

where $\mathbf{L}$ denotes the weighted Laplacian, using the TraceMin scheme for obtaining the Fiedler vector. The basic TraceMin algorithm can be summarized as follows. Let $\mathbf{X}_k$ be an approximation of the eigenvectors corresponding to the p smallest eigenvalues such that $\mathbf{X}_k^T \mathbf{L} \mathbf{X}_k = \boldsymbol{\Sigma}_k$ and $\mathbf{X}_k^T \mathbf{X}_k = \mathbf{I}$, where $\boldsymbol{\Sigma}_k = \mathrm{DIAG}(\rho_1^{(k)}, \rho_2^{(k)}, \ldots, \rho_p^{(k)})$. The updated approximation is obtained by solving the minimization problem

$$\min \operatorname{tr}(\mathbf{X}_k - \boldsymbol{\Delta}_k)^T \mathbf{L} (\mathbf{X}_k - \boldsymbol{\Delta}_k), \quad \text{subject to } \boldsymbol{\Delta}_k^T \mathbf{X}_k = 0. \tag{8.7}$$

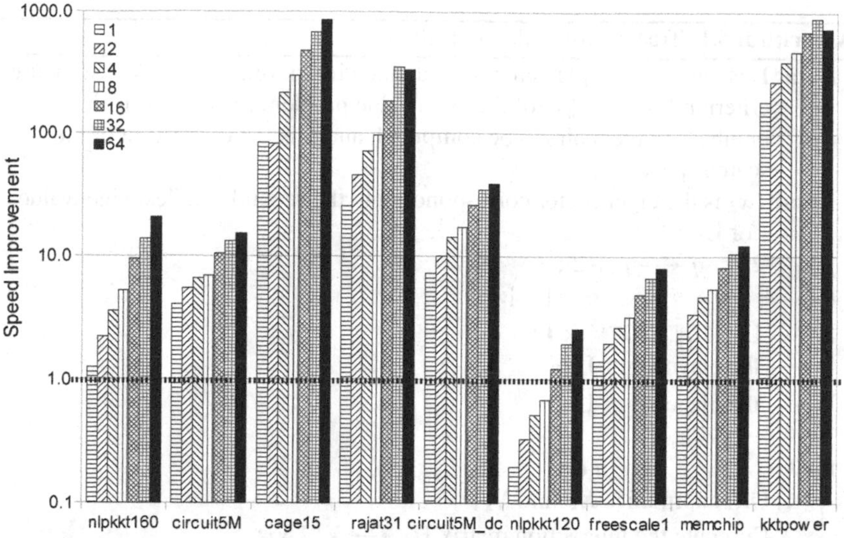

Fig. 8.7 Speed improvement: Time (MC73_Fiedler)/Time (TraceMin-Fiedler) for 9 test problems

This in turn leads to the need for solving a saddle point problem, in each iteration of the TraceMin algorithm, of the form

$$\begin{bmatrix} \mathbf{L} & \mathbf{X}_k \\ \mathbf{X}_k^T & 0 \end{bmatrix} \begin{bmatrix} \boldsymbol{\Delta}_k \\ \mathbf{N}_k \end{bmatrix} = \begin{bmatrix} \mathbf{L}\mathbf{X}_k \\ 0 \end{bmatrix}. \tag{8.8}$$

We solve first the Schur complement system $(\mathbf{X}_k^T \mathbf{L}^{-1} \mathbf{X}_k)\mathbf{N}_k = \mathbf{X}_k^T \mathbf{X}_k$ for obtaining $\mathbf{N}_k$. After $\boldsymbol{\Delta}_k$ is retrieved, $(\mathbf{X}_k - \boldsymbol{\Delta}_k)$ is then used to obtain $\mathbf{X}_{k+1}$ which forms the section

$$\mathbf{X}_{k+1}^T \mathbf{L} \mathbf{X}_{k+1} = \boldsymbol{\Sigma}_{k+1}, \qquad \mathbf{X}_{k+1}^T \mathbf{X}_{k+1} = \mathbf{I}. \tag{8.9}$$

The TraceMin-Fiedler algorithm, which is based on the basic TraceMin algorithm, is given in Algorithm 8.1.

Using the above algorithm, speed improvements over the uniprocessor MC73_Fiedler using TraceMin-Fiedler on 1, 8, 16, 32, and 64 cores are shown in Fig. 8.7 for matrices obtained from the University of Florida Sparse Matrix Collection [8]. The platform we use is a cluster with Infiniband interconnection where each node consists of two six-core Intel Xeon CPUs (Westmere X5670) running at 2.93 GHz (12 cores per node).

Reordering using the Fiedler vector provides matrices in which the large elements are clustered around the main diagonal as shown in Fig. 8.8 for a matrix obtained from the University of Florida Sparse Matrix Collection.

Once the reordered system is obtained one can extract a banded preconditioner and solve the general system using a preconditioned iterative method. Systems involving the preconditioner are solved at each iteration using the **DS** factorization. In which systems involving the diagonal blocks in **D** can be solved by using (i) dense

Algorithm 8.1: TraceMin-Fiedler algorithm

Data: $\mathbf{L}$ is the $n \times n$ Laplacian matrix defined in Eq. (8.5), ε_{out} is the stopping criterion for the $\|.\|_\infty$ of the eigenvalue problem residual, p is the number of eigenpairs to be computed, and q is the dimension of the search space

Result: x_2 is the eigenvector corresponding to the second smallest eigenvalue of $\mathbf{L}$

$p \longleftarrow 2$; $q \longleftarrow p+2$;

$n_{\text{conv}} \longleftarrow 0$; $\mathbf{X}_{\text{conv}} \longleftarrow [\ \ ]$;

$\hat{\mathbf{L}} \longleftarrow \mathbf{L} + \|\mathbf{L}\|_\infty 10^{-12} \times \mathbf{I}$;

$\mathbf{D} \longleftarrow$ the diagonal of $\mathbf{L}$;

$\hat{\mathbf{D}} \longleftarrow$ the diagonal of $\hat{\mathbf{L}}$;

$\mathbf{X}_1 \longleftarrow rand(n, q)$;

for $k = 1, 2, \ldots$ *max_it* **do**

- 1. Orthonormalize $\mathbf{X}_k$ into $\mathbf{V}_k$;
- 2. Compute the interaction matrix $\mathbf{H}_k \longleftarrow \mathbf{V}_k^T \mathbf{L} \mathbf{V}_k$;
- 3. Compute the eigendecomposition $\mathbf{H}_k \mathbf{Y}_k = \mathbf{Y}_k \boldsymbol{\Sigma}_k$ of $\mathbf{H}_k$. The eigenvalues $\boldsymbol{\Sigma}_k$ are arranged in ascending order and the eigenvectors are chosen to be orthogonal;
- 4. Compute the corresponding Ritz vectors $\mathbf{X}_k \longleftarrow \mathbf{V}_k \mathbf{Y}_k$;
- Note that $\mathbf{X}_k$ is a section, i.e. $\mathbf{X}_k^T \mathbf{L} \mathbf{X}_k = \boldsymbol{\Sigma}_k, \mathbf{X}_k^T \mathbf{X}_k = \mathbf{I}$;
- 5. Compute the relative residual $\|\mathbf{L}\mathbf{X}_k - \mathbf{X}_k \boldsymbol{\Sigma}_k\|_\infty / \|\mathbf{L}\|_\infty$;
- 6. Test for convergence: If the relative residual of an approximate eigenvector is less than ε_{out}, move that vector from $\mathbf{X}_k$ to $\mathbf{X}_{\text{conv}}$ and replace n_{conv} by $n_{\text{conv}} + 1$ increment. If $n_{\text{conv}} \geq p$, stop;
- 7. Deflate: If $n_{\text{conv}} > 0$, $\mathbf{X}_k \longleftarrow \mathbf{X}_k - \mathbf{X}_{\text{conv}}(\mathbf{X}_{\text{conv}}^T \mathbf{X}_k)$;
- 8. **if** $n_{\text{conv}} = 0$ **then**
  - Solve the linear system $\hat{\mathbf{L}} \mathbf{W}_k = \mathbf{X}_k$ approximately with relative residual ε_{in} via the PCG scheme using the diagonal preconditioner $\hat{\mathbf{D}}$;
- **else**
  - Solve the linear system $\mathbf{L} \mathbf{W}_k = \mathbf{X}_k$ approximately with relative residual ε_{in} via the PCG scheme using the diagonal preconditioner $\mathbf{D}$;
- 9. Form the Schur complement $\mathbf{S}_k \longleftarrow \mathbf{X}_k^T \mathbf{W}_k$;
- 10. Solve the linear system $\mathbf{S}_k \mathbf{N}_k = \mathbf{X}_k^T \mathbf{X}_k$ for $\mathbf{N}_k$;
- 11. Update $\mathbf{X}_{k+1} \longleftarrow \mathbf{X}_k - \boldsymbol{\Delta}_k = \mathbf{W}_k \mathbf{N}_k$;

sequential or multithreaded banded solvers [26] or (ii) a sparse sequential or multithreaded direct solver. Second variation has been implemented in [23, 27, 41] and is called PSPIKE.

We obtained the g3_circuit (1,585,478 unknowns and 7,660,826 nonzeros) matrix from the University of Florida Sparse Matrix Collection. In Fig. 8.9, speed

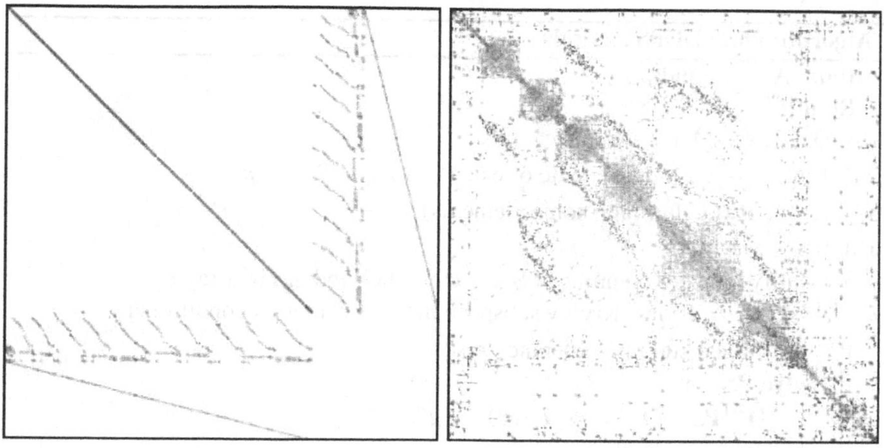

(a) original matrix (b) TraceMin-Fiedler

Fig. 8.8 Sparsity plots of *eurqsa*; red and blue indicates the largest and the smallest elements, respectively

Fig. 8.9 The speed improvement for g3_circuit compared to Pardiso using one core (73.5 seconds)

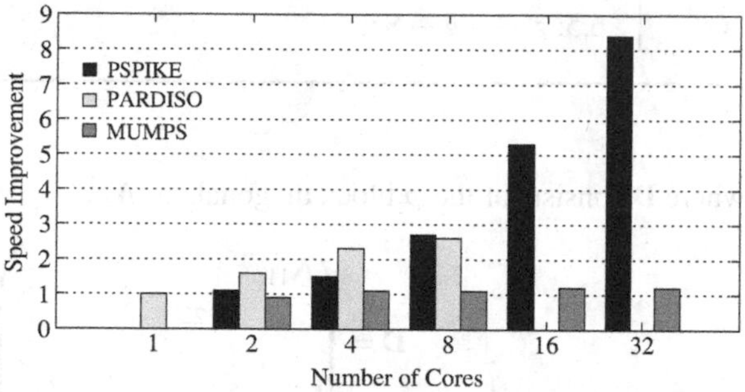

improvements of PSPIKE, MUMPS and Pardiso are presented for g3_circuit system is presented. The platform we use is an Intel Xeon (X5560@2.8GHz) cluster with Infini-band interconnection and 16 GB memory per node. In PSPIKE, BiCGStab [43] is used as the outer iterative solver and the iterations are stopped when $\|f - \mathbf{A}x\|_\infty / \|f\|_\infty \leq 10^{-6}$. The reduced system is truncated to enhance parallelism and solved directly.

8.3.2 Domain Decomposing Parallel Sparse Solver

Given a general sparse linear system $\mathbf{A}x = f$, we partition $\mathbf{A} \in R^{n \times n}$ into p block rows $\mathbf{A} = [\mathbf{A}_1, \mathbf{A}_2, \ldots, \mathbf{A}_p]^T$. Let

$$\mathbf{A} = \mathbf{D} + \mathbf{R}, \tag{8.10}$$

Algorithm 8.2: DDPS algorithm

Data: $\mathbf{A}x = f$ and p
Result: x
1. $\mathbf{D} + \mathbf{R} \longleftarrow \mathbf{A}$ for a given p;
2. $\tilde{\mathbf{L}}_i \tilde{\mathbf{U}}_i \longleftarrow \mathbf{A}_{ii}$ (approximate or exact) for $i = 1, 2, \ldots, p$;
3. $\tilde{\mathbf{R}} \longleftarrow \mathbf{R}$ (by dropping some elements);
4. $\mathbf{S} \longleftarrow \tilde{\mathbf{D}}^{-1}\tilde{\mathbf{R}}$;
5. identify nonzero columns of $\mathbf{S}$ and store their indices in array c;
6. **solve** $\mathbf{A}x = f$ via a Krylov subspace method with a preconditioner $\mathbf{P} = \tilde{\mathbf{D}} + \tilde{\mathbf{R}}$ and stopping tolerance ε_{out}
    solve $\mathbf{P}z = y$
        $(\tilde{\mathbf{D}}^{-1}\mathbf{P}z = \tilde{\mathbf{D}}^{-1}y \Rightarrow (\mathbf{I} + \mathbf{S})z = g)$;
        6.1. $g \longleftarrow \tilde{\mathbf{D}}^{-1}y$;
        6.2. $\hat{\mathbf{S}} \longleftarrow (\mathbf{I}(c,c) + \mathbf{S}(c,c))$; $\hat{z} \longleftarrow z(c)$; $\hat{g} \longleftarrow g(c)$;
        6.3. solve the smaller independent system: $\hat{\mathbf{S}}\hat{z} = \hat{g}$ (directly or iteratively with stopping tolerance ε_{in});
        6.4. $z(c) \longleftarrow \hat{z}$;
        6.5. $z \longleftarrow g - \mathbf{S}z$;

where $\mathbf{D}$ consists of the p block diagonals of $\mathbf{A}$,

$$\mathbf{D} = \begin{pmatrix} \mathbf{A}_{11} & & & \\ & \mathbf{A}_{22} & & \\ & & \ddots & \\ & & & \mathbf{A}_{pp} \end{pmatrix}, \tag{8.11}$$

and $\mathbf{R}$ consists of the remaining elements (i.e. $\mathbf{R} = \mathbf{A} - \mathbf{D}$). We note that the process can be viewed as an algebraic domain decomposition. Therefore, we will call the method described in this section *Domain Decomposing Parallel Sparse Solver* (DDPS). The **DS** factorization in DDPS is for solving the systems involving the preconditioner.

Let $\tilde{\mathbf{L}}_i$ and $\tilde{\mathbf{U}}_i$ be the incomplete or complete LU factorizations of A_{ii}, where $i = 1, 2, \ldots, p$. We define

$$\tilde{\mathbf{D}} = \begin{pmatrix} \tilde{\mathbf{A}}_{11} & & & \\ & \tilde{\mathbf{A}}_{22} & & \\ & & \ddots & \\ & & & \tilde{\mathbf{A}}_{pp} \end{pmatrix} \tag{8.12}$$

in which $\tilde{\mathbf{A}}_{ii} = \tilde{\mathbf{L}}_i \tilde{\mathbf{U}}_i$. The DDPS algorithm is shown in Algorithm 8.2.

Stages 1–5 are preprocessing stages where the right-hand-side is not required. After preprocessing we solve the system via a Krylov subspace method and using a preconditioner. The major operations in a Krylov subspace method are: (i) matrix–vector multiplications, (ii) inner products, and (iii) preconditioning operations in the form of $\mathbf{P}z = y$. Details of the preconditioning operations for DDPS are given in Algorithm 8.2.

Each stage, with the exception of Stage 6.3, can be executed with perfect parallelism, requiring no interprocessor communication. In Stage 6.3, the solution of the smaller system $\hat{\mathbf{S}}\hat{z} = \hat{g}$ is the only part of the algorithm that requires communication. We solve this smaller reduced system iteratively via BiCGStab without preconditioning. The size of $\hat{\mathbf{S}}$, which is determined by the nonzero columns it has, is problem dependent and is expected to have an influence on the overall scalability of the algorithm. We employ several techniques to reduce the dimension of $\hat{\mathbf{S}}$. First, we use METIS [19] reordering to minimize the total communication volume, hence reducing the size of $\hat{\mathbf{S}}$. We also use the following dropping strategy: if for any column j in $\mathbf{R}_i \|\mathbf{R}(:,j)_i\|_\infty \le \delta \times \max_l \|\mathbf{R}(:,l)_i\|_\infty$ $(i = 1, 2, \ldots, p)$ we do not consider that column when forming $\hat{\mathbf{S}}$. Here $\mathbf{R}_i$ is the block row partition of $\mathbf{R}$ (i.e. $\mathbf{R} = [\mathbf{R}_1, \mathbf{R}_2, \ldots, \mathbf{R}_p]^T$). We call this dropping strategy *a priori dropping* and use it for obtaining the results in this section. Another possibility is to drop elements after computing $\mathbf{S}$ *a posteriori dropping*.

We note that dropping elements from $\mathbf{R}$ in Stage 3 to reduce the size of $\hat{\mathbf{S}}$ results in an approximation of the solution. Furthermore, we can use approximate LU factorization of the diagonal blocks in Stage 2 and solve $\hat{\mathbf{S}}\hat{z} = \hat{g}$ iteratively in Stage 6.3. Therefore, we place an outer iterative layer (e.g. BiCGStab) where we use the above algorithm as a solver for systems involving the preconditioner $\mathbf{P} = \tilde{\mathbf{D}} + \tilde{\mathbf{R}}$, where $\tilde{\mathbf{R}}$ consists only of the columns that are not dropped. We stop the outer iterations when the relative residual at the kth iteration $\|r_k\|_\infty / \|r_0\|_\infty \le \varepsilon_{\text{out}}$.

DDPS is a direct solver if (i) nothing is dropped from $\mathbf{R}$, (ii) exact LU factorization of $\mathbf{A}_{ii}$ is computed, and (iii) $\hat{\mathbf{S}}\hat{z} = \hat{g}$ is solved exactly. In the case of using DDPS as a direct solver, an outer iterative scheme is not required. The choices we make in Stages 2, 3 and 6.3, result in a solver that can be as robust as a direct solver or as scalable as an iterative solver, or anything in between. We note that the outer iterative layer also benefits from our partitioning strategy as METIS minimizes the total communication volume in parallel sparse matrix vector multiplications. We further note that $\hat{\mathbf{S}}$ consists of dense columns, which we store as a two-dimensional array in memory, and as a result matrix–vector multiplications can be done via BLAS2 (or BLAS3 in case of multiple right hand sides).

We obtained the torso3 matrix (259, 156 unknowns and 4, 429, 042 nonzeros) from the University of Florida sparse matrix collection. In Fig. 8.10 we present for this matrix the speed improvements of DDPS and MUMPS solvers compared to Pardiso on a single core of an Intel Xeon (X5560@2.8GHz) cluster with Infini-band interconnection and 16 GB memory per node, and with $\varepsilon_{\text{out}} = 10^{-6}$. In this example, DDPS uses Pardiso for solving systems involving the diagonal blocks of $\mathbf{D}$.

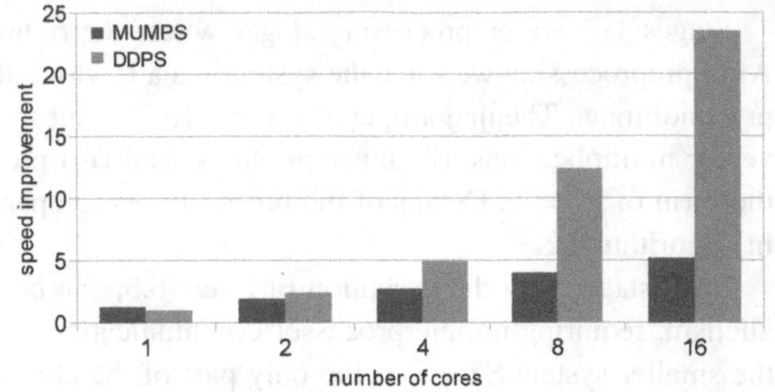

Fig. 8.10 The speed improvement for torso3 compared to Pardiso using one core (49.4 seconds)

8.4 Conclusions

We have presented two alternative formulations of **DS** factorization based methods for solution of sparse linear systems on parallel computing platforms. Performance comparisons to traditional **LU** factorization based parallel solvers show that combining iterative methods with direct solvers and using **DS** factorization, one can achieve better scalability and shorter time to solution.

Acknowledgements I would like to thank Ahmed Sameh, Ananth Grama, Faisal Saied, Eric Cox, Kenji Takizawa, Madan Sathe, Mehmet Koyuturk, Olaf Schenk, and Tayfun Tezduyar for their help and many useful discussions.

References

1. Amestoy, P.R., Duff, I.S.: Multifrontal parallel distributed symmetric and unsymmetric solvers. Comput. Methods Appl. Mech. Eng. **184**, 501–520 (2000)
2. Amestoy, P.R., Duff, I.S., L'Excellent, J.Y., Koster, J.: A fully asynchronous multifrontal solver using distributed dynamic scheduling. SIAM J. Matrix Anal. Appl. **23**(1), 15–41 (2001)
3. Amestoy, P.R., Guermouche, A., L'Excellent, J.Y., Pralet, S.: Hybrid scheduling for the parallel solution of linear systems. Parallel Comput. **32**(2), 136–156 (2006)
4. Anderson, E., Bai, Z., Bischof, C., Blackford, S., Demmel, J., Dongarra, J., Croz, J.D., Greenbaum, A., Hammerling, S., McKenney, A., Sorensen, D.: LAPACK Users' Guide, 3rd edn. SIAM, Philadelphia (1999)
5. Barnard, S.T., Pothen, A., Simon, H.: A spectral algorithm for envelope reduction of sparse matrices. Numer. Linear Algebra Appl. **2**(4), 317–334 (1995)
6. Berry, M.W., Sameh, A.: Multiprocessor schemes for solving block tridiagonal linear systems. Int. J. Supercomput. Appl. **1**(3), 37–57 (1988)
7. Blackford, L., Choi, J., Cleary, A., D'Azevedo, E., Demmel, J., Dhillon, I., Dongarra, J., Hammarling, S., Henry, G., Petitet, A., Stanley, K., Walker, D., Whaley, R.: ScaLAPACK User's Guide. SIAM, Philadelphia (1997). See also www.netlib.org/scalapack
8. Davis, T.A.: University of Florida sparse matrix collection. NA Digest (1997)
9. Dongarra, J.J., Sameh, A.H.: On some parallel banded system solvers. Parallel Comput. **1**(3), 223–235 (1984)
10. Duff, I., Koster, J.: On algorithms for permuting large entries to the diagonal of a sparse matrix (1999). citeseer.comp.nus.edu.sg/duff99algorithms.html
11. Duff, I.S., Koster, J.: The design and use of algorithms for permuting large entries to the diagonal of sparse matrices. SIAM J. Matrix Anal. Appl. **20**(4), 889–901 (1999). citeseer.ist.psu.edu/duff97design.html

12. Fiedler, M.: Algebraic connectivity of graphs. Czechoslov. Math. J. **23**(2), 298–305 (1973)
13. Gupta, A.: Recent advances in direct methods for solving unsymmetric sparse systems of linear equations. ACM Trans. Math. Softw. **28**(3), 301–324 (2002). http://doi.acm.org/10.1145/569147.569149
14. Gupta, A., Koric, S., George, T.: Sparse matrix factorization on massively parallel computers. In: SC'09 USB Key. ACM/IEEE, Portland, p. A1 (2009)
15. He, X., Zha, H., Ding, C.H., Simon, H.D.: Web document clustering using hyperlink structures. Comput. Stat. Data Anal. **41**(1), 19–45 (2002)
16. Higham, D.J., Kalna, G., Kibble, M.: Spectral clustering and its use in bioinformatics. J. Comput. Appl. Math. **204**(1), 25–37 (2007). http://www.sciencedirect.com/science/article/B6TYH-4K3D33V-3/2/204acbb72a44113bdd062272c3513921. Special issue dedicated to Professor Shinnosuke Oharu on the occasion of his 65th birthday
17. HSL: A collection of Fortran codes for large-scale scientific computation (2004). See http://www.cse.scitech.ac.uk/nag/hsl/
18. Hu, Y., Scott, J.: HSL_MC73: a fast multilevel Fiedler and profile reduction code. Technical Report RAL-TR-2003-036 (2003)
19. Karypis, G., Kumar, V.: Multilevel k-way partitioning scheme for irregular graphs. J. Parallel Distrib. Comput. **48**, 96–129 (1998)
20. Kundu, S., Sorensen, D., Philiphsi, G.N.J.: Automatic domain decomposition of proteins by a Gaussian network model. Proteins **57**(4), 725–733 (2004)
21. Lawrie, D.H., Sameh, A.H.: The computation and communication complexity of a parallel banded system solver. ACM Trans. Math. Softw. **10**(2), 185–195 (1984)
22. Li, X., Demmel, J.W.: SuperLU DIST: A scalable distributed-memory sparse direct solver for unsymmetric linear systems. ACM Trans. Math. Softw. **29**, 110–140 (2003)
23. Manguoglu, M.: A parallel hybrid sparse linear system solver. In: Computational Electromagnetics International Workshop, 2009. CEM 2009, pp. 38–43 (2009)
24. Manguoglu, M.: A domain-decomposing parallel sparse linear system solver. J. Comput. Appl. Math. **236**(3), 319–325 (2011)
25. Manguoglu, M., Cox, E., Saied, F., Sameh, A.: TRACEMIN-fiedler: A parallel algorithm for computing the Fiedler vector. In: High Perf. Comput. Comput. Sci.–VECPAR 2010, pp. 449–455 (2011)
26. Manguoglu, M., Koyuturk, M., Sameh, A., Grama, A.: Weighted matrix ordering and parallel banded preconditioners for iterative linear system solvers. SIAM J. Sci. Comput. **32**(3), 1201–1216 (2010)
27. Manguoglu, M., Sameh, A., Schenk, O.: Pspike: A parallel hybrid sparse linear system solver. In: Euro-Par 2009 Parallel Processing, pp. 797–808 (2009)
28. Manguoglu, M., Takizawa, K., Sameh, A., Tezduyar, T.: Nested and parallel sparse algorithms for arterial fluid mechanics computations with boundary layer mesh refinement. Int. J. Numer. Methods Fluids **65**, 135–149 (2011). doi:10.1002/fld.2415
29. Ng, A.Y., Jordan, M.I., Weiss, Y.: On spectral clustering: Analysis and an algorithm. In: Advances in Neural Information Processing Systems, vol. 14, pp. 849–856. MIT Press, Cambridge (2001)
30. Polizzi, E., Sameh, A.H.: A parallel hybrid banded system solver: the SPIKE algorithm. Parallel Comput. **32**(2), 177–194 (2006)
31. Polizzi, E., Sameh, A.H.: SPIKE: A parallel environment for solving banded linear systems. Comput. Fluids **36**(1), 113–120 (2007)
32. Pothen, A., Simon, H.D., Liou, K.P.: Partitioning sparse matrices with eigenvectors of graphs. SIAM J. Matrix Anal. Appl. **11**(3), 430–452 (1990)
33. Qiu, H., Hancock, E.R.: Graph matching and clustering using spectral partitions. Pattern Recognit. **39**(1), 22–34 (2006)
34. Chen, S.C., Kuck, D.J., Sameh, A.H.: Practical parallel band triangular system solvers. ACM Trans. Math. Softw. **4**(3), 270–277 (1978)
35. Sameh, A., Tong, Z.: The trace minimization method for the symmetric generalized eigenvalue problem. J. Comput. Appl. Math. **123**(1-2), 155–175 (2000)

36. Sameh, A.H., Kuck, D.J.: On stable parallel linear system solvers. J. ACM **25**(1), 81–91 (1978)
37. Sameh, A.H., Sarin, V.: Hybrid parallel linear system solvers. Int. J. Comput. Fluid Dyn. **12**, 213–223 (1999)
38. Sameh, A.H., Wisniewski, J.A.: A trace minimization algorithm for the generalized eigenvalue problem. SIAM J. Numer. Anal. **19**(6), 1243–1259 (1982)
39. Schenk, O., Gärtner, K.: Solving unsymmetric sparse systems of linear equations with PARDISO. Future Gener. Comput. Syst. **20**(3), 475–487 (2004)
40. Schenk, O., Gärtner, K.: On fast factorization pivoting methods for sparse symmetric indefinite systems. Electron. Trans. Numer. Anal. **23**, 158–179 (2006)
41. Schenk, O., Manguoglu, M., Sameh, A., Christian, M., Sathe, M.: Parallel scalable PDE-constrained optimization: antenna identification in hyperthermia cancer treatment planning. Comput. Sci. Res. Dev. **23**(3–4) (2009)
42. Shepherd, S.J., Beggs, C.B., Jones, S.: Amino acid partitioning using a Fiedler vector model. Eur. Biophys. J. **37**(1), 105–109 (2007)
43. van der Vorst, H.A.: BI-CGSTAB: a fast and smoothly converging variant of BI-CG for the solution of nonsymmetric linear systems. SIAM J. Sci. Stat. Comput. **13**(2), 631–644 (1992)

Chapter 9
Parallel Block-Jacobi SVD Methods

Martin Bečka, Gabriel Okša, and Marián Vajteršic

Abstract The serial Jacobi algorithm (either one-sided or two-sided) for the computation of a singular value decomposition (SVD) of a general matrix has excellent numerical properties and parallelization potential, but it is considered to be the slowest method for computing the SVD. Even its parallelization with some parallel cyclic (static) ordering of subproblems does not lead to much improvement when comparing with parallel methods based on the matrix bi-diagonalization principle. However, in the last 10 years some progress has been achieved in increasing the efficiency of the parallel block-Jacobi SVD method by using two new ideas: (i) the new parallel dynamic ordering of subproblems, and, (ii) the matrix pre-processing by QR iterations. For the parallel two-sided block-Jacobi algorithm, these ideas were already thoroughly tested on various parallel platforms, and our implementation can be faster than the ScaLAPACK routine `PDGESVD` for some distributions of singular values. With respect to the one-sided variant, the new parallel dynamic ordering, when compared to parallel cyclic ordering, can substantially decrease the number of parallel iteration steps needed for the convergence. However, its more scalable implementation is desirable because currently it occupies a relatively high portion of the total parallel execution time.

9.1 Background

This section contains a very brief introduction into the mathematical background behind the Singular Value Decomposition (SVD) of a matrix. Full SVD theory can be

M. Bečka (✉) · G. Okša · M. Vajteršic
Mathematical Institute, Department of Informatics, Slovak Academy of Sciences, Bratislava, Slovak Republic
e-mail: Martin.Becka@savba.sk

G. Okša
e-mail: Gabriel.Oksa@savba.sk

M. Vajteršic
Department of Computer Sciences, University of Salzburg, Salzburg, Austria
e-mail: marian@cosy.sbg.ac.at

M.W. Berry et al. (eds.), *High-Performance Scientific Computing*,
DOI 10.1007/978-1-4471-2437-5_9, © Springer-Verlag London Limited 2012

found in many excellent books, e.g. [7, 13, 16]. Afterwards, some serial algorithms for the SVD computation, other than the Jacobi method, are briefly mentioned. We concentrate on dense matrices, so some projection methods that are well-suited for sparse matrices are omitted.

In what follows, A^H denotes the Hermitian operation over the elements of matrix A, i.e., their complex conjugation and transposition. Further, $\|A\|_F$ and $\|A\|_2$ are the Frobenius and spectral norms of matrix A, respectively.

9.1.1 Singular Value Decomposition

The SVD of a complex matrix A of size $m \times n$, $(m \geq n)$, is defined by

$$A = U \Sigma V^H, \tag{9.1}$$

where U and V are unitary matrices of orders m and n, respectively, and Σ is an $m \times n$ diagonal matrix. The real, nonnegative diagonal elements $\sigma_1 \geq \sigma_2 \geq \cdots \geq \sigma_n \geq 0$ of Σ are the singular values of A, and the columns of U and V are the left and right singular vectors, respectively. When $m > n$, the matrix Σ contains the zero block of size $(m - n) \times n$ at the bottom.

The decomposition $A = U \Sigma V^H$ can be also written as $AV = U\Sigma$ or $Av_i = \sigma_i u_i$ for $i = 1, 2, \ldots, n$. The alternative way of saying the same thing is $A^H U = V \Sigma^H$ or $A^H u_i = \sigma_i v_i$ for $i = 1, 2, \ldots, n$ and $A^H u_i = 0$ for $i = n + 1, \ldots, m$. When $m > n$, the so called *thin* (or *economy-sized*) SVD is often computed in the form $A = U_n \Sigma_n V^H$, where $U_n = [u_1, u_2, \ldots, u_n]$ (i.e., only first n left singular vectors are computed), and $\Sigma_n = \mathrm{diag}(\sigma_1, \ldots, \sigma_n)$.

If $\mathrm{rank}(A) = r$ with $r < n$ then last $n - r$ singular values are zero and $A = U_r \Sigma_r V_r^H$ where $V_r = [v_1, v_2, \ldots, v_r]$. This is the so-called *compact* SVD of A where only first r left and right singular vectors play a role.

Taking only first t, $t < r$, left and right singular vectors and singular values, one obtains the rank-t approximation $A_t = U_t \Sigma_t V_t^H$, which is called the *rank-t truncated (partial)* SVD of A. Among all rank-t matrices B, $B = A_t$ is the unique minimizer of $\|A - B\|_F$. The truncated SVD is much smaller to store and cheaper to compute than the compact SVD when $t \ll r$ and it is the most often used form of the SVD in applications where the small singular values are of no interest to the user (e.g. in signal and image filtration, data retrieval computations, etc.).

9.1.2 Serial SVD Algorithms

As is well known, there exists a connection between the SVD of A and the Eigenvalue Decomposition (EVD) of Hermitian matrices $A^H A$, AA^H and $H(A) = \begin{pmatrix} 0 & A \\ A^H & 0 \end{pmatrix}$. There are special algorithms for the EVD of Hermitian matrices; see [1, 7, 13]. Hence, the straightforward approach for the SVD computation of A is to work with one of these matrices and compute its EVD. However, the explicit computation of $A^H A$ (or AA^H) is not advisable from the numerical point of view. When A has very small singular values, they are squared, become even smaller and can be computed with much less accuracy than required. On the other hand, squaring makes gaps between consecutive singular values even larger, so that the largest eigenvalues are well separated from the rest of the spectrum and are computed faster. Thus, EVD methods working with $A^H A$ (or AA^H) may be well-suited for the computation of largest singular values of A only.

In the case of $H(A)$, the dimension of the EVD problem is $m+n$, which can be prohibitive in the case of large matrices. Also small singular values of A become eigenvalues of $H(A)$ in the middle of its spectrum; to compute them accurately requires to use the shift-and-invert method or the Jacobi–Davidson method [1]. Both methods are rather expensive, because they require the solution of large linear systems in each iteration step. Also they are used only to compute a limited number of singular values near some *target* τ and are not well-suited for the computation of the whole SVD (or thin SVD).

Therefore, it is common practice to compute the SVD of A using directly matrix A. Most serial SVD algorithms apply some transformations to compute the thin SVD of A in three phases:

1. Find an $n \times n$ unitary matrix V_1 and $m \times n$ matrix U_1 with orthonormal columns such that $U_1^H A V_1 = B$ is $n \times n$ *bi-diagonal* matrix (i.e., only main diagonal and first superdiagonal are nonzero).
2. Compute the SVD of B: $B = U_2 \Sigma V_2^H$.
3. Multiply $U = U_1 U_2$ and $V = V_1 V_2$ to obtain the thin SVD of A: $A = U \Sigma V^H$.

Phase 1, reduction to the bidiagonal form, is computed using a sequence of unitary Householder reflections from the right and left. Its approximate cost is $O(mn^2)$ floating point operations. If singular values only are required, phase 2 costs just $O(n^2)$ floating point operations and phase 3 is omitted. If all left and right singular vectors are desired, the cost depends strongly on how the SVD of a bidiagonal matrix B is computed as described below.

Computing the SVD of a bidiagonal matrix B is in some sense similar to the EVD of a Hermitian, tridiagonal matrix [1, 7, 13]. Hence, modified EVD procedures are used including:

- *The QR algorithm* (not to be confused with the QR decomposition): This algorithm computes all singular values and optionally all the left and right singular vectors of a bidiagonal matrix [8]. The cost is $O(n^2)$ for singular values only and $O(n^3)$ for the whole thin SVD.
- *Divide-and-conquer algorithm*: It divides the matrix in two halves, computes the SVD of each half, and then "glues" the solutions together by solving a special rational equation [7]. Halving the matrix can be recursively repeated until a small submatrices are obtained (say, of order 20), for which the QR algorithm can be used. The cost depends on the number of halving steps, but in general it is comparable with the QR algorithm [1].
- *Bisection and inverse iteration*: This algorithm is used to find only singular values and vectors of interest (i.e., singular values are restricted to some interval).

It works in time $O(n)$ per singular triplet, but has big problems in the case of clustered singular values when the work can be as high as $O(n^3)$, and the orthogonality of computed singular vectors may not be guaranteed [10].

The Golub–Kahan–Lanczos algorithm applies the Lanczos algorithm to the Jordan–Wielandt matrix $H(A)$ and computes the set of left and right singular vectors of A, and, simultaneously, the bi-diagonalization of A [13]. Singular values are computed separately from the bidiagonal form. This algorithm requires matrix-vector multiplications both with A and A^H and is well-suited especially for large, sparse matrices. The reduction of the bidiagonal form to the diagonal matrix together with the update of all left and right singular vectors requires $O(m^2n + n^3)$ flops.

In summary, SVD methods based on bi-diagonalization can suffer from the loss of accuracy in computing the smallest singular values [9], since the high *relative* accuracy is not guaranteed.

9.2 Two-Sided Block Jacobi SVD Method

Next we shortly discuss the progress achieved in the two-sided block Jacobi SVD method, when a matrix $A \in \mathbb{C}^{m\times n}$, $m \geq n$, is cut row-wise and column-wise into an $\ell \times \ell$ block structure. Details can be found in our published papers [2–4, 15].

Having p processors, each processor contains exactly two block columns of A, U and V, so that the blocking factor is $\ell = 2p$. Notice that this is a rather natural partition of matrices, because each processor has to solve one 2×2 SVD subproblem in each parallel iteration step.

The rate of convergence measured by the decrease of the off-diagonal norm depends on the ordering of p subproblems that are solved in parallel. To achieve faster convergence, one should maximize the off-diagonal norm that is nullified in each parallel iteration step. Jacobi's approach [14] is optimal for the scalar case, because it annihilates the element with a maximum absolute value in each serial iteration step. We extended his idea to the parallel, block formulation in paper [4] using the solution of the *maximum-weight perfect matching problem*, and we obtained the new *dynamic ordering*. As opposed to any fixed, static list of subproblems, the dynamic ordering takes into account the actual status of the matrix w.r.t. the distribution of its off-diagonal norm.

Another way, how to further decrease the number of outer parallel iteration steps, can be based on applying an appropriate *preconditioner* to the original matrix A at the beginning of iteration process. Ideally, such a preconditioner should concentrate the Frobenius norm of A toward diagonal as much as possible.

The connection between diagonal elements of the R- or L-factor of a general matrix A and its singular values (SVs) was studied by Stewart in [16]. He has shown experimentally that after the QR factorization with column pivoting, followed optionally by the LQ factorization of the R-factor with or without column pivoting, the absolute values of diagonal elements in the resulting upper or lower triangular matrix (so called *R-values* or *L-values*) are, in general, very good approximations of SVs of A.

In paper [15], we have extended this serial preconditioner to the parallel case and shown that its combination with dynamic ordering can lead to a substantial decrease of the number of parallel iteration steps. The best results were achieved for well-conditioned matrices with a multiple minimal SV, where the reduction can be as large as two orders of magnitude.

In paper [2], the pre-processing step was extended to the method of *QR iterations* (QRI) using the optimal data layout for the QR (LQ) factorization. In general, the use of about six QRI steps can be recommended in the pre-processing, followed by a (quite limited) number of parallel iterations in the Jacobi algorithm with dynamic ordering. Such a strategy usually leads to a significant reduction of the total parallel execution time of the whole algorithm for almost all tested distributions of SVs.

9.3 One-Sided Block Jacobi SVD Algorithm

The one-sided block Jacobi SVD algorithm is suited for the SVD computation of a general complex matrix A of order $m \times n$, $m \geq n$. However, we will restrict ourselves to real matrices with obvious modifications in the complex case.

We start with the block-column partitioning of A in the form

$$A = [A_1, A_2, \ldots, A_\ell],$$

where the width of A_i is n_i, $1 \leq i \leq \ell$, so that $n_1 + n_2 + \cdots + n_\ell = n$.

The serial algorithm can be written as an iterative process:

$$\begin{aligned} A^{(0)} &= A, & V^{(0)} &= I_n, \\ A^{(k+1)} &= A^{(k)} U^{(k)}, & V^{(k+1)} &= V^{(k)} U^{(k)}, \quad k \geq 0. \end{aligned} \tag{9.2}$$

Here the $n \times n$ orthogonal matrix $U^{(k)}$ is the so-called *block rotation* of the form

$$U^{(k)} = \begin{pmatrix} I & & & & \\ & U_{ii}^{(k)} & & U_{ij}^{(k)} & \\ & & I & & \\ & U_{ji}^{(k)} & & U_{jj}^{(k)} & \\ & & & & I \end{pmatrix}, \tag{9.3}$$

where the unidentified matrix blocks are zero. The purpose of matrix multiplication $A^{(k)}U^{(k)}$ in Eq. (9.2) is to mutually orthogonalize the columns between column-blocks i and j of $A^{(k)}$. The matrix blocks $U_{ii}^{(k)}$ and $U_{jj}^{(k)}$ are square of order n_i and n_j, respectively, while the first, middle and last identity matrix is of order $\sum_{s=1}^{i-1} n_s$, $\sum_{s=i+1}^{j-1} n_s$ and $\sum_{s=j+1}^{\ell} n_s$, respectively. The orthogonal matrix

$$\hat{U}^{(k)} = \begin{pmatrix} U_{ii}^{(k)} & U_{ij}^{(k)} \\ U_{ji}^{(k)} & U_{jj}^{(k)} \end{pmatrix} \tag{9.4}$$

of order $n_i + n_j$ is called the *pivot submatrix* of $U^{(k)}$ at step k. During the iterative process Eq. (9.2), two index functions are defined: $i = i(k)$, $j = j(k)$ whereby

$1 \le i < j \le \ell$. At each step k, the pivot pair (i, j) is chosen according to a given *pivot strategy* that can be identified with a function $\mathcal{F}: \{0, 1, \ldots\} \to \mathbf{P}_r = \{(l, m) : 1 \le l < m \le \ell\}$. If $\mathbf{O} = \{(l_1, m_1), (l_2, m_2), \ldots, (l_{N(\ell)}, m_{N(\ell)})\}$ is some ordering of $\mathbf{P}_\ell$ with $N(\ell) = \ell\,(\ell - 1)/2$, then the *cyclic* strategy is defined by

$$\text{If } k \equiv \ell - 1 \mod N(\ell) \text{ then } (i(k), j(k)) = (l_s, m_s) \text{ for } 1 \le s \le N(\ell).$$

The most common cyclic strategies are the *row-cyclic* one and the *column-cyclic* one, where the orderings are given row-wise and column-wise, respectively, with regard to the upper triangle of A. The first $N(\ell)$ iterations constitute the first *sweep*. When the first sweep is completed, the pivot pairs (i, j) are repeated during the second sweep, and so on, up to the convergence of the entire algorithm.

Notice that in Eq. (9.2) only the matrix of right singular vectors $V^{(k)}$ is iteratively computed by orthogonal updates. If the process ends at iteration t, say, then $A^{(t)}$ has mutually highly orthogonal columns. Their norms are the singular values of A, and the normalized columns (with unit 2-norm) constitute the matrix of left singular vectors.

The parallel version of the one-sided block Jacobi SVD algorithm implemented on p processors with the blocking factor $\ell = 2p$ is given in the form of Algorithm 9.1.

Algorithm 9.1: Parallel one-sided block Jacobi SVD algorithm

$V = I_n,\ \ell = 2 * p$
▷ each processor has 2 block columns of $A : A_L$ and A_R
$G = \begin{pmatrix} G_{LL} & G_{LR} \\ G_{LR}^T & G_{RR} \end{pmatrix} = \begin{pmatrix} A_L^T A_L & A_L^T A_R \\ A_R^T A_L & A_R^T A_R \end{pmatrix}$
▷ *global convergence criterion with a constant* $\varepsilon,\ 0 < \varepsilon \ll 1$
while $(F(A, \ell) \ge \varepsilon)$ **do**
  ▷ *local convergence criterion with a constant* $\delta,\ 0 < \delta \ll 1$
  if $(F(G, \ell) \ge \delta)$ **then**
    ▷ *diagonalization of G*
    `EVD`(G, X)
    ▷ *update of block columns*
    $(A_L, A_R) = (A_L, A_R) * X$
    $(V_L, V_R) = (V_L, V_R) * X$
  end if
  ▷ *parallel ordering–choice of p independent pairs* (i, j) *of block columns*
  `ReOrderingComp`(p)
  `Send-Receive`(A_k, V_k, G_{kk}), where k is either L or R
end while
sv_L : square roots of diagonal elements of G_{LL}
sv_R : square roots of diagonal elements of G_{RR}
▷ *two block columns of left singular vectors*
$U_L = A_L * \text{diag}(1/sv_L),\ U_R = A_R * \text{diag}(1/sv_R)$

Note that the diagonalization of the auxiliary matrix G is equivalent to the mutual orthogonalization of block columns A_L and A_R of matrix A. Some parallel ordering is required in the procedure `ReOrderingComp` that defines p independent pairs of block columns of A which are simultaneously mutually orthogonalized in a given parallel iteration step by computing p eigenvalue decompositions $\texttt{EVD}(G, X)$ of p auxiliary matrices G. Up to now, some cyclic (static) parallel ordering (see [5, 6]) has been used. In next subsection, we describe a new *dynamic* ordering that takes into account the actual status of matrix A with respect to the mutual inclination of its block columns.

9.3.1 Dynamic Ordering

A big disadvantage of any fixed ordering is the fact that the actual status of orthogonality is usually checked only after a whole sweep and one has no information about the quality of this process at the beginning of a parallel iteration step. In other words, in a given parallel iteration step one can try to orthogonalize some mutually 'almost orthogonal' block columns while neglecting pairs that are far from being orthogonal. It is clear, at least intuitively, that orthogonalizing block columns with small mutual angles first would mean to eliminate the 'worst' pairs first, and this would mean (hopefully) the faster convergence of the whole algorithm as compared with any fixed, cyclic ordering.

Hence, the main question is how to choose p pairs of block columns with smallest principal angles among all $\ell(\ell-1)/2 = p(2p-1)$ pairs. The obvious, but very naive way is to compute, for each column block X, all possible matrix products $X^T Y$, then to compute the SVD of $X^T Y$ and look at the singular values, which are the cosines of acute principal angles (the smaller angle, the larger cosine). When the block columns are distributed in processors, to compute matrix products $X^T Y$ for each two different block columns X and Y means to move block columns across processors, i.e., it leads to heavy communication at the beginning of each parallel iteration step. Besides that, one needs to compute many matrix products and SVDs. Moreover, when p pairs of column blocks with smallest principal angles are chosen, they must meet in processors, which means yet another communication.

Our idea is different. After the first parallel iteration step, the block columns inside contain mutually orthogonal columns. Suppose that each processor contains exactly two block columns (this is not substantial for the following discussion). Moreover, suppose that $k \equiv n/2p$ columns in each block column are *normalized* so that each has the unit Euclidean norm. Hence, each column block is the *orthonormal basis* of the k-dimensional subspace which is spanned by the column vectors of a given block column.

Now take two block columns A_i, A_j which should be orthogonalized in a given parallel iteration step. Having p processors, our goal is to choose p pairs of those block columns that are maximally *inclined* to each other, i.e., their mutual position differs maximally from the orthogonal one.

This vague description can be made mathematically correct using the notion of *principal angles* between two k-dimensional subspaces spanned by two block columns A_i, A_j. Since A_i and A_j are orthonormal bases of two subspaces with the equal dimension, the cosines of principal angles are defined as the singular values of the matrix $A_i^T A_j$. Let $\sigma_1 \geq \sigma_2 \geq \cdots \geq \sigma_k$ be k singular values of the $k \times k$ matrix $A_i^T A_j$. Then the principal angles $\theta_1 \leq \theta_2 \leq \cdots \leq \theta_k$, $\theta_i \in [0, \pi/2]$, $1 \leq i \leq k$, are defined as

$$\theta_i = \arccos(\sigma_i), \quad 1 \leq i \leq k. \tag{9.5}$$

Since A_i and A_j have orthonormal columns, all singular values of $A_i^T A_j$ are in the interval [0, 1], so that the relation (9.5) is well defined.

We are interested in, say, L *smallest* principal angles, i.e., in L *largest* cosines (largest singular values) $\sigma_1 \geq \sigma_2 \geq \cdots \geq \sigma_L$. When $\sigma_1 = 0$, then all $\sigma_i = 0$, $2 \leq i \leq k$, and two block columns A_i and A_j are perfectly orthogonal; we do not need to orthogonalize them explicitly. On the other hand, when all σ_k are significantly greater than 0, column blocks A_i and A_j are certainly far from the mutual orthogonality.

However, this approach means that we must explicitly compute the matrix $A_i^T A_j$. When two block columns A_i and A_j are placed in two different processors, we can either compute this matrix product in parallel (but for each pair of block columns), or store both blocks in one processor and compute the matrix product locally using the LAPACK library. Afterwards, we must compute (or at least somehow estimate) the largest L singular values and afterwards compute some function of them (e.g., the sum of their squares) to get our weight w_{ij} for the maximal perfect matching. In both cases we need again too much communication at the beginning of each parallel iteration step to construct the actual parallel ordering for that step.

To estimate L largest singular value of the $k \times k$ matrix $A_i^T A_j$, we suggest to use the Lanczos process applied to the symmetric Jordan–Wielandt matrix C,

$$C \equiv \begin{pmatrix} 0 & A_i^T A_j \\ A_j^T A_i & 0 \end{pmatrix}. \tag{9.6}$$

It is well known that the eigenvalues of the $2k \times 2k$ matrix C are $\pm\sigma_1, \pm\sigma_2, \ldots, \pm\sigma_k$. Notice that there are k pairs of eigenvalues with the same absolute value.

It follows from the theory of Krylov space methods that the Lanzcos algorithm applied to a symmetric matrix is the good iterative method for estimating its largest (in absolute value) eigenvalues. This algorithm, applied to the symmetric Jordan–Wielandt matrix C, is listed as Algorithm 9.2 for a *fixed* number of iteration steps L. Steps 2–9 constitute an adaptation of the Arnoldi method for a symmetric matrix. Due to the special structure of C (see Eq. (9.6)), the matrix-vector product in step 2 is applied in two substeps: $w_s^1 = A_i^T A_j v_s^1$, $w_s^2 = A_j^T A_i v_s^2$, where $v_s = (v_s^{1T}, v_s^{2T})^T$ and $w_s = (w_s^{1T}, w_s^{2T})^T$.

The result is the orthonormal basis of the Krylov subspace $\mathcal{K}_L(C, x_0)$ formed by vectors v_s, $1 \leq s \leq L$. Besides that, the coefficients α_s and β_s are computed that are stored in the symmetric, tri-diagonal matrix T_L (step 10).

In our application, the orthonormal vectors v_s are not important (they are used, for example, in the solution of a linear system of equations). What is most important,

Algorithm 9.2: Lanczos algorithm for the symmetric Jordan–Wielandt matrix C

Choose an even integer L and the vector x_0 of length $2k$, and compute $\beta_1 = \|x_0\|$; $v_1 = x_0/\beta$
for $s = 1$ to L **do**
  $w_s = Cv_s$
  if $(s \neq 1)$ **then**
    $w_s = w_s - \beta_s v_{s-1}$
  end if
  $\alpha_s = w_s^T v_s$
  $w_s = w_s - \alpha_s v_s$
  $\beta_{s+1} = \|w_s\|$
  if $(\beta_{s+1} \neq 0)$ **then**
    $v_{s+1} = w_s/\beta_{s+1}$
  end if
  if $(\beta_{s+1} == 0)$ **then**
    $s = L$
  end if
end for
Set: $T_L = \text{tridiag}(\beta_i, \alpha_i, \beta_{i+1})$
Compute the Frobenius norm of T_L.

is the square of the Frobenius norm of T_L written in terms of its eigenvalues ω_s, $1 \leq s \leq L$ (they are known as Ritz values):

$$\|T_L\|_F^2 = \sum_{s=1}^{L} \omega_s^2.$$

As already mentioned, the L Ritz values approximate reasonably well L largest (in the absolute value) eigenvalues λ_s of the Jordan–Wielandt matrix C. However, in our application, there are exactly two eigenvalues of C with the same absolute value (with opposite signs) and they are related to the squares of singular values of $A_i^T A_j$. Therefore,

$$\|T_L\|_F^2 = \sum_{s=1}^{L} \omega_s^2 \approx \sum_{s=1}^{L} \lambda_s^2 = 2\sum_{s=1}^{L/2} \sigma_s^2 = 2\sum_{s=1}^{L/2} \cos^2(\theta_s),$$

i.e., the Frobenius norm of T_L can be used as the (good) approximation for the sum of $L/2$ largest cosines defining $L/2$ *smallest* principal angles between subspaces span(A_i) and span(A_j). In other words, we have found an easily computable weight w_{ij} for the maximum perfect matching in the one-sided block Jacobi method. We stress that we do *not* need to compute the Ritz values (i.e., the EVD of T_L)—the Frobenius norm squared is enough.

Moreover, note that in our application T_L is not needed in its explicit form. All that is needed is the square of its Frobenius norm. Since

$$w_{ij} = \|T_L\|_F^2 = \sum_{s=1}^{L} \alpha_s^2 + 2\sum_{s=2}^{L} \beta_s^2,$$

$\|T_L\|_F^2$ can be updated recursively immediately after computing α_s and β_{s+1} in the sth iteration step of the Lanczos algorithm.

Note that the weight w_{ij} takes into account the *actual* mutual position of two subspaces span(A_i) and span(A_j). Therefore, we can simply choose the 'worst' p pairs of column blocks for their parallel orthogonalization by choosing the pairs with highest values of w_{ij}. This is an analogy to the two-sided dynamic ordering where the actual Frobenius norm of the off-diagonal blocks was taken into account. Therefore, the above described ordering can be defined as the *one-sided dynamic ordering*. To choose the p 'worst' block columns for the parallel orthogonalization, the same maximum-weight perfect matching algorithm on the complete graph with ℓ vertices and weights w_{ij} can be used as in the two-sided case (see [4]).

We have just described how we can quite cheaply compute the weight w_{ij} that is the function of $L/2$ (estimated) largest cosines of principal angles between subspaces span(A_i) and span(A_j). The larger the weight, the lower the degree of mutual orthogonality between these two subspaces. However, at the beginning of each parallel iteration step we have to compute those weights for *all* pairs of block columns of matrix A. Next we describe how this computation can be done in parallel *without* sending/receiving whole block columns and *without* computing explicitly the matrix products $A_i^T A_j$.

In a parallel environment with p processors and the blocking factor $\ell = 2p$, these computations must be done for all $2(p-1)$ Lanczos processes for which each processor P_j is the master and this work is serialized inside processors. Each processor stores the information about two block columns that it currently overviews, and about all Lanczos processes for which it serves as the master. Therefore, each processor can read/write from/to the data structure the data/results of its own computations for all Lanczos processes for which it is the master (matrix-vector products, updates of Frobenius norms). To communicate data between all processors, the MPI collective communication `ALLTOALL` is used. Two such communications are needed per one parallel iteration step, i.e., together $2L$ collective communications are needed. These communications serve also like the global synchronization steps in the whole computation.

At the end of computation with Lanczos processes, all processors contain all weights w_{ij} for all block column pairs (excluding those residing in p processors), which are simply the squares of Frobenius norms of all matrices T_L produced in all Lanczos processes. Therefore, each processor can compute the maximum-weight perfect matching and the resulting parallel ordering; the algorithm is the same as for the parallel two-sided block Jacobi method (see [4]). For transferring the chosen pairs in processors, the optimal parallel scheduling is used (see [3]).

The global stopping criterion of the iteration process is based on the maximum value of currently computed weights w_{ij}. When using a computer with machine precision ε, the convergence is reached when

$$\max_{i,j} w_{ij} < n\, L\, \varepsilon, \tag{9.7}$$

where n is the matrix order and L is the number of steps in Lanczos processes. In other words, the computation is finished when the cosines of $L/2$ largest principal angles between all column blocks are 'sufficiently' small. The local stopping criterion is similar: A given pair (i, j) of block columns is *not* orthogonalized when

$$w_{ij} < n\, L\, \varepsilon. \tag{9.8}$$

In the following tables, we present first numerical results comparing the behavior of the parallel one-sided block Jacobi SVD algorithm with dynamic ordering with two different cyclic (static) orderings, `static1` (the odd-even ordering CO(0), see [5]) and `static2` (the robin-round ordering DO(0), see [5]). Computations were performed on the Woodcrest Cluster at Nuernberg-Erlangen University for random matrices with six various distributions of SVs defined by the variable `mode`. `mode` = 1 corresponds to a multiple minimal singular value, `mode` = 2 to a multiple maximal SV, `mode` = 3 describes a geometric sequence of SVs, `mode` = 4 defines an arithmetic sequence of SVs, `mode` = 5 defines the SVs as random numbers such that their logarithms are uniformly distributed, and, finally, `mode` = 6 sets the SVs to random numbers from the same distribution as the rest of a matrix (i.e., in our case they were normally distributed).

Table 9.1 contains the results for the SVD of well-conditioned matrices (with the condition number $\kappa = 10^1$) of order $n = 4000$ with a variable number of Lanczos steps L. For both static cyclic orderings, the number of sweeps is given by $n_{\text{it}}/15$ where n_{it} is the number of parallel iterations needed for the convergence of the whole algorithm. The total parallel execution time T_{p} is given in seconds. For `mode` = 1 and 2, our dynamic ordering needs about *five times less* parallel iterations than a static ordering. For harder cases, with `mode` ≥ 3, the ratio is about 2–3. But notice, that the decrease of T_{p} is much less. The dynamic ordering is about 2.5 times faster for `mode` = 1 and 2, but only about 1.5 faster for other modes. Also, T_{p} increases with L, the number of Lanczos steps, suggesting that the estimation of weights at the beginning of each parallel iteration step is quite time-demanding.

This conclusion is confirmed in Table 9.2 with results for ill-conditioned matrices (with $\kappa = 10^8$) where the last row depicts the average time T_{WC} of weight computations for a given number of Lanzcos steps for `mode` = 5. With respect to n_{it}, the situation is similar to well-conditioned matrices. However, it is clearly seen that our current implementation of the dynamic ordering is not very efficient. For example, in the case of $L = 6$ Lanczos steps the time spent in the computation of weights is 60 per cent of T_{p}. If this portion of algorithm were faster, one would substantially decrease T_{p} and be even more efficient as compared to the static ordering.

Table 9.1 Performance for $n = 4000$, $p = 8$, $\kappa = 10^1$

mode		$L=1$	$L=2$	$L=4$	$L=6$	static1	static2
1	n_{it}	4	4	4	4	30	30
	T_{p} [s]	5	6	9	11	13	13
2	n_{it}	4	4	4	4	30	30
	T_{p} [s]	5	6	9	11	13	12
3	n_{it}	108	99	98	99	240	270
	T_{p} [s]	225	234	292	354	354	390
4	n_{it}	103	97	98	97	225	240
	T_{p} [s]	212	228	289	345	327	354
5	n_{it}	109	103	99	100	255	285
	T_{p} [s]	230	242	293	360	367	409
6	n_{it}	108	107	106	103	270	285
	T_{p} [s]	226	252	315	371	395	420

Table 9.2 Performance for $n = 4000$, $p = 8$, $\kappa = 10^8$

mode		$L=1$	$L=2$	$L=4$	$L=6$	static1	static2
2	n_{it}	19	19	19	19	45	45
	T_{p} [s]	21	26	38	50	25	24
3	n_{it}	226	205	169	183	780	795
	T_{p} [s]	515	535	552	696	1233	1252
4	n_{it}	111	105	103	101	240	270
	T_{p} [s]	225	241	303	358	347	390
5	n_{it}	219	208	184	177	795	795
	T_{p} [s]	501	538	597	688	1243	1266
	T_{WC} [s]	159	250	345	413		
6	n_{it}	219	208	184	177	795	795
	T_{p} [s]	501	538	597	688	1243	1266

9.4 Conclusions

Recent progress in the parallel block Jacobi SVD algorithm has been achieved by applying two ideas: (i) the new parallel dynamic ordering of subproblems, and (ii) the matrix pre-processing by QR iterations. For the parallel two-sided block Jacobi method, these ideas were implemented and tested on various parallel platforms during last 10 years and results were published in papers [2–4, 15]. In the case of the one-sided variant, the results presented here using the new parallel dynamic ordering are the most recent ones. They are quite promising, but a more efficient implementation of the estimation of principal angles between any two block matrix columns is needed. In other words, one should spend much less portion of the parallel execution time in the computation and distribution of weights for the dynamic ordering.

It should be stressed, however, that the new dynamic ordering alone cannot make the parallel one-sided block Jacobi SVD algorithm competitive to the ScaLAPACK routine `PDGESVD`. Again, some sort of matrix pre-processing has to be included similarly as was the case in the two-sided variant [15]. A concentration of the Frobenius norm near the main matrix diagonal is not enough. It has to be coupled with a special ordering inside EVDs of 2×2 subproblems computed in each processor within a given parallel iteration step (see [11, 12]). We plan to investigate and implement these ideas in the near future.

Acknowledgements The authors were supported by the VEGA grant no. 2/0003/11 from the Scientific Grant Agency of the Ministry of Education and Slovak Academy of Sciences, Slovakia.

References

1. Barrett, R., Berry, M., Chan, T., Demmell, J., Donato, J., Dongarra, J., Eijkhout, V., Pozo, R., Romine, C., van der Vorst, H.: Templates for the Solution of Linear Systems: Building Blocks for Iterative Methods. SIAM, Philadelphia (1993)
2. Bečka, M., Okša, G., Vajteršic, M., Grigori, L.: On iterative QR pre-processing in the parallel block-Jacobi SVD algorithm. Parallel Comput. **36**, 297–307 (2010)
3. Bečka, M., Okša, G.: On variable blocking factor in a parallel dynamic block-Jacobi SVD algorithm. Parallel Comput. **28**, 1153–1174 (2003)
4. Bečka, M., Okša, G., Vajteršic, M.: Dynamic ordering for a parallel block-Jacobi SVD algorithm. Parallel Comput. **28**, 243–262 (2002)
5. Bečka, M., Vajteršic, M.: Block-Jacobi SVD algorithms for distributed memory systems: I. hypercubes and rings. Parallel Algorithms Appl. **13**, 265–287 (1999)
6. Bečka, M., Vajteršic, M.: Block-Jacobi SVD algorithms for distributed memory systems: II. meshes. Parallel Algorithms Appl. **14**, 37–56 (1999)
7. Demmel, J.: Applied Numerical Linear Algebra, 1st edn. Philadelphia, SIAM (1997)
8. Demmel, J., Kahan, W.: Accurate singular values of bidiagonal matrices. SIAM J. Sci. Stat. Comput. **11**, 873–912 (1990)
9. Demmel, J., Veselić, K.: Jacobi's method is more accurate than QR. SIAM J. Matrix Anal. Appl. **13**, 1204–1245 (1992)
10. Dhillon, I.: Current inverse iteration software can fail. BIT Numer. Math. **38**, 685–704 (1998)
11. Drmač, Z., Veselić, K.: New fast and accurate Jacobi SVD algorithm: I. Tech. rep. (2005)
12. Drmač, Z., Veselić, K.: New fast and accurate Jacobi SVD algorithm: II. Tech. rep. (2005)
13. Golub, G., Van Loan, C.: Matrix Computations, 3rd edn. Johns Hopkins University Press, Baltimore (1996)
14. Jacobi, C.: Über ein leichtes verfahren die in der theorie der säculärstörungen vorkom menden gleichungen numerisch aufzulösen. Crelle's J. für reine und angewandte Mathematik **30**, 51–94 (1846)
15. Okša, G., Vajteršic, M.: Efficient preprocessing in the parallel block-Jacobi SVD algorithm. Parallel Comput. **31**, 166–176 (2005)
16. Stewart, G.: Matrix Algorithms, vol. II: Eigensystems, 1st edn. SIAM, Philadelphia (2001)

Chapter 10
Robust and Efficient Multifrontal Solver for Large Discretized PDEs

Jianlin Xia

Abstract This paper presents a robust structured multifrontal factorization method for large symmetric positive definite sparse matrices arising from the discretization of partial differential equations (PDEs). For PDEs such as 2D and 3D elliptic equations, the method costs roughly $O(n)$ and $O(n^{4/3})$ flops, respectively. The algorithm takes advantage of a low-rank property in the direct factorization of some discretized matrices. We organize the factorization with a supernodal multifrontal method after the nested dissection ordering of the matrix. Dense intermediate matrices in the factorization are approximately factorized into hierarchically semiseparable (HSS) forms, so that a data-sparse Cholesky factor is computed and is guaranteed to exist, regardless of the accuracy of the approximation. We also use an idea of rank relaxation for HSS methods so as to achieve similar performance with flexible structures in broader types of PDE. Due to the structures and the rank relaxation, the performance of the method is relatively insensitive to parameters such as frequencies and sizes of discontinuities. Our method is also much simpler than similar structured multifrontal methods, and is more generally applicable (to PDEs on irregular meshes and to general sparse matrices as a black-box direct solver). The method also has the potential to work as a robust and effective preconditioner even if the low-rank property is insignificant. We demonstrate the efficiency and effectiveness of the method with several important PDEs. Various comparisons with other similar methods are given.

10.1 Introduction

Large sparse linear systems arise frequently from numerical and engineering problems, in particular, the discretization of partial differential equations (PDEs). Typically, there are two types of linear system solver, direct methods and iterative methods. Direct methods are reliable and are efficient for multiple right-hand sides, but are often expensive due to the generation of *fill-in* or loss of sparsity. Iterative methods take good advantage of sparsity and require less storage, but may diverge or

J. Xia (✉)
Department of Mathematics, Purdue University, West Lafayette, IN, USA
e-mail: xiaj@math.purdue.edu

M.W. Berry et al. (eds.), *High-Performance Scientific Computing*,
DOI 10.1007/978-1-4471-2437-5_10,

converge slowly if no effective preconditioners are available. Also, classical ILU preconditioners may suffer from breakdown.

Assume we have a system

$$Ax = b, \tag{10.1}$$

where A is an $n \times n$ symmetric positive definite (SPD) matrix. If A arises from the discretization of some PDEs. It may be associated with a mesh. In the direct solution of the system, A or the mesh points can be reordered so as to reduce fill-in. For example, the nested dissection ordering [12] and its generalizations can be used to get nearly optimal exact factorization complexity, which is generally $O(n^{3/2})$ in 2D or $O(n^2)$ in 3D [17]. In nested dissection, a mesh is recursively divided with separators (small sets of mesh points). However, notice that some iterative methods such as multigrid converge with $O(n)$ complexity for some PDEs.

In the recent years, nearly linear complexity structured approximate factorization methods have been developed based on a *low-rank property*. It has been noticed that, during the direct solution of some PDEs such as elliptic equations, certain off-diagonal blocks of the intermediate dense matrices or fill-in have small numerical ranks [1, 2, 4, 19, 33, etc.]. This property is closely related to the idea of the fast multipole method [14] and the property of certain Green's functions which are smooth away from the diagonal singularity under certain conditions. This property can be used to improve the computational efficiency, with dense intermediate matrices approximated by rank structured matrices such as quasiseparable, semiseparable, or hierarchical matrices [2, 10, 15, 16, 28, etc.]. This idea is widely used in the development of new fast algorithms. Related techniques have also been shown very useful in high performance scientific computing [23, 24, 29].

Rank structured methods can be fully integrated into sparse matrix techniques to provide new fast solvers. In [33] and [25, 26], structured sparse factorization algorithms are proposed based on the multifrontal method [9, 20] and hierarchically semiseparable (HSS) matrices [3, 5, 34] or hierarchical matrices. The algorithms have nearly linear complexity and linear storage requirement for some problems. The method in [33] involves complicated HSS operations, and are mainly applicable to regular meshes. Later, more general structured multifrontal methods have been discussed in [30, 31] and [25]. The method in [25] also requires the mesh to be nearly regular (or the location and layout of the separators in nested dissection follow the patterns of those in a regular mesh). Both methods in [25, 33] only work for 2D problems. The 3D method in [26] only works for regular meshes. All these methods may suffer from the problem of breakdown, especially when a low accuracy is used, say, in preconditioning. In addition, these methods generally require bounded off-diagonal ranks in the low-rank property.

In this paper, we propose a more robust and more general structured multifrontal algorithm, following the preliminary discussions in the report [30]. We use a flexible nested dissection algorithm that works for irregular meshes in both 2D and 3D. In the meantime, a robust HSS Cholesky factorization algorithm in [35] is generalized to the context of the multifrontal method, so that an approximate multifrontal factorization can always be computed without breakdown, in general. We also simplify the process by preserving certain dense operations, which keeps the performance to be similar to the fully structured version.

An optimization step is used in the multifrontal scheme (Theorem 10.1 below), so that the complexity can be lower than similar methods in [11, 25] by up to a factor of $O(\log n)$. Moreover, we relax the classical rank requirement in [25, 33], so that the structured sparse solution is fast even if the related numerical ranks are not bounded. Traditionally, HSS operations require the off-diagonal (numerical) ranks of a dense matrix to be bounded in order to achieve linear complexity. Here, the rank relaxation idea in [32] indicates that similar complexity can be achieved without this requirement. That is, the ranks are actually allowed to increase along the block sizes. This is then generalized to the rank relaxation in our robust sparse solution. It enhances the flexibility and applicability of structured multifrontal solvers, and is especially useful for difficult problems such as Helmholtz equations with high frequencies and 3D equations.

With the relaxed rank requirement, this new method has complexity similar to the one in [33], but applies to more general sparse matrices including 3D discretized ones. The factorization costs for some 2D and 3D discretized equations (elliptic, Helmholtz, etc.) are roughly $O(n)$ and $O(n^{4/3})$ flops, respectively (see Theorem 10.1 and Remark 10.1). In contrast, the exact factorization generally costs at least $O(n^{3/2})$ in 2D and $O(n^2)$ in 3D. We point out that, after the factorization, the solution cost and the storage requirement are both nearly $O(n)$, including for 3D. Furthermore, the rank structures and the rank relaxation idea indicate that the performance of the method is relatively insensitive to parameters such as frequencies in some problems.

Our method is especially useful for direct solutions of sparse linear systems with multiple right-hand sides, involving some parameters, and/or with only modest accuracy desired. It also has the potential to be used as a robust and effective preconditioner when the rank property is insignificant. The method uses two layers of tree structures, an outer one for the multifrontal method, and an inner one for each intermediate HSS matrix. It is thus suitable for parallel implementations. Several numerical examples are shown, including a Poisson equation, an interface problem, and a linear elasticity equation. The later two are ill conditioned, but our method (as a solver or a preconditioner) has similar performance for a large range of parameters. Both analytical and numerical comparisons with other similar methods are given.

The remaining sections are organized as follows. Section 10.2 reviews a dense HSS Cholesky factorization method. New structured multifrontal factorization and solution algorithms are developed in Sect. 10.4. Section 10.5 shows the algorithm and its complexity analysis. The numerical experiments are given in Sect. 10.6.

10.2 Review of HSS Cholesky Factorization of a Dense Matrix

10.2.1 Hierarchically Semiseparable Structures

HSS structures are very useful in handling dense matrices with the low-rank property. The definition of a postordering HSS form is as follows [34].

Definition 10.1 Assume F is an $N \times N$ (real) matrix, and $\mathcal{I} = \{1, 2, \ldots, N\}$. Let T be a binary tree with k nodes, and $t_i \subset \mathcal{I}$ be an index set associated with each node i of T. Let $F|_{t_i \times t_j}$ denote the submatrix of F with a row index set t_i and a column index set t_j in $\mathcal{I}$. We say F is in an *HSS form* with the corresponding *HSS tree* T if:

1. T is a postordered full binary tree: each node i is either a leaf or is a non-leaf node with two children c_1 and c_2 which are ordered as $c_1 < c_2 < i$.
2. For each non-leaf node i, $t_{c_1} \cup t_{c_2} = t_i$, $t_{c_1} \cap t_{c_2} = \phi$, and $t_{2k-1} = \mathcal{I}$.
3. There exists matrices $D_i, U_i, V_i, R_i, W_i, B_i$ (called *HSS generators*) associated with each node i satisfying

$$D_i = \begin{pmatrix} D_{c_1} & U_{c_1} B_{c_1} V_{c_2}^T \\ U_{c_2} B_{c_2} V_{c_1}^T & D_{c_2} \end{pmatrix}, \qquad U_i = \begin{pmatrix} U_{c_1} & R_{c_1} \\ U_{c_2} & R_{c_2} \end{pmatrix}, \tag{10.2}$$

$$V_i = \begin{pmatrix} V_{c_1} & W_{c_1} \\ V_{c_2} & W_{c_2} \end{pmatrix},$$

so that $D_i \equiv F|_{t_i \times t_i}$. Here, the generators associated with the root k are empty matrices except $D_k \equiv F$.

The HSS form of F is given by the generators. For a non-leaf node i, the generators D_i, U_i, V_i are recursively defined and are not explicitly stored. Clearly, U_i is a basis for the column space of $F_i^- = F|_{t_i \times (\mathcal{I} \backslash t_i)}$, and V_i^T is a basis for the row space of $F_i^| = F|_{(\mathcal{I} \backslash t_i) \times t_i}$. These off-diagonal blocks F_i^- and $F_i^|$ are called *HSS blocks*. The maximum (numerical) rank of all the HSS blocks is called the *HSS rank* of F. If F is symmetric, we can set [34]

$$D_i = D_i^T, \quad V_i = U_i, \quad B_j = B_i^T \quad (j\text{: sibling of } i).$$

10.2.2 Robust HSS Cholesky Factorization

Given a dense real SPD matrix F, we can use the method in [35] to compute an approximate Cholesky factorization $F \approx LL^T$, where L is a lower triangular HSS matrix. LL^T generally exists for any given approximation accuracy. The fundamental idea can be illustrated in terms of a block 2×2 SPD matrix. Factorize the $(1, 1)$

block of the following matrix:

$$F \equiv \begin{pmatrix} F_{1,1} & F_{2,1}^T \\ F_{2,1} & F_{2,2} \end{pmatrix} = \begin{pmatrix} D_1 & \\ F_{2,1}D_1^{-T} & I \end{pmatrix} \begin{pmatrix} D_1^T & D_1^{-1}F_{2,1}^T \\ & S \end{pmatrix},$$

where $F_{1,1} = D_1 D_1^T$ is the Cholesky factorization of $F_{1,1}$, and $S = F_{22} - (F_{2,1}D_1^{-T}) \cdot (D_1^{-1}F_{2,1}^T)^T$ is the Schur complement. Compute an SVD $F_{2,1}D_1^{-T} = U_2 B_1^T U_1^T + \hat{U}_2 \hat{B}_1^T \hat{U}_1^T$, where all the singular values greater than a tolerance τ are in B_1. (The number of singular values in B_1 is the off-diagonal *numerical rank* r.) Then

$$F \approx \begin{pmatrix} D_1 & \\ U_2 B_1^T U_1^T & I \end{pmatrix} \begin{pmatrix} D_1^T & U_1 B_1 U_2^T \\ & \tilde{S} \end{pmatrix},$$

where $\tilde{S}$ is an approximate Schur complement given by

$$\tilde{S} = F_{2,2} - U_2 B_1^2 U_2^T = S + O(\tau^2).$$

That is, a positive semidefinite term is implicitly added to the Schur complement. Then a Cholesky factorization $\tilde{S} = D_2 D_2^T$ yields

$$F \approx LL^T, \qquad L = \begin{pmatrix} D_1 & \\ U_2 B_1^T U_1^T & D_2 \end{pmatrix}.$$

Therefore, we obtain an approximate Cholesky factor L which is a block 2×2 HSS form. It is also shown in [35] that, with certain modifications, L can work as an effective preconditioner when the low-rank property is insignificant. That is, if the HSS rank of A for a small tolerance is large, L can be obtained by manually choosing a small rank r (and a large tolerance). The idea can be generalized to multiple blocks so that L is a general lower-triangular HSS matrix.

10.3 Nested Dissection for General Graphs

Before the numerical factorization of a sparse SPD matrix A, it is often reordered so as to reduce fill-in. Nested dissection generally leads to the optimal complexity for 2D and 3D discretized matrices [17].

In the following discussions, we focus on discretized matrices. For general sparse matrices, we can similarly consider the adjacency graph. Treat the mesh in the discretization as an undirected graph $(\mathcal{V}, \mathcal{E})$. Each mesh point $i \in \mathcal{V}$ corresponds to a row and a column of A, and each edge $(i, j) \in \mathcal{E}$ corresponds to the entries $A_{ij} = A_{ji} \neq 0$. A separator in $\mathcal{V}$ is found to divide the entire mesh into to two subregions, which are further divided recursively. Unlike the method in [33] which uses coordinates of mesh points, graph partition tools can be employed to handle more general meshes. Here, we use `METIS` [18], and follow the basic ideas in `Meshpart` [13]. See Figs. 10.1, 10.2 for some examples.

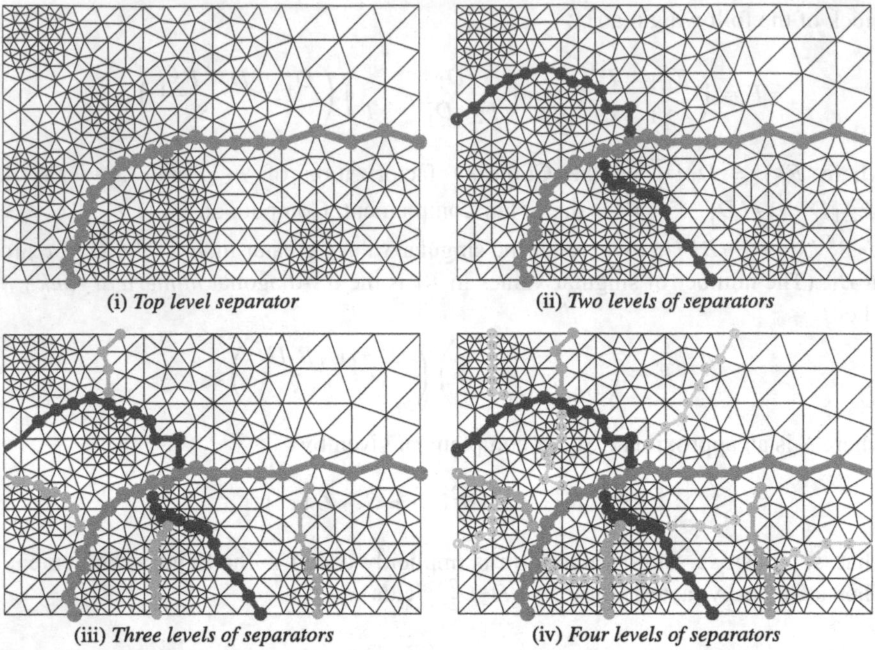

Fig. 10.1 Multiple levels of separators in nested dissection for an irregular mesh from `Meshpart` [13]

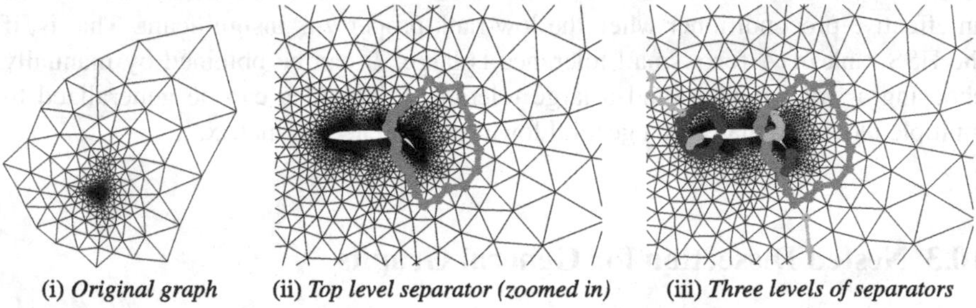

(i) Original graph
(ii) Top level separator (zoomed in)
(iii) Three levels of separators

Fig. 10.2 Multiple levels of separators in nested dissection for an irregular mesh with a missing piece, where the matrix is from the University of Florida sparse matrix collection [7]

Lower level separators are ordered before upper level ones. For example, Fig. 10.3 shows the nonzero pattern of a discretized matrix A after the reordering of a 3D mesh. During the factorization of A, the elimination of a mesh point mutually connects points which are previously connected to it [22, 27]. This creates fill-in.

As compared with the methods in [25, 33], our work has more flexibility:

1. The domains can be in any shape, such as with missing pieces (Fig. 10.2).
2. Both 2D and 3D domains can be handled (Fig. 10.3).

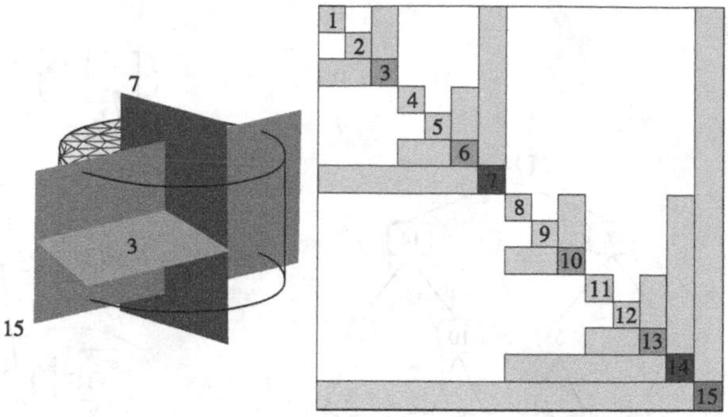

Fig. 10.3 Three levels of partition of a 3D mesh and the corresponding nonzero pattern of A after the nested dissection ordering

3. The mesh points and separators can be arbitrarily located, and the separators can be arbitrarily connected to each other.
4. Our method can also work as a black box for general sparse matrices.

10.4 Robust Structured Multifrontal Factorization

In this section, we consider the direct factorization of sparse SPD matrices with the multifrontal method [9, 20], which is one of the most important sparse factorization algorithms. During the factorization, if the dense intermediate matrices have the low-rank property, we use fast robust HSS methods to replace the dense operations.

10.4.1 Multifrontal Method

The multifrontal method [9, 20] reorganizes the sparse Cholesky factorization $A = \mathcal{L}\mathcal{L}^T$ into local factorizations of intermediate dense matrices, where $\mathcal{L}$ is lower triangular. The factorization is conducted following a tree called *elimination tree*, or more generally, an *assembly tree*. In general, an elimination tree $\mathcal{T}$ has n nodes and a node p is the parent of i if and only if

$$p = \min\{j > i | \mathcal{L}|_{j\times i} \neq 0\},$$

where $\mathcal{L}|_{j\times i}$ represents the (j,i) entry of $\mathcal{L}$. Use $\mathcal{T}[i]$ to denote the subtree of $\mathcal{T}$ with root i. Let $\mathcal{N}_i \equiv \{j_1, j_2, \ldots, j_d\}$ be the set of row indices of nonzeros in $\mathcal{L}_{:,i}$ (the ith column of $\mathcal{L}$) with i excluded. The ith *frontal matrix* is defined to be

$$\mathcal{F}_i = \begin{pmatrix} A|_{i\times i} & (A|_{\mathcal{N}_i\times i})^T \\ A|_{\mathcal{N}_i\times i} & 0 \end{pmatrix} - \sum_{j\in\mathcal{T}[i]\backslash i} \mathcal{L}|_{(i\cup\mathcal{N}_i)\times j} (\mathcal{L}|_{(i\cup\mathcal{N}_i)\times j})^T .$$

One step of elimination applied to $\mathcal{F}_i$ provides the column $\mathcal{L}_{(i\cup\mathcal{N}_i)\times i}$:

$$\mathcal{F}_i = \begin{pmatrix} \mathcal{L}|_{i\times i} & 0 \\ \mathcal{L}|_{\mathcal{N}_i\times i} & I \end{pmatrix} \begin{pmatrix} (\mathcal{L}|_{i\times i})^T & (\mathcal{L}|_{\mathcal{N}_i\times i})^T \\ 0 & \mathcal{U}_i \end{pmatrix},$$

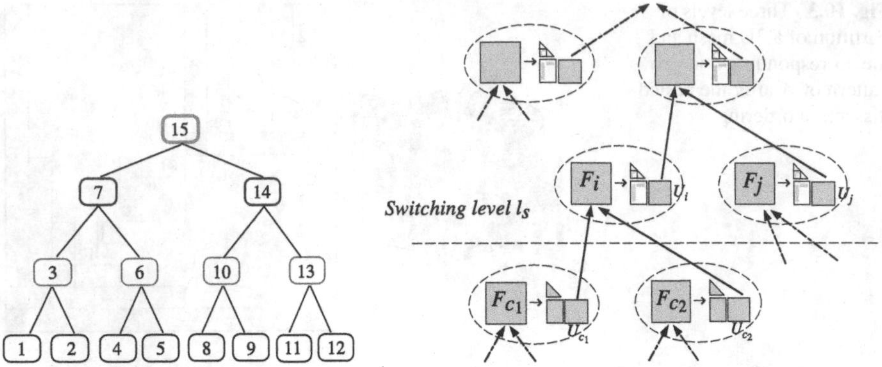

(i) *Supernodal elimination tree* $\mathcal{T}$ (ii) *Robust structured multifrontal factorization*

Fig. 10.4 Supernodal version elimination tree for the problem in Fig. 10.3, and also a general pattern of triangular structured multifrontal factorization, where a switching level $\mathbf{l}_s$ is marked

where $\mathcal{U}_i$ is the contribution from $\mathcal{T}[i]$ to p and is called the ith *update matrix*:

$$\mathcal{U}_i = -\sum_{j\in\mathcal{T}[i]} \mathcal{L}|_{\mathcal{N}_i\times j}(\mathcal{L}|_{\mathcal{N}_i\times j})^T.$$

Update matrices are used to form upper level frontal matrices. This process is called an *extend–add* operation, which matches indices and add entries, denoted

$$\mathcal{F}_i = \begin{pmatrix} A|_{j\times j} & (A|_{\mathcal{N}_i\times i})^T \\ A|_{\mathcal{N}_i\times i} & 0 \end{pmatrix} \Leftrightarrow \mathcal{U}_{c_1} \Leftrightarrow \mathcal{U}_{c_2} \Leftrightarrow \cdots \Leftrightarrow \mathcal{U}_{c_q},$$

where nodes $c_1, c_2, \ldots, c_q$ are the children of i in the elimination tree. The elimination process then repeats along the elimination tree.

10.4.2 Structured Supernodal Multifrontal Factorization

Here, we use nested dissection to reorder A and to produce a binary tree $\mathcal{T}$ as the assembly tree in a supernodal version of the multifrontal method, where each separator is treated as a node in the tree. Figure 10.4(i) shows the assembly tree for the mesh in Fig. 10.3.

Assume the root of $\mathcal{T}$ is at level 0, and the leaves are at the largest level. For a separator i, let $\mathcal{N}_i \equiv \{j_1, j_2, \ldots, j_d\}$ be the set of neighbor separators of i at the same or upper levels of i in $\mathcal{T}$. Also let t_j denote the index set of the neighbor j in A, and let $\hat{t}_j$ denote the subset of t_j that is connected to i due to lower level eliminations. The frontal matrix $\mathcal{F}_i$ is formed by the block form extend–add operation

$$\mathcal{F}_i = \mathcal{F}_i^0 \Leftrightarrow \mathcal{U}_{c_1} \Leftrightarrow \mathcal{U}_{c_2}, \quad \mathcal{F}_i^0 \equiv \begin{pmatrix} A|_{t_i\times t_i} & \left(A|_{(\cup_{j=1}^d \hat{t}_j)\times t_i}\right)^T \\ A|_{(\cup_{j=1}^d \hat{t}_j)\times t_i} & 0 \end{pmatrix}, \quad (10.3)$$

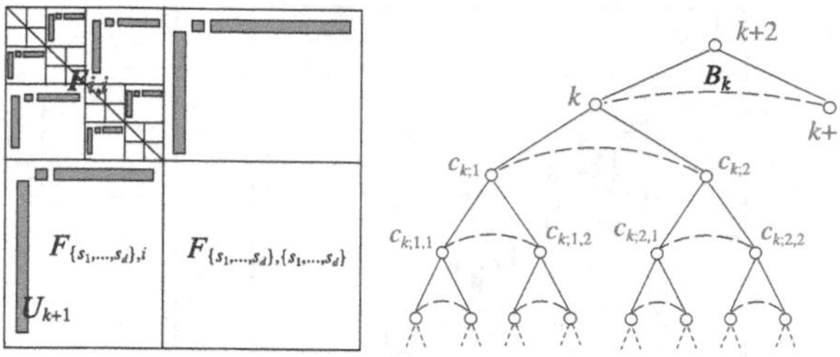

Fig. 10.5 Partial factorization of a frontal matrix $\mathcal{F}_i$ and the HSS tree used. $\mathcal{F}_i$ is shown with its leading block $F_{i,i}$ represented by structured factors. The subtree $T[k]$ is the HSS tree for $F_{i,i}$

where c_1 and c_2 are the children of i. For notational convenience, rewrite $\mathcal{F}_i$ as

$$\mathcal{F}_i \equiv \begin{pmatrix} F_{i,i} & F_{\mathcal{N}_i,i}^T \\ F_{\mathcal{N}_i,i} & F_{\mathcal{N}_i,\mathcal{N}_i} \end{pmatrix}. \tag{10.4}$$

In our structured multifrontal method, we set a *switching level* l_s so that if a separator i is at level l of $\mathcal{T}$ and $l > l_s$, we use exact Cholesky factorizations, and otherwise, we use HSS Cholesky factorizations. Similar to [33], we can show that this can help minimize the cost, which is smaller than using structured factorizations at all levels as in [25] by a factor up to $O(\log n)$. This is justified by Theorem 10.1 below. See Fig. 10.4(ii).

Here, we only need to describe the structured factorization part, which includes:

1. Factorizing $\mathcal{F}_i$ with the robust HSS method in Sect. 10.2.2, except that the last diagonal block $F_{\mathcal{N}_i,\mathcal{N}_i}$ is not factorized. Then $F_{i,i} \approx L_{i,i}L_{i,i}^T$.
2. In the meantime, $L_{i,i}^{-1}F_{\mathcal{N}_i,i}$ is compressed into a low-rank form.
3. Computing the update matrix or Schur complement $\mathcal{U}_i$ with a low-rank update.

Then $\mathcal{U}_i$ participates in the extend–add operation to form the parent frontal matrix. The details are elaborated as follows.

In order to perform the partial factorization of $\mathcal{F}_i$, we use a full HSS tree T with $k+2$ nodes, where the left and right children of the root are k and $k+1$ respectively. See Fig. 10.5. The subtree $T[k]$ is used as the HSS tree for completely factorizing $F_{i,i}$, and the single node $k+1$ is for the unfactorized part $F_{\mathcal{N}_i,\mathcal{N}_i}$. The algorithm in [35] is applied to $\mathcal{F}_i$ following the postordering traversal of the nodes of T. The factorization stops after the entire $T[k]$ is visited (or after the Schur complement $\mathcal{U}_i$ is computed). At the point, we have an approximate HSS Cholesky factorization

$$F_{i,i} \approx L_{i,i}L_{i,i}^T, \quad L_{i,i} = \left(\begin{array}{c|c} \left(\begin{array}{c|c} \ddots & 0 \\ \hline U_{c_{k,1,2}}B_{c_{k,1,2}}U_{c_{k,1,1}}^T & \ddots \end{array}\right) & 0 \\ \hline U_{c_{k,2}}B_{c_{k,2}}U_{c_{k,1}}^T & \left(\begin{array}{c|c} \ddots & 0 \\ \hline U_{c_{k,2,2}}B_{c_{k,2,2}}U_{c_{k,2,1}}^T & \ddots \end{array}\right) \end{array}\right), \tag{10.5}$$

where $L_{i,i}$ is a lower-triangular HSS matrix and $c_{k,1}, c_{k,2}, \ldots$ are the children of appropriate nodes as shown in Fig. 10.5.

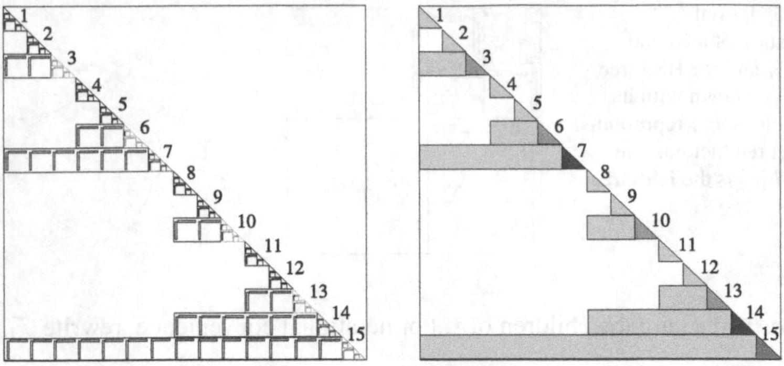

(i) *Factor from our structured factorization* (ii) *Factor from exact factorization*

Fig. 10.6 The nonzero pattern of a structured multifrontal factor of the matrix A in Fig. 10.3, as compared with the factor from the exact factorization

Then the frontal matrix $\mathcal{F}_j$ in Eq. (10.3) is (approximately) factorized as

$$\mathcal{F}_i \approx \begin{pmatrix} L_{i,i} & \\ L_{\mathcal{N}_i,i} & I \end{pmatrix} \begin{pmatrix} I & \\ & \mathcal{U}_i \end{pmatrix} \begin{pmatrix} L_{i,i}^T & L_{\mathcal{N}_i,i}^T \\ & I \end{pmatrix}, \tag{10.6}$$

where

$$L_{\mathcal{N}_i,i} = U_{k+1} B_k^T U_k^T, \tag{10.7}$$

Note that U_{k+1} and B_k are explicitly available, and U_k is implicitly represented by lower level U and R generators and has orthonormal columns (see Eq. (10.2)). Thus, $\mathcal{U}_i$ can be formed explicitly with a low-rank update

$$\begin{aligned} \mathcal{U}_i &= F_{\mathcal{N}_i,\mathcal{N}_i} - \left(U_{k+1} B_k^T U_k^T\right)\left(U_{k+1} B_k^T U_k^T\right)^T \\ &= F_{\mathcal{N}_i,\mathcal{N}_i} - \left(U_{k+1} B_k^T\right)\left(U_{k+1} B_k^T\right)^T. \end{aligned} \tag{10.8}$$

Here, for simplicity, we keep $\mathcal{U}_i$ as a dense matrix so that the extend–add operation is the same as in the standard supernodal multifrontal method (see Remark 10.2 below for more explanations). According to the idea of Schur compensation in [35], $\mathcal{U}_i$ is roughly equal to the exact Schur complement of $F_{i,i}$ plus a positive semidefinite term, and is always positive definite, in general. $\mathcal{U}_i$ then participates in the construction of the parent frontal matrix just like in the standard multifrontal method. Similarly, the parent frontal matrix is also guaranteed to be positive definite.

This process then proceeds along the assembly tree $\mathcal{T}$. After the elimination, we have an approximate factorization

$$A \approx \mathcal{L}\mathcal{L}^T,$$

where $\mathcal{L}$ always exists and is called a *triangular structured multifrontal factor*. See Fig. 10.6 for an example. $\mathcal{L}$ is associated with two layers of postordering trees, the outer layer assembly tree $\mathcal{T}$, and an inner layer HSS tree T for each node of $\mathcal{T}$. For a node i of $\mathcal{T}$, we store a triangular HSS form $L_{i,i}$ and also the U and B generators in Eq. (10.8). These are used in the structured multifrontal solution.

10.4.3 Structured Multifrontal Solution

Next, we consider the solution of Eq. (10.1) with the structured multifrontal factor. We solve two structured triangular systems

$$\mathcal{L}y = b, \tag{10.9}$$

$$\mathcal{L}^T x = y. \tag{10.10}$$

For convenience, assume b, x, y are partitioned conformably according to the sizes of the separators. For example, $b = (b_1^T, b_2^T, \ldots, b_K^T)^T$ with the length of b_i equal to the number of mesh points in separator i. Since the situation for a node i at a level l greater than the switching level l_s is trivial (with regular dense solutions), we only focus on structured solutions when describing the algorithm.

The solution of Eq. (10.9) with forward substitution involves forward (or post-ordering) traversal of the assembly tree $\mathcal{T}$. For a node i of $\mathcal{T}$, according to Eq. (10.6), we need to solve a system of the following form for y_i:

$$\begin{pmatrix} L_{i,i} & \\ L_{\mathcal{N}_i,i} & I \end{pmatrix} \begin{pmatrix} y_i \\ \tilde{b}_{\mathcal{N}_i} \end{pmatrix} = \begin{pmatrix} b_i \\ b_{\mathcal{N}_i} \end{pmatrix}, \tag{10.11}$$

where $b_{\mathcal{N}_i}$ is related to $b_{j_1}, b_{j_2}, \ldots, b_{j_d}$ (the separators $j_1, j_2, \ldots, j_d \in \mathcal{N}_i$ are connected to i and partially contribute to $b_{\mathcal{N}_i}$). Here, b_i is either from b (when i is a leaf), or is an updated vector due to the solution steps associated with lower level nodes. (We still use b_i for notational convenience. See Eq. (10.12).)

We first solve $L_{i,i} y_i = b_i$ with a lower triangular HSS solver in [21]. Then $L_{\mathcal{N}_i,i} y_i$ is the contribution of separator i to its neighbors. That is, we update $b_{\mathcal{N}_i}$ by

$$b_{\mathcal{N}_i} \leftarrow b_{\mathcal{N}_i} - L_{\mathcal{N}_i,i} y_i = b_{\mathcal{N}_i} - U_{k+1}\big(B_k^T (U_k^T y_i)\big), \tag{10.12}$$

where Eq. (10.7) is used. Again, U_{k+1} and B_k are explicitly available, and $U_k^T y_i$ can be quickly computed since it is partially formed in the HSS solution of $L_{i,i} y_i = b_i$ [21]. Thus, $b_{\mathcal{N}_i}$ can be convenient computed, and is then used to update $b_{j_1}, b_{j_2}, \ldots, b_{j_d}$.

In the backward substitution stage for solving Eq. (10.10), we traverse the elimination tree top-down. Similarly for each node i, according to Eqs. (10.6)–(10.7), we need to solve a system of the following form for x_i:

$$\begin{pmatrix} L_{i,i}^T & U_k B_k U_{k+1}^T \\ & I \end{pmatrix} \begin{pmatrix} x_i \\ x_{\mathcal{N}_i} \end{pmatrix} = \begin{pmatrix} y_i \\ x_{\mathcal{N}_i} \end{pmatrix},$$

Algorithm 10.1: Robust structured multifrontal factorization (RSMF)

```
for nodes (separators) i = 1, 2, ... of T do
    if i is a leaf then
        form F_i ≡ F_i^0, where F_i^0 is given in Eq. (10.3)
    if i is at level l > l_s then
        Compute traditional Cholesky factorization F_{i,i} = L̂_i L̂_i^T of F_{i,i} in Eq. (10.4);
        Compute the Schur complement U_i
    else
        Apply the robust HSS Cholesky factorization to F_i so that Eqs. (10.5) and (10.7) in Eq. (10.6) are computed;
        Compute U_i with a low-rank update as in Eq. (10.8)
    if i is a left node then
        push U_i onto the update matrix stack
    else
        Pop U_j from the update matrix stack;
        F_p = F_p^0 ⊕ U_i ⊕ U_j, where p is the parent of i
```

where $x_{\mathcal{N}_i}$ is already available from the solution steps associated with the upper level separators. This just needs the solution of an upper triangular HSS system $L_{i,i}^T x_i = y_i - U_k(B_k(U_{k+1}^T x_{\mathcal{N}_i}))$.

Note that the space of b can be used to store y and then x. After all the updates and solutions are performed, b is transformed into x.

10.5 Algorithm, Complexity, and Rank Relaxation

The structured multifrontal factorization algorithm is summarized as follows.

In a parallel implementation, we can traverse the assembly tree levelwise. The algorithm can be applied to general sparse SPD matrices. But we only consider its complexity in terms of sparse matrices arising from 2D and 3D discretized PDEs. The detailed flop count uses an idea of rank relaxation. The following lemma is a simple extension of the results in [32].

Lemma 10.1 (Dense rank relaxation) *Suppose an order N matrix F is hierarchically partitioned into $O(\log N)$ levels of HSS blocks following a perfect binary tree. Let $N_l = O(N/2^l)$ be the row dimension of the HSS block rows at level l, and r_l be their maximum numerical rank. Then for a given r_l, a triangular HSS factorization of A can be computed in ξ_{fact} flops, the HSS system can be solved in ξ_{sol} flops, and the HSS form needs memory size σ_{mem}, where the values are given in Table* 10.1.

Table 10.1 Costs and storage of dense-to-triangular-HSS factorization and solution with rank relaxation, where $p \in \mathbb{N}$, and $r = \max r_l$ is the HSS rank

r_l		$r = \max r_l$	ξ_{fact}	ξ_{sol}	σ_{mem}
$O(1)$		$O(1)$			
$O((\log_2 N_l)^p),\ p \ge 0$		$O((\log_2 N)^p)$	$O(N^2)$	$O(N)$	$O(N)$
$O(N_l^{1/p})$	$p > 3$	$O(N^{1/p})$			
	$p = 3$	$O(N^{1/3})$	$O(N^2)$	$O(N \log N)$	$O(N)$
	$p = 2$	$O(N^{1/2})$	$O(N^2 \log N)$	$O(N^{3/2})$	$O(N \log N)$

Table 10.2 Factorization cost ξ_{fact}, solution cost ξ_{sol}, and storage σ_{mem} of the structured multifrontal method applied to a discretized matrix A of order n on a 2D $n^{1/2} \times n^{1/2}$ mesh, where $p \in \mathbb{N}$

r_l		$r = \max_i \max r_l$	ξ_{fact}	ξ_{sol}	σ_{mem}
$O(1)$		$O(1)$	$O(n \log n)$		
$O((\log N_l)^p),\ p \ge 0$		$O((\log N)^p)$		$O(n \log \log n)$	$O(n \log \log n)$
$O(N_l^{1/p})$	$p \ge 3$	$O(N^{1/p})$			
	$p = 2$	$O(N^{1/2})$	$O(n \log^2 n)$		

Lemma 10.1 and an extension of the derivations in [31] yield the following results.

Theorem 10.1 (Sparse rank relaxation) *Suppose the robust structured multifrontal factorization method (Algorithm* 10.1*) and the solution method (Sect.* 10.4.3*) are applied to a discretized matrix A of order n on a regular mesh. Assume each frontal matrix $\mathcal{F}_i$ has order $O(N)$ and is treated as F in Lemma* 10.1 *so that the HSS blocks at level l of the HSS tree of $\mathcal{F}_i$ has row dimension N_l and rank r_l. Let the factorization cost, solution cost, and memory size of the structured multifrontal method be ξ_{fact}, ξ_{sol}, and σ_{mem}, respectively. Then if r_l satisfies the patterns as in Lemma* 10.1,

- *If A is obtained from a 2D $n^{1/2} \times n^{1/2}$ mesh, the results are given in Table* 10.2. *The switching level $l_s = O(\log n^{1/2})$ is chosen so that the* factorization costs *before and after the switching level are the same.*
- *If A is obtained from a 3D $n^{1/3} \times n^{1/3} \times n^{1/3}$ mesh, the results are given in Table* 10.3. *The switching level $l_s = O(\log n^{1/3})$ is chosen so that the* solution costs *before and after the switching level are the same.*

As an example, for problems such as 2D discrete Poisson's equations, it is shown that the maximum rank bound for all nodes i of $\mathcal{T}$ is $r = \max_i \max_l = O(1)$ [4]. Thus, Table 10.2 applies and our solver has nearly linear complexity and nearly linear storage. In contrast, the factorization method in [11] costs $O(n \log^2 n)$. Moreover, Theorem 10.1 indicates that we can relax the rank requirement to get similar complexity. For 3D discrete Poisson's equations, it is shown that $r = O(n^{1/3})$ [4].

Table 10.3 Factorization cost ξ_{fact}, solution cost ξ_{sol}, and storage σ_{mem} of the structured multifrontal method applied to a discretized matrix A of order n on a 3D $n^{1/3} \times n^{1/3} \times n^{1/3}$ mesh, where $p \in \mathbb{N}$

r_l		$r = \max_i \max r_l$	ξ_{fact}	ξ_{sol}	σ_{mem}
$O(1)$		$O(1)$			
$O((\log_2 N_l)^p)$, $p \geq 0$		$O((\log_2 N)^p)$	$O(n^{4/3})$	$O(n)$	$O(n)$
$O(N_l^{1/p})$, $p > 3$		$O(N^{1/p})$			
$O(N_l^{1/p})$	$p = 3$	$O(N^{1/3})$	$O(n^{4/3})$	$O(n \log^{1/2} n)$	$O(n \log^{1/2} n)$
	$p = 2$	$O(N^{1/2})$	$O(n^{4/3} \log n)$	$O(n \log n)$	$O(n \log n)$

In some numerical tests, the pattern of r_l is observed to follow the last row of Table 10.3.

Remark 10.1 For 2D Helmholtz equations, the rank bound is $r = O(\log n)$ with certain assumptions [11], which only depends on the logarithm of the frequency. Thus, not only our method has nearly $O(n)$ complexity, but also its performance is relatively insensitive to the frequency, because of the rank bound and the rank relaxation. Similar results are also observed for other problems with parameters such as sizes of discontinuities and Poisson's ratios. See Sect. 10.6 for some examples.

Remark 10.2 In our discussions, we keep $\mathcal{U}_i$ as a dense matrix. It turns out that this is at most $O(\log n)$ times slower than a fully structured version where an HSS form of $\mathcal{U}_i$ is used, but it significantly simplifies the descriptions and implementations. Moreover, our solution cost is very close to $O(n)$ (such as $O(n \log \log n) \sim O(n \log^{1/2} n)$ in Theorem 10.1). The storage for a stack needed for the update matrices is about the same as the factor size. A fully structured robust multifrontal solver with HSS form extend–add operations will appear in our future work.

Remark 10.3 Moreover, our method has various other advantages:

- Our method applies to PDEs on irregular grids and general sparse problems. The method in [26, 33] is mainly designed for regular grids, and the one in [25] requires that the mesh is nearly regular, or the separators in the partition roughly follow the layout in a regular grid.
- Our method applies to both 2D and 3D PDEs, while it is not clear how the ones in [11, 25, 33] perform in 3D.
- We use a switching level to optimize the cost, and the factorization cost in 2D is faster than the one in [11] by a factor of $O(\log n)$.
- We incorporate robustness enhancement so that, for an SPD matrix A, the structured factor $\mathcal{L}$ always exists and $\mathcal{L}\mathcal{L}^T$ is positive definite, in general. This does not hold for the methods in [11, 25, 26, 33].
- The algorithm is parallelizable, while the one in [11] is sequential.

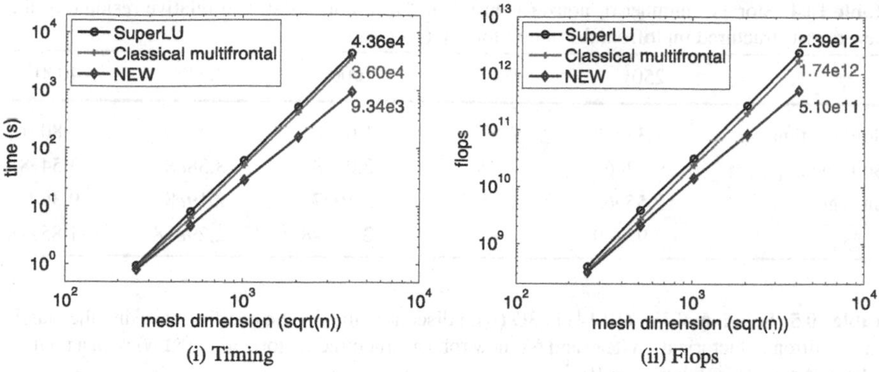

Fig. 10.7 Performance of the new robust structured multifrontal method (NEW) for Eq. (10.13), as compared with SuperLU [8] and the classical multifrontal method, where NEW uses a relative tolerance $\tau = 10^{-6}$, the total number of levels in $\mathcal{T}$ increases from 14 to 21 when n increases, and there are about nine levels below the switching level l_s

10.6 Numerical Experiments

The method is implemented in Fortran 90, and can work as a fast direct solver. If the low-rank property is insignificant, the method can serve as an efficient and effective preconditioner. We test it on various important discretized PDE examples.

Example 10.1 We first demonstrate the efficiency of the solver for the standard five-point discretized Laplacian from the 2D Poisson equation with a Dirichlet boundary condition:

$$-\Delta \mathbf{u} = \mathbf{f}, \quad \mathbf{u} \in \mathbb{R}^2. \tag{10.13}$$

Here, we let the matrix size n range from 255^2 to 4095^2. Every time n nearly quadruples. See Fig. 10.7 for the timing and flops of the factorizations. We see that the robust structured method is much faster than both SuperLU [8] and the exact multifrontal method when n is large.

The results of the structured solution are shown in Table 10.4. We observe that both the storage and the solution cost scale nearly linearly in terms of n. The accuracy is also well controlled. In addition, with few steps of iterative refinement, the full computer precision is reached.

Example 10.2 Next, we solve a 3D interface problem with jumps in the coefficient:

$$-\nabla \cdot (c(\delta)\nabla \mathbf{u}) = \mathbf{f}, \quad \mathbf{u} \in \mathbb{R}^3,$$
$$c(\delta) = 1 \text{ or } \delta. \tag{10.14}$$

We follow the choice of $c(\delta)$ and the boundary condition in *i* FEM [6]. The smaller δ is, the more ill conditioned the problem is. As compared with the exact

Table 10.4 Storage (number of nonzero entries in $\mathcal{L}$), solution cost, and relative residual of the new robust structured multifrontal solution for Eq. (10.13)

n	250^2	500^2	1000^3	2000^3	4000^2
Solution time (s)	7.37*e*–2	2.52*e*–1	1.06*e*0	4.38*e*0	1.80*e*1
Solution flops	9.63*e*6	4.62*e*7	2.00*e*8	8.38*e*8	3.54*e*8
Storage	2.55*e*6	1.25*e*7	5.30*e*7	2.19*e*8	9.36*e*8
$\frac{\Vert Ax-b\Vert_2}{\Vert b\Vert_2}$	7.95*e*–9	1.74*e*–8	2.31*e*–8	2.29*e*–8	1.85*e*–8

Table 10.5 Solution of Eq. (10.14) in 3D (with discontinuities in the coefficient) using the classical multifrontal factorization (MF) and our new robust structured factorization (NEW) with a relative tolerance $\tau = 10^{-3}$, where $\delta = 10^{-8}$

n	Flops				Storage (Number of nonzeros in L)			
	1.70*e*5	2.75*e*5	5.37*e*5	12.7*e*5	1.70*e*5	2.75*e*5	5.37*e*5	12.7*e*5
MF	0.91*e*11	4.21*e*11	12.5*e*11	41.0*e*11	1.11*e*8	2.01*e*8	4.20*e*8	9.21*e*8
NEW	0.80*e*11	2.44*e*11	6.44*e*11	17.2*e*11	1.05*e*8	1.39*e*8	2.34*e*8	4.60*e*8

Table 10.6 Solution of Eq. (10.14) in 3D (with discontinuities in the coefficient) using our new robust structured factorization with a relative tolerance $\tau = 10^{-3}$ for different δ, where the storage is measured by the number of nonzero entries in $\mathcal{L}$

δ	10^{-2}	10^{-4}	10^{-6}	10^{-8}
Flops	9.84*e*11	8.90*e*11	8.15*e*11	6.44*e*11
Storage	2.71*e*8	2.60*e*8	2.50*e*8	2.34*e*8

multifrontal method, our robust structured factorization attains satisfactory speedup with modest accuracy τ. See Table 10.5. We also test the structured method for different δ. Table 10.6 indicates that the performance is relatively insensitive to δ.

Example 10.3 Finally, we consider the preconditioning of a linear elasticity equation

$$-(\mu\Delta\mathbf{u} + (\lambda+\mu)\nabla\nabla\cdot\mathbf{u}) = \mathbf{f} \quad \text{in } \Omega = (0,1)\times(0,1). \qquad (10.15)$$

This equation is frequently solved in structural mechanics. Standard solvers including multigrid often suffer from the deterioration of the convergence rate for large Poisson's ratios λ/μ or near the incompressible limit. When λ/μ grows, the condition number of the discretized matrix A grows quickly. Here, we demonstrate the effectiveness of our structured solver as a preconditioner (although direct factorizations may cost less). For A with size $n \approx 1.28 \times 10^6$, we manually specify a numerical rank $r = 40$ in the structured multifrontal preconditioner. The convergence results for λ/μ varying from 1 to 10^6 is shown in Table 10.7. The convergence behavior is illustrated in Fig. 10.8. We observe that the preconditioned conjugate gra-

Table 10.7 Convergence of direct CG and preconditioned CG with our robust structured multifrontal solver (CG-RSMF) as a preconditioner for solving Eq. (10.15), where direct CG is set to stop when certain number of iterations is reached, and N_{iter} is the number of iterations.

	λ/μ	1	10^2	10^4	10^6
Direct CG	N_{iter}	3917	7997	14950	31004
	Flops	$1.28e16$	$2.61e16$	$4.87e16$	$1.01e17$
	$\frac{\|Ax-b\|_2}{\|b\|_2}$	$1.00e{-}12$	$1.33e{-}9$	$3.92e{-}9$	$4.89e{-}10$
CG-RSMF	N_{iter}	40	47	72	141
	Flops	$1.31e14$	$1.54e14$	$2.36e14$	$4.61e14$
	$\frac{\|Ax-b\|_2}{\|b\|_2}$	$8.17e{-}16$	$6.54e{-}16$	$7.73e{-}16$	$4.66e{-}16$

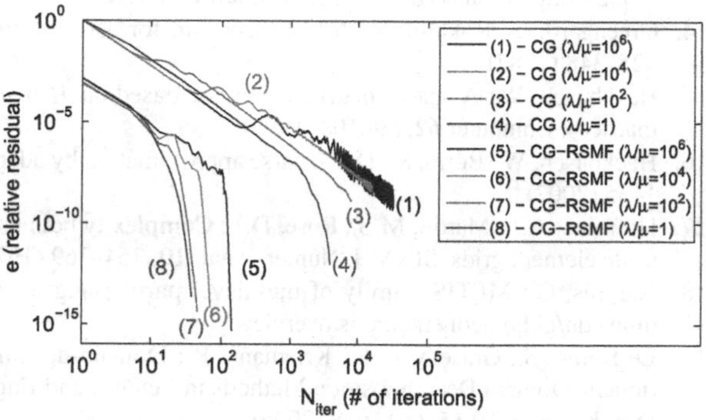

Fig. 10.8 Convergence of direct CG and preconditioned CG with our robust structured multifrontal solver (CG-RSMF) as a preconditioner for solving Eq. (10.15)

dient (CG) method converges quickly for all λ/μ. In comparison, direct CG costs more and has difficulty converging for large λ/μ.

Acknowledgements The author thanks Ming Gu, Xiaoye S. Li, and Jie Shen for some useful discussions, and thanks Zhiqiang Cai and Long Chen for providing two test examples. This research was supported in part by NSF grant CHE-0957024.

References

1. Bebendorf, M.: Efficient inversion of the Galerkin matrix of general second-order elliptic operators with nonsmooth coefficients. Math. Comput. **74**(251), 1179–1199 (2005)
2. Bebendorf, M., Hackbusch, W.: Existence of $\mathcal{H}$-matrix approximants to the inverse FE-matrix of elliptic operators with l^∞-coefficients. Numer. Math. **95**, 1–28 (2003)
3. Chandrasekaran, S., Dewilde, P., Gu, M., Lyons, W., Pals, T.: A fast solver for HSS representations via sparse matrices. SIAM J. Matrix Anal. Appl. **29**, 67–81 (2006)
4. Chandrasekaran, S., Dewilde, P., Gu, M., Somasunderam, N.: On the numerical rank of the off-diagonal blocks of Schur complements of discretized elliptic PDEs. SIAM J. Matrix Anal. Appl. **31**, 2261–2290 (2010)
5. Chandrasekaran, S., Gu, M., Pals, T.: A fast ULV decomposition solver for hierarchically semiseparable representations. SIAM J. Matrix Anal. Appl. **28**, 603–622 (2006)

6. Chen, L.: *i*FEM: An innovative finite element methods package in MATLAB. Technical Report (2008). http://math.uci.edu/~chenlong/Papers/iFEMpaper.pdf
7. Davis, T.A., Hu, Y.: The university of Florida sparse matrix collection. ACM Trans. Math. Softw.
8. Demmel, J.W., Gilbert, J.R., Li, X.S.: SuperLU Users' Guide (2003). http://crd.lbl.gov/~xiaoye/SuperLU/
9. Duff, I.S., Reid, J.K.: The multifrontal solution of indefinite sparse symmetric linear. ACM Trans. Math. Softw. **9**, 302–325 (1983)
10. Eidelman, Y., Gohberg, I.C.: On a new class of structured matrices. Integral Equ. Oper. Theory **34**, 293–324 (1999)
11. Engquist, B., Ying, L.: Sweeping preconditioner for the Helmholtz equation: Hierarchical matrix representation. Commun. Pure Appl. Math. **64**, 697–735 (2011)
12. George, A.: Nested dissection of a regular finite element mesh. SIAM J. Numer. Anal. **10**, 345–363 (1973)
13. Gilbert, J.R., Teng, S.H.: MESHPART, a Matlab mesh partitioning and graph separator toolbox (2002). http://aton.cerfacs.fr/algor/Softs/MESHPART/
14. Greengard, L., Rokhlin, V.: A fast algorithm for particle simulations. J. Comp. Physiol. **73**, 325–348 (1987)
15. Hackbusch, W.: A sparse matrix arithmetic based on $\mathcal{H}$-matrices. Part I: Introduction to $\mathcal{H}$-matrices. Computer **62**, 89–108 (1999)
16. Hackbusch, W., Börm, S.: Data-sparse approximation by adaptive $\mathcal{H}^2$-matrices. Computer **69**, 1–35 (2002)
17. Hoffman, A.J., Martin, M.S., Rose, D.J.: Complexity bounds for regular finite difference and finite element grids. SIAM J. Numer. Anal. **10**, 364–369 (1973)
18. Karypis, G.: METIS: family of multilevel partitioning algorithms (1998). http://glaros.dtc.umn.edu/gkhome/metis/metis/overview
19. Le Borne, S., Grasedyck, L., Kriemann, R.: Domain-decomposition based $\mathcal{H}$-LU preconditioners. Domain Decomposition Methods in Science and Engineering XVI. in O.B. Widlund, D.E. Keyes (Eds) **55**, 661–668 (2006)
20. Liu, J.W.H.: The multifrontal method for sparse matrix solution: Theory and practice. SIAM Rev. **34**, 82–109 (1992)
21. Lyons, W.: Fast algorithms with applications to PDEs. Ph.D. Thesis, UCSB (2005)
22. Parter, S.: The use of linear graphs in Gauss elimination. SIAM Rev. **3**, 119–130 (1961)
23. Polizzi, E., Sameh, A.H.: A parallel hybrid banded system solver: the SPIKE algorithm. Parallel Comput. **32**, 177–194 (2006)
24. Sambavaram, S.R., Sarin, V., Sameh, A.H., Grama, A.: Multipole-based preconditioners for large sparse linear systems. Parallel Comput. **29**, 1261–1273 (2003)
25. Schmitz, P.G., Ying, L.: A fast direct solver for elliptic problems on general meshes in 2D. J. Comput. Phys. (2011). doi:10.1016/j.jcp.2011.10.013
26. Schmitz, P.G., Ying, L.: A fast direct solver for elliptic problems on Cartesian meshes in 3D (2011, submitted). http://www.math.utexas.edu/users/lexing/publications/direct3d.pdf
27. Tewarson, R.P.: On the product form of inverses of sparse matrices. SIAM Rev. **8** (1966)
28. Vandebril, R., Barel, M.V., Golub, G., Mastronardi, N.: A bibliography on semiseparable matrices. Calcolo **42**, 249–270 (2005)
29. Wang, S., Li, X.S., Xia, J., Situ, Y., de Hoop, V.M.: Efficient scalable algorithms for hierarchically semiseparable matrices. Preprint (2011). http://www.math.purdue.edu/~xiaj/work/parhss.pdf
30. Xia, J.: Robust structured multifrontal factorization and preconditioning for discretized PDEs. Preprint (2008). http://www.math.purdue.edu/~xiaj/work/mfprec.pdf
31. Xia, J.: Efficient structured multifrontal factorization for large sparse matrices. Preprint (2010). http://www.math.purdue.edu/~xiaj/work/mfhss.pdf
32. Xia, J.: On the complexity of some hierarchical structured matrices. SIAM J. Matrix Anal. Appl. (2011, submitted). http://www.math.purdue.edu/~xiaj/work/hsscost.pdf

33. Xia, J., Chandrasekaran, S., Gu, M., Li, X.S.: Superfast multifrontal method for large structured linear systems of equations. SIAM J. Matrix Anal. Appl. **31**, 1382–1411 (2009)
34. Xia, J., Chandrasekaran, S., Gu, M., Li, X.S.: Fast algorithms for hierarchically semiseparable matrices. Numer. Linear Algebra Appl. **17**, 953–976 (2010)
35. Xia, J., Gu, M.: Robust approximate Cholesky factorization of rank-structured symmetric positive definite matrices. SIAM J. Matrix Anal. Appl. **31**, 2899–2920 (2010)

Chapter 11
A Preconditioned Scheme for Nonsymmetric Saddle-Point Problems

Abdelkader Baggag

Abstract In this paper, we present an effective preconditioning technique for solving nonsymmetric saddle-point problems. In particular, we consider those saddle-point problems that arise in the numerical simulation of particulate flows—flow of solid particles in incompressible fluids, using mixed finite element discretization of the Navier–Stokes equations.

These indefinite linear systems are solved using a preconditioned Krylov subspace method with an indefinite preconditioner. This creates an inner–outer iteration, in which the inner iteration is handled via a preconditioned Richardson scheme. We provide an analysis of our approach that relates the convergence properties of the inner to the outer iterations. Also "optimal" approaches are proposed for the implicit construction of the Richardson's iteration preconditioner. The analysis is validated by numerical experiments that demonstrate the robustness of our scheme, its lack of sensitivity to changes in the fluid–particle system, and its "scalability".

11.1 Introduction

Many scientific applications require the solution of saddle-point problems of the form

$$\begin{bmatrix} A & B \\ B^T & 0 \end{bmatrix} \begin{bmatrix} \mathbf{u} \\ \mathbf{p} \end{bmatrix} = \begin{bmatrix} \mathbf{a} \\ \mathbf{b} \end{bmatrix}, \tag{11.1}$$

where $A \in \mathbb{R}^{n \times n}$ and $B \in \mathbb{R}^{n \times m}$ with $m \leq n$, and where the $(n+m) \times (n+m)$ coefficient matrix

$$\mathscr{A} = \begin{bmatrix} A & B \\ B^T & 0 \end{bmatrix},$$

is assumed to be nonsingular. Such systems are typically obtained when "Lagrange multipliers" or mixed finite element discretization techniques are employed. Examples of these include, but are not limited to, the equality-constrained quadratic pro-

A. Baggag (✉)
College of Science and Engineering, Université Laval, Quebec City, Canada
e-mail: abdelkader.baggag@gci.ulaval.ca

M.W. Berry et al. (eds.), *High-Performance Scientific Computing*,
DOI 10.1007/978-1-4471-2437-5_11, © Springer-Verlag London Limited 2012

gramming problems, the discrete equations which result from the approximation of elasticity problems, Stokes equations, and the linearization of Navier–Stokes equations [2, 15, 16, 29, 39, 47]. When the matrix A is symmetric and positive definite, the problem (11.1) has n positive and m negative eigenvalues, with well defined bounds [57]. If the matrix A is symmetric indefinite or nonsymmetric, little can be said about the spectrum of the indefinite matrix $\mathscr{A}$.

Much attention has been paid to the case when A is symmetric positive definite, e.g. see [1, 6, 8, 9, 11–13, 18, 22, 25–28, 32, 40, 46, 54–56, 64, 71–75], and more recently to the case when A is nonsymmetric [3, 4, 10, 14, 17, 19–21, 23, 24, 31, 41, 44, 45, 61–63, 65]. In this paper, A is assumed to be nonsymmetric and B of full column rank. Here, we adopt one of the symmetric indefinite preconditioners studied, among others, by Golub and Wathen [31], for solving Eq. (11.1) via a preconditioned Krylov subspace method, such as GMRES, with the preconditioner given by

$$\mathscr{M} = \frac{1}{2}\left(\mathscr{A} + \mathscr{A}^T\right) = \begin{bmatrix} A_s & B \\ B^T & 0 \end{bmatrix}. \tag{11.2}$$

Here, A_s is the symmetric part of A, i.e., $A_s = (A + A^T)/2$. The motivating application in our paper produces a block diagonal matrix A_s in which each block has the following properties:

1. positive definite and irreducibly diagonally dominant, i.e., for each diagonal block $A_s^{(k)} = [a_{ij}^{(s)}]$ is irreducible, and $a_{ii}^{(s)} \geq \sum_{j \neq i} |a_{ij}^{(s)}|$ with strict inequality holding for at least one i, and
2. $\| A_s \|_F \geq \| A_{ss} \|_F$ where $\| \cdot \|_F$ denotes the Frobenius norm, and $A_{ss} = (A - A^T)/2$ is the skew symmetric part of A.

Thus, the preconditioner $\mathscr{M}$ is nonsingular, and the Schur complement, $-(B^T A_s^{-1} B)$, is symmetric negative definite.

The application of the preconditioner $\mathscr{M}$ in each Krylov iteration requires the solution of a linear system of the form

$$\begin{bmatrix} A_s & B \\ B^T & 0 \end{bmatrix} \begin{bmatrix} \mathbf{x} \\ \mathbf{y} \end{bmatrix} = \begin{bmatrix} \mathbf{f} \\ \mathbf{g} \end{bmatrix}. \tag{11.3}$$

The focus of our study is the development of a preconditioned Richardson iterative scheme for solving the above symmetric indefinite system (11.3) in a nested iterations setting that ensures the convergence of the inner iterations.

This system can be reformulated as

$$A_s \mathbf{x} = \mathbf{f} - B\mathbf{y}, \tag{11.4}$$

$$\left(B^T A_s^{-1} B\right)\mathbf{y} = B^T A_s^{-1}\mathbf{f} - \mathbf{g}. \tag{11.5}$$

Thus, one may first solve Eq. (11.5) to obtain $\mathbf{y}$, then solve Eq. (11.4) to get $\mathbf{x}$. Using a conjugate gradient algorithm for solving Eqs. (11.5), and (11.4), one creates an inner–outer iterative scheme [11]. This is the approach used in the classical

Uzawa scheme [1]. It turns out that in order to ensure convergence of the outer iteration, it is necessary to solve systems in the inner iteration with relatively high accuracy [13, 22]. For large-scale applications, such as the numerical simulation of particulate flows, solving linear systems involving A_s or $(B^T A_s^{-1} B)$ is not practical, as the action of A_s^{-1} must be computed on various vectors. Consequently, the approach we adopt here is to replace the cost of computing the action of A_s^{-1} by the cost of evaluating the action of some other "more economical" symmetric positive definite operator $\hat{A}^{-1}$ which approximates A_s^{-1} in some sense. Thus, the linear system (11.4) is solved via the iteration

$$\mathbf{x}_{k+1} = \left(I - \hat{A}^{-1} A_s\right) \mathbf{x}_k + \hat{A}^{-1} \underline{\mathbf{f}}, \tag{11.6}$$

where $\underline{\mathbf{f}} = \mathbf{f} - B\,\mathbf{y}$ and $\hat{A}$ is an appropriate symmetric positive definite splitting that assures convergence, i.e., $\alpha = \rho(I - \hat{A}^{-1} A_s) < 1$, where $\rho(\cdot)$ is the spectral radius.

Similarly, we replace A_s by $\hat{A}$ in (11.5) and solve the resulting "inexact" system,

$$\left(B^T \hat{A}^{-1} B\right) \mathbf{y} = B^T \hat{A}^{-1} \mathbf{f} - \mathbf{g}, \tag{11.7}$$

instead of the original system Eq. (11.5), via the iteration

$$\mathbf{y}_{k+1} = \left[I - \hat{G}^{-1}\left(B^T \hat{A}^{-1} B\right)\right] \mathbf{y}_k + \hat{G}^{-1} \hat{\mathbf{s}}, \tag{11.8}$$

where $\hat{\mathbf{s}} = B^T \hat{A}^{-1} \mathbf{f} - \mathbf{g}$, and $\hat{G}^{-1}$ is an inexpensive symmetric positive definite approximation of the inverse of the inexact Schur complement $(B^T \hat{A}^{-1} B)^{-1}$ that assures convergence of Eq. (11.8), i.e., $\beta = \rho(I - \hat{G}^{-1}(B^T \hat{A}^{-1} B)) < 1$. Moreover, $\hat{G}^{-1}$ is chosen such that $(I - \hat{G}^{-\frac{1}{2}}(B^T \hat{A}^{-1} B)\, \hat{G}^{-\frac{1}{2}})$ is positive definite.

Similarly, if we define the symmetric preconditioner $\hat{\mathscr{M}}$ to the system (11.3) as

$$\hat{\mathscr{M}} = \begin{bmatrix} \hat{A} & B \\ B^T & -\hat{G} + (B^T \hat{A}^{-1} B) \end{bmatrix}, \tag{11.9}$$

we obtain the following preconditioned Richardson iterative scheme for solving Eq. (11.3):

$$\begin{bmatrix} \mathbf{x}_{k+1} \\ \mathbf{y}_{k+1} \end{bmatrix} = \begin{bmatrix} \mathbf{x}_k \\ \mathbf{y}_k \end{bmatrix} + \begin{bmatrix} \hat{A} & B \\ B^T & -\hat{G} + B^T \hat{A}^{-1} B \end{bmatrix}^{-1} \left\{ \begin{bmatrix} \mathbf{f} \\ \mathbf{g} \end{bmatrix} - \begin{bmatrix} A_s & B \\ B^T & 0 \end{bmatrix} \begin{bmatrix} \mathbf{x}_k \\ \mathbf{y}_k \end{bmatrix} \right\}, \tag{11.10}$$

that is convergent if and only if $\rho(\mathscr{I} - \hat{\mathscr{M}}^{-1} \mathscr{M}) < 1$.

Thus, our proposed nested iterative scheme is shown in Fig. 11.1 in which the outermost iteration is that of a Krylov subspace method (we use *restarted* GMRES throughout this paper), and the preconditioning operation itself is doubly nested. Our focus here is the development of an algorithm for the most inner iteration, i.e. solving systems involving the symmetric indefinite preconditioner (11.2) using the preconditioned Richardson iteration (11.10).

Fig. 11.1 A nested iterative scheme

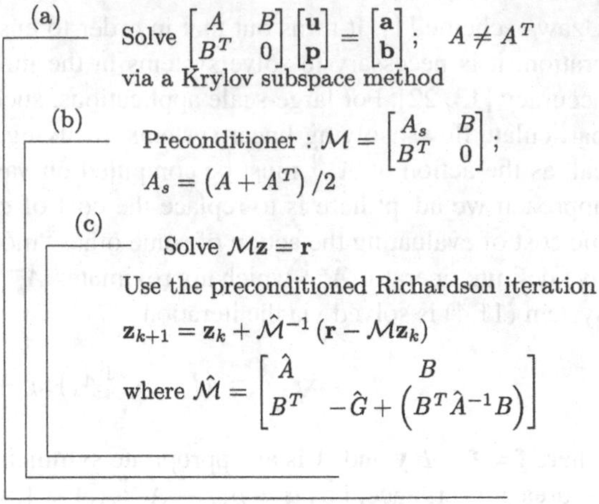

(a)
Solve $\begin{bmatrix} A & B \\ B^T & 0 \end{bmatrix} \begin{bmatrix} \mathbf{u} \\ \mathbf{p} \end{bmatrix} = \begin{bmatrix} \mathbf{a} \\ \mathbf{b} \end{bmatrix}$; $A \neq A^T$
via a Krylov subspace method
(b)
Preconditioner : $\mathcal{M} = \begin{bmatrix} A_s & B \\ B^T & 0 \end{bmatrix}$;
$A_s = (A + A^T)/2$
(c)
Solve $\mathcal{M}\mathbf{z} = \mathbf{r}$
Use the preconditioned Richardson iteration
$\mathbf{z}_{k+1} = \mathbf{z}_k + \hat{\mathcal{M}}^{-1}(\mathbf{r} - \mathcal{M}\mathbf{z}_k)$
where $\hat{\mathcal{M}} = \begin{bmatrix} \hat{A} & B \\ B^T & -\hat{G} + \left(B^T \hat{A}^{-1} B\right) \end{bmatrix}$

In the Golub–Wathen study [31], an iteration of the form

$$\begin{bmatrix} A_s & B \\ B^T & 0 \end{bmatrix} \begin{bmatrix} \mathbf{u}_{k+1} \\ \mathbf{p}_{k+1} \end{bmatrix} = \begin{bmatrix} (A_s - A) & 0 \\ 0 & 0 \end{bmatrix} \begin{bmatrix} \mathbf{u}_k \\ \mathbf{p}_k \end{bmatrix} + \begin{bmatrix} \mathbf{a} \\ \mathbf{b} \end{bmatrix},$$

is used, which does not always converge, and when used as an inner iteration within *full* GMRES, systems of the form $\mathcal{M}\mathbf{z} = \mathbf{r}$, in the inner-most loop $\{c\}$ of Fig. 11.1, are solved using a direct scheme. This could be as time-consuming as solving directly the nonsymmetric system (11.1), especially for very large systems.

In our study, the *monotone* convergence of our inner iteration (11.10) is guaranteed, and the performance of our nested scheme in Fig. 11.1 does not degrade as the mesh size decreases. Moreover, the construction of the preconditioner $\hat{\mathcal{M}}$ of the Richardson iteration is simple and economical.

In this paper, we analyze the iterative scheme (11.10) and show that a sufficient condition for *monotone* convergence is $\max\{\alpha, \beta\} < (\sqrt{5} - 1)/2$, and thus relating the rate of convergence of the inner iterations to the outer iteration, even though Eq. (11.8) is not the iteration that corresponds to the exact system (11.5) to be solved but to a modified one, (11.7), which, we will show, is not required to be solved accurately.

We use a simple explicit approximate inverse A_0^{-1} of A_s^{-1} for which $\alpha_0 = \rho(I - A_0^{-1} A_s) < 1$ and obtain an iteration for improving the convergence rate of Eq. (11.6). The matrix $\hat{G}^{-1}$ is not formed explicitly and the solution of systems involving $\hat{G}$ is achieved via the CG scheme, thus the only operations involved in the proposed nested iterative scheme (11.10) are matrix-vector multiplications and vector operations.

Our preconditioning strategy of the inner Richardson iteration is motivated by the study of Bank, Welfert and Yserentant [6] on a class of iterative methods for solving saddle-point problems. We extend it in this paper with some new results and a new analysis that relates the proposed iterative scheme to Uzawa's method. Further, we use our scheme for solving those *indefinite* linear systems that arise from the mixed finite element discretization of 2D particulate flow problems, using P2-P1 type elements.

In what follows, we introduce the motivating application, the proposed nested iterative scheme, and analyze its convergence properties. We propose "optimal" approaches for the construction of $\hat{A}^{-1}$ and $\hat{G}^{-1}$ approximating A_s^{-1} and $(B^T \hat{A}^{-1} B)^{-1}$, or their actions on vectors, so as to assure convergence of our scheme. We also demonstrate the robustness of our nested iterative scheme as a preconditioner, its lack of sensitivity to changes in the fluid–particle systems, and its "scalability".

11.2 Motivating Application

Direct numerical simulation of particulate flows is of great value in a wide range of industrial applications such as enhancing productivity of oil reservoirs and the manufacturing process of polymers. From the numerical point of view, there are three classes of algorithms to handle such direct simulations, namely the space-time technique [35–38, 69], the "fictitious domain" formulation [30], and the "Arbitrary Lagrangian Eulerian" formulation, e.g. see [34, 42, 43, 48–51, 70]. All use finite elements for spatial discretization, and are based on a combined *weak* formulation, in which fluid and particle equations of motion are combined into a single *weak* equation of motion from which the hydrodynamic forces and torques on the particles have been eliminated, e.g. see [3, 4, 42] for details.

The particulate flow system is represented via the use of projection matrices that describe the constraints imposed on the system by the boundary conditions on the particle surfaces, where the vector velocity is reordered as $[\mathbf{u}_I^T, \mathbf{u}_\Gamma^T]^T$, in which $\mathbf{u}_\Gamma$ contains the components of the velocity field associated with the vertices on the particle boundaries, with the projection matrix applied to the decoupled system leading to a *nonsymmetric* (indefinite) saddle-point matrix with "borders".

For the direct numerical simulation of particulate flows, one must simultaneously integrate the Navier–Stokes equations, which govern the motion of the fluid, and the equations of rigid-body motion. These equations are coupled through the no-slip condition on the particle boundaries, and through the hydrodynamic forces and torques which appear in the equations of the rigid-body motion, e.g. see [3, 4, 42].

To establish a structurally symmetric matrix formulation of the coupled fluid–particle system, e.g. see [51] or [42], the first step is to assemble the matrices corresponding to the decoupled problem, where the no-slip condition is not taken into consideration. The Jacobian $\widetilde{J}$ of the decoupled fluid–particle system has the following algebraic form:

$$\widetilde{J} = \left[\begin{array}{cc|c} A & B & \\ B^T & 0 & \\ \hline & & M_p \end{array}\right],$$

in which case, the variable unknowns are ordered as follows

$$\begin{bmatrix} \mathbf{u} \\ \mathbf{p} \\ \mathbf{U} \end{bmatrix} \quad \begin{array}{l} \mathbf{u}: \text{fluid velocity at each node} \\ \mathbf{p}: \text{fluid pressure} \\ \mathbf{U}: \text{particles velocity vector} \end{array}$$

and where M_p denotes the mass matrix of the n_p particles. M_p is block-diagonal and its size is $3\,n_p$ for $2D$ motion.

Since the approximate solution of the particulate flow problem is to be found in the subspace satisfying the no-slip condition, the constraints can be described in terms of a projection matrix. To clarify this further, the velocity unknowns may be divided into two categories, $\mathbf{u}_I$ for interior velocity unknowns and $\mathbf{u}_\Gamma$ for velocity unknowns on the surface of the particles. The Jacobian of the decoupled fluid–particle system is reordered accordingly, and hence the corresponding linear system is expressed in the following form:

$$\left[\begin{array}{ccc|c} A_{II} & A_{I\Gamma} & B_I & \\ A_{\Gamma I} & A_{\Gamma\Gamma} & B_\Gamma & \\ B_I^T & B_\Gamma^T & 0 & \\ \hline & & & M_p \end{array}\right] \begin{bmatrix} \mathbf{u}_I \\ \mathbf{u}_\Gamma \\ \mathbf{p} \\ \mathbf{U} \end{bmatrix} = \begin{bmatrix} \mathbf{f}_I \\ \mathbf{f}_\Gamma \\ \mathbf{g} \\ \mathbf{f}_p \end{bmatrix}.$$

The no-slip condition on the surface of the particles requires that $\mathbf{u}_\Gamma = Q\mathbf{U}$, where Q is the projection matrix from the space of the surface unknowns onto the particle unknowns. Hence,

$$\begin{bmatrix} \mathbf{u}_I \\ \mathbf{u}_\Gamma \\ \mathbf{p} \\ \mathbf{U} \end{bmatrix} = \left[\begin{array}{cc|c} I_{n\times n} & 0 & 0 \\ 0 & 0 & Q \\ 0 & I_{m\times m} & 0 \\ \hline 0 & 0 & I_{3n_p\times 3n_p} \end{array}\right] \begin{bmatrix} \mathbf{u}_I \\ \mathbf{p} \\ \mathbf{U} \end{bmatrix} = \widetilde{Q} \begin{bmatrix} \mathbf{u}_I \\ \mathbf{p} \\ \mathbf{U} \end{bmatrix}.$$

Finally, the Jacobian of the nonlinear coupled fluid–particle system can be written as $J = \widetilde{Q}^T\, \widetilde{J}\, \widetilde{Q}$, and we obtain the *nonsymmetric* bordered "saddle-point" problem,

$$\left[\begin{array}{cc|c} A_{II} & B_I & A_{I\Gamma}Q \\ B_I^T & 0 & B_\Gamma^T Q \\ \hline Q^T A_{\Gamma I} & Q^T B_\Gamma & Q^T A_{\Gamma\Gamma} Q + M_p \end{array}\right] \begin{bmatrix} \mathbf{u}_I \\ \mathbf{p} \\ \mathbf{U} \end{bmatrix} = \begin{bmatrix} \mathbf{f}_I \\ \mathbf{g} \\ \bar{\mathbf{f}}_p \end{bmatrix}, \tag{11.11}$$

where the last block-column has a size equal to $3\,n_p$ for $2D$ motion.

Writing the Jacobian as

$$J = \left[\begin{array}{c|c} \mathscr{A} & \mathscr{B} \\ \hline \mathscr{C}^T & \mathscr{D} \end{array}\right], \tag{11.12}$$

Fig. 11.2 Field ordering

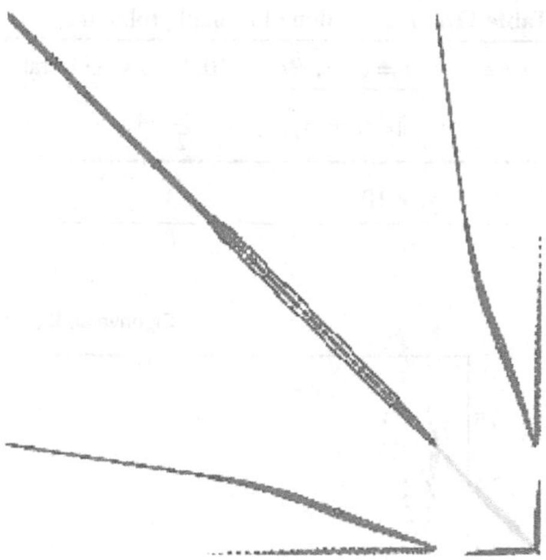

in which case, the variable unknowns are (always) ordered as follows, see Fig. 11.2

$$\begin{bmatrix} \mathbf{u}_I \\ \mathbf{p} \\ \mathbf{U} \end{bmatrix} \qquad \begin{array}{l} \mathbf{u}_I: \text{fluid velocity for the interior nodes} \\ \mathbf{p}: \text{ fluid pressure} \\ \mathbf{U}: \text{ particles velocity vector} \end{array}$$

and where the different block matrices are given by

$$\mathscr{A} = \begin{bmatrix} A_{II} & B_I \\ B_I^T & 0 \end{bmatrix} \in \mathbb{R}^{(n+m)\times(n+m)}, \qquad \mathscr{B} = \begin{bmatrix} A_{I\Gamma} \\ B_\Gamma^T \end{bmatrix} Q \in \mathbb{R}^{(n+m)\times 3n_p},$$

$$\mathscr{C}^T = Q^T \begin{bmatrix} A_{\Gamma I} & B_\Gamma \end{bmatrix} \in \mathbb{R}^{3n_p\times(n+m)}, \qquad \mathscr{D} = Q^T A_{\Gamma\Gamma} Q + M_p \in \mathbb{R}^{3n_p\times 3n_p}.$$

It can be shown, e.g. see [3, 4], that

1. $\mathscr{A}$ and $\mathscr{D}$ are nonsingular, with $(\mathscr{D} + \mathscr{D}^T)/2$ symmetric positive definite,
2. $\mathscr{B}$ and $\mathscr{C}$ are of full-column rank,
3. $A_s = (A_{II} + A_{II}^T)/2$ is symmetric positive definite, and that A_{II} and $A_{\Gamma\Gamma}$ are positive stable, and
4. for the application considered here, Reynolds number ≤ 100, our choice of an effective time step, $\Delta t = 0.01$, and the discretization scheme adopted, the block diagonal matrix A_s is assured of having irreducibly diagonally dominant blocks.

Example 11.1 To verify numerically the above observations, we have conducted the simulation of a sedimentation experiment with 20 circular particles of diameter 1.0 in a channel of width 12.8 and length 124.0. Some information about the associated linear systems are displayed in Table 11.1, and the eigenvalue distribution of A_{II} is shown in Fig. 11.3. We clearly see that all the eigenvalues of A_{II} are on the right half of the complex plane, i.e., they all have positive real parts.

Table 11.1 Description of a small problem

Time	$\Delta t = 0.01$, $Re = 100.0$, Newton Iteration 5, 20 Particles				
	$\frac{1}{2}\|A_{II} + A_{II}^T\|_F$	$\frac{1}{2}\|A_{II} - A_{II}^T\|_F$	size(A_{II})	size($\mathscr{A}$)	cond($\mathscr{A}$)
$5\Delta t$	4×10^3	13	3994	4733	10^8

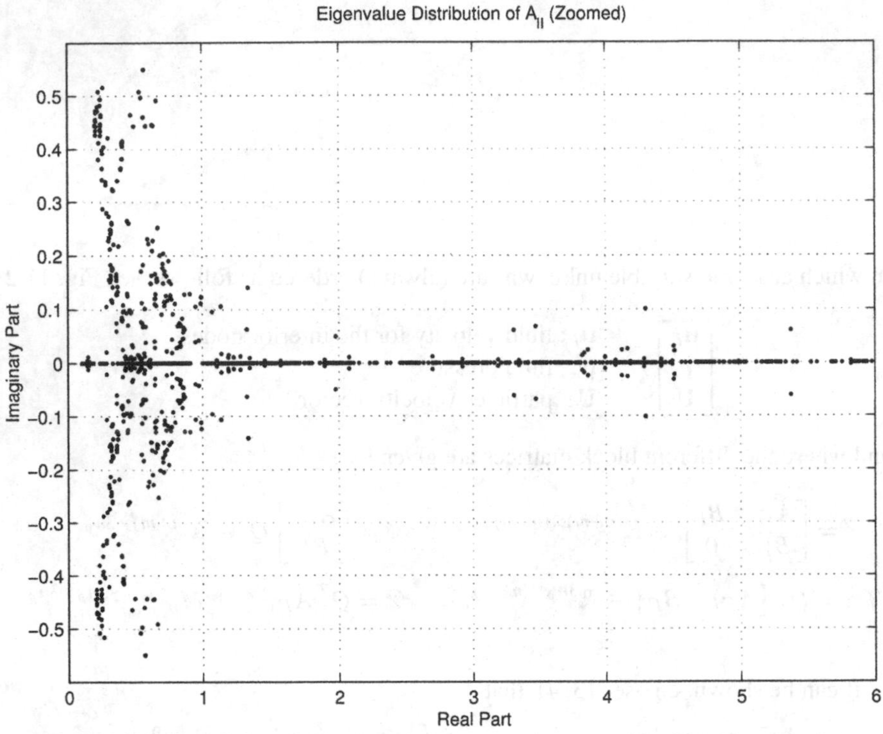

Fig. 11.3 Eigenvalue distribution of A_{II} at time step 5 (zoomed)

11.2.1 Properties of the Matrices

As the simulation time progresses, the structure of $\mathscr{A}$, its size and bandwidth vary, and its condition number increases. Generally, the flow simulation is characterized by three stages: the beginning, middle, and end of the simulation. Throughout the beginning and end stages, $\|A_s\|_F \gg \|A_{ss}\|_F$. In the middle stage, however, as the particulate flow becomes fully coupled, the Frobenius norm of the skew-symmetric part, $\|A_{ss}\|_F$, increases to approach $\|A_s\|_F$. Our experience indicates that Krylov subspace methods fail in solving Eq. (11.11) with classical ("black-box") preconditioners, even after only a few time steps, e.g., see [34, 42].

11.3 Solution Strategy

Since $\mathscr{D}$ is of much smaller dimension than $\mathscr{A}$ in Eq. (11.12), we solve Eq. (11.11) using the Schur complement approach by solving,

$$\left[\begin{array}{c|c} \mathscr{A} & \mathscr{B} \\ \hline & \mathscr{S}_1 \end{array}\right] \left[\begin{array}{c} \widetilde{\mathbf{x}} \\ \hline \widetilde{\mathbf{y}} \end{array}\right] = \left[\begin{array}{c} \mathbf{f} \\ \hline \hat{\mathbf{f}}_p \end{array}\right].$$

We first solve the Schur complement system $\mathscr{S}_1 \widetilde{\mathbf{y}} = \hat{\mathbf{f}}_p$ for $\widetilde{\mathbf{y}}$, where

$$\mathscr{S}_1 = \left[\mathscr{D} - \mathscr{C}^T \mathscr{A}^{-1} \mathscr{B}\right] \quad \text{and} \quad \hat{\mathbf{f}}_p = \left[\bar{\mathbf{f}}_p - \mathscr{C}^T \mathscr{A}^{-1} \mathbf{f}\right],$$

and once $\widetilde{\mathbf{y}}$ is obtained, $\widetilde{\mathbf{x}}$ is recovered by solving,

$$\mathscr{A}\widetilde{\mathbf{x}} = \mathbf{f} - \mathscr{B}\widetilde{\mathbf{y}},$$

via a preconditioned Krylov subspace method, such as GMRES [59]. In any case, the major task in solving Eq. (11.11) is the solution of a *nonsymmetric* saddle-point problem.

In the remainder of this paper, we concentrate on solving saddle-point systems of the form $\mathscr{A}\widetilde{\mathbf{x}} = \mathbf{b}$ using a Krylov subspace method such as GMRES, see Algorithm 11.1, with the indefinite preconditioner,

$$\mathscr{M} = \frac{1}{2}\left(\mathscr{A} + \mathscr{A}^T\right) = \begin{bmatrix} A_s & B_I \\ B_I^T & 0 \end{bmatrix}.$$

The inclusion of the exact representation of the (1,2) and (2,1) blocks of the preconditioner $\mathscr{M}$ leads one to hope for a more favorable distribution of the eigenvalues of the (left-)preconditioned linear system. The eigenvalues of the preconditioned coefficient matrix $\mathscr{M}^{-1}\mathscr{A}$ may be derived by considering the generalized eigenvalue problem

$$\begin{bmatrix} A_{II} & B_I \\ B_I^T & 0 \end{bmatrix} \begin{bmatrix} \mathbf{x} \\ \mathbf{y} \end{bmatrix} = \lambda \begin{bmatrix} A_s & B_I \\ B_I^T & 0 \end{bmatrix} \begin{bmatrix} \mathbf{x} \\ \mathbf{y} \end{bmatrix},$$

which has an eigenvalue at 1 with multiplicity $2m$, and $(n - m)$ eigenvalues which are defined by the generalized eigenvalue problem, e.g. see [39]

$$Q_2^T A_{II} Q_2 \mathbf{z} = \lambda\, Q_2^T A_s Q_2 \mathbf{z}, \tag{11.13}$$

where $B_I = [Q_1 Q_2][\,{}^R_0\,]$. Thus, since $A_{II} = A_s + A_{ss}$, Eq. (11.13) is equivalent to

$$(\lambda - 1) = \frac{\mathbf{z}^\star Q_2^T A_{ss} Q_2 \mathbf{z}}{\mathbf{z}^\star Q_2^T A_s Q_2 \mathbf{z}}, \quad \forall \mathbf{z} \neq \mathbf{0},$$

Algorithm 11.1: Generalized Minimum RESidual (GMRES) [59]

1: Compute $\mathbf{r}_0 = \mathscr{M}^{-1}(\mathbf{b} - \mathscr{A}\,\widetilde{\mathbf{x}}_0), \beta = \| \mathbf{r}_0 \|_2,\ \mathbf{v}_1 = \mathbf{r}_0/\beta$.
2: **for** $j = 1, \ldots, m$ **do**
3: Compute $\mathbf{w} = \mathscr{M}^{-1}(\mathscr{A}\,\mathbf{v}_j)$.
4: **for** $i = 1, \ldots, j$ **do**
5: $h_{i,j} = \langle \mathbf{w}, \mathbf{v}_i \rangle$.
6: $\mathbf{w} = \mathbf{w} - h_{i,j}\,\mathbf{v}_i$.
7: **end for**
8: Compute $h_{j+1,j} = \| \mathbf{w} \|_2$;
and$\mathbf{v}_{j+1} = \mathbf{w}/h_{j+1,j}$.
9: **end for**
10: Define $V_m = \left[\mathbf{v}_1 \cdots \mathbf{v}_m\right],\ \bar{H}_m = \{h_{i,j}\}_{1 \le i \le j+1; 1 \le j \le m}$
11: Compute $\bar{\mathbf{y}}_m = \text{argmin}_{\bar{\mathbf{y}}} \| \beta\,\mathbf{e}_1 - \bar{H}_m\,\bar{\mathbf{y}} \|_2$;
and ;
$\widetilde{\mathbf{x}}_m = \widetilde{\mathbf{x}}_0 + V_m\,\bar{\mathbf{y}}_m$.
12: If satisfied Stop, else set $\widetilde{\mathbf{x}}_0 = \widetilde{\mathbf{x}}_m$;
and GOTO 1.

which means that if A_s is dominant, the eigenvalues λ are clustered around $(1 \pm i\gamma)$. Hence, the solution procedure is as follows. *Solve linear systems involving the matrix $\mathscr{A}$, which is the (1,1) (indefinite) saddle-point block of the Jacobian given in Eq. (11.11), by a preconditioned Krylov subspace method. Choosing an indefinite preconditioner $\mathscr{M}$ of the form Eq. (11.2), Step 3 of Algorithm 11.1, i.e., operations of the form $\mathbf{w} = \mathscr{M}^{-1}(\mathscr{A}\mathbf{v})$ are handled via the proposed nested iterative scheme.* Thus the algorithms presented in this paper are for solving the systems in the innermost loop of Fig. 11.1.

In what follows, we drop the subscript "I".

11.4 Proposed Nested Iterative Scheme

The matrix $\mathscr{M}$ can be factored as

$$\mathscr{M} = \begin{bmatrix} A_s & 0 \\ B^T & I \end{bmatrix} \begin{bmatrix} A_s^{-1} & 0 \\ 0 & -G \end{bmatrix} \begin{bmatrix} A_s & B \\ 0 & I \end{bmatrix}, \tag{11.14}$$

where $G = (B^T\ A_s^{-1}\ B)$. For many practical problems, an important feature of the system (11.14) is that the action of the matrices A_s^{-1} and G^{-1} can be approximated by "simple" matrices $\hat{A}^{-1}$ and $\hat{G}^{-1}$, in the sense that even though the computational cost of solving linear systems with the coefficient matrices $\hat{A}$ and $\hat{G}$ is low, the overall behavior of the algorithm lends itself to fast convergence. Other methods, with optimal order of computational complexity, are available for solving linear systems involving A_s, such as multigrid methods, e.g. see [5, 18, 71, 72].

The factorization (11.14) suggests an approximation of $\mathscr{M}$ given by

$$\hat{\mathscr{M}} = \begin{bmatrix} \hat{A} & 0 \\ B^T & I \end{bmatrix} \begin{bmatrix} \hat{A}^{-1} & 0 \\ 0 & -\hat{G} \end{bmatrix} \begin{bmatrix} \hat{A} & B \\ 0 & I \end{bmatrix}, \tag{11.15}$$

where $\hat{A}^{-1}$ and $\hat{G}^{-1}$ are approximations of A_s^{-1} and $(B^T\,\hat{A}^{-1}\,B)^{-1}$, respectively, and are assumed to be symmetric and positive definite.

Observing that the symmetric indefinite preconditioner $\hat{\mathscr{M}}$ in Eq. (11.15) is nonsingular, with exactly n positive and m negative eigenvalues, the nested iterative scheme for solving the symmetric saddle-point problem (11.3) consists of the preconditioned Richardson iteration (11.10),

$$\begin{bmatrix} \mathbf{x}_{k+1} \\ \mathbf{y}_{k+1} \end{bmatrix} = \begin{bmatrix} \mathbf{x}_k \\ \mathbf{y}_k \end{bmatrix} + \begin{bmatrix} \hat{A} & B \\ B^T & -\hat{G} + B^T\hat{A}^{-1}B \end{bmatrix}^{-1} \left\{ \begin{bmatrix} \mathbf{f} \\ \mathbf{g} \end{bmatrix} - \begin{bmatrix} A_s & B \\ B^T & 0 \end{bmatrix} \begin{bmatrix} \mathbf{x}_k \\ \mathbf{y}_k \end{bmatrix} \right\},$$

where we consider the splitting $\mathscr{M} = \hat{\mathscr{M}} - \mathscr{N}$, with $\mathscr{N}$ being the defect matrix of the splitting.

This is equivalent to solving the following set of equations, which may be regarded as a version of a preconditioned inexact Uzawa algorithm with an additional correction step for $\mathbf{x}$, e.g. see [6, 74, 75]:

$$\begin{aligned} \hat{A}\left(\hat{\mathbf{x}}_{k+1} - \mathbf{x}_k\right) &= \mathbf{f} - [A_s\,\mathbf{x}_k + B\,\mathbf{y}_k], \\ \hat{G}\,(\mathbf{y}_{k+1} - \mathbf{y}_k) &= B^T\hat{\mathbf{x}}_{k+1} - \mathbf{g}, \\ \hat{A}\left(\mathbf{x}_{k+1} - \hat{\mathbf{x}}_{k+1}\right) &= -B\,(\mathbf{y}_{k+1} - \mathbf{y}_k). \end{aligned}$$

The corresponding algorithm is outlined by the following steps in Algorithm 11.2.

Algorithm 11.2: Nested iterative scheme

1: Initialize: $\mathbf{x} = \mathbf{x}_0$, $\mathbf{y} = \mathbf{y}_0$.
2: **for** $k = 0, 1, \ldots,$ until convergence **do**
3: Compute $\mathbf{r}_k = \mathbf{f} - \left[A_s\,\mathbf{x}_k + B\,\mathbf{y}_k\right]$.
4: Compute $\mathbf{s}_k = \mathbf{g} - B^T\mathbf{x}_k$.
5: Solve $\hat{A}\,\mathbf{c}_k = \mathbf{r}_k$.
6: Solve $\hat{G}\,\mathbf{d}_k = B^T\mathbf{c}_k - \mathbf{s}_k$.
7: Solve $\hat{A}\,\mathbf{c}_k = \mathbf{r}_k - B\,\mathbf{d}_k$.
8: Update $\begin{bmatrix} \mathbf{x}_{k+1} \\ \mathbf{y}_{k+1} \end{bmatrix} = \begin{bmatrix} \mathbf{x}_k \\ \mathbf{y}_k \end{bmatrix} + \begin{bmatrix} \mathbf{c}_k \\ \mathbf{d}_k \end{bmatrix}$.
9: **end for**

Remark 11.1 Step 7 in Algorithm 11.2 may be rearranged as

$$\mathbf{c}_k := \mathbf{c}_k - \hat{A}^{-1}(B\,\mathbf{d}_k),$$

where $\hat{A}^{-1}(B\,\mathbf{d}_k)$ is obtained as a byproduct of Step 6. This can save the application of $\hat{A}^{-1}$ at the end of every outer iteration, and thus improves the efficiency of the algorithm.

In each iteration step k of the nested iterative algorithm, five matrix-vector multiplications are required, namely $A_s\,\mathbf{x}_k$, $B\,\mathbf{y}_k$, $B^T\mathbf{x}_k$, $B^T\mathbf{c}_k$, and $B\,\mathbf{d}_k$; in addition to solving three systems, two of them involve $\hat{A}$ and the third involves $\hat{G}$.

11.5 Convergence Analysis of the Nested Iterative Scheme

The iteration matrix of the preconditioned Richardson iteration (11.10) is given by

$$\mathscr{K} = \left(\mathscr{I} - \hat{\mathscr{M}}^{-1}\,\mathscr{M}\right) = \hat{\mathscr{M}}^{-1}(\hat{\mathscr{M}} - \mathscr{M}),$$

where $\mathscr{M}$ and $\hat{\mathscr{M}}$ are given by Eqs. (11.2) and (11.9), respectively. Observing that

$$\hat{\mathscr{M}} - \mathscr{M} = \begin{bmatrix} \hat{A} - A_s & 0 \\ 0 & -\hat{G} + (B^T\hat{A}^{-1}B) \end{bmatrix},$$

and assuming that $\hat{A}$ and $\hat{G}$ are symmetric positive definite, then

$$\bar{\mathscr{K}} = \begin{bmatrix} \hat{A}^{\frac{1}{2}} & 0 \\ 0 & \hat{G}^{\frac{1}{2}} \end{bmatrix} \mathscr{K} \begin{bmatrix} \hat{A}^{-\frac{1}{2}} & 0 \\ 0 & \hat{G}^{-\frac{1}{2}} \end{bmatrix},$$

has the same eigenvalues as $\mathscr{K}$, and is given by

$$\bar{\mathscr{K}} = \begin{bmatrix} I - \bar{B}\bar{B}^T & \bar{B} \\ \bar{B}^T & -I \end{bmatrix} \begin{bmatrix} (I - \bar{A}) & 0 \\ 0 & -(I - \bar{G}) \end{bmatrix}, \tag{11.16}$$

in which

$$\bar{A} = \hat{A}^{-\frac{1}{2}} A_s\,\hat{A}^{-\frac{1}{2}} \qquad \in \mathbb{R}^{n\times n}, \tag{11.17}$$

$$\bar{B} = \hat{A}^{-\frac{1}{2}} B\,\hat{G}^{-\frac{1}{2}} \qquad \in \mathbb{R}^{n\times m}, \tag{11.18}$$

$$\bar{G} = \hat{G}^{-\frac{1}{2}}\left(B^T\hat{A}^{-1}B\right)\hat{G}^{-\frac{1}{2}} = \bar{B}^T\bar{B} \qquad \in \mathbb{R}^{m\times m}. \tag{11.19}$$

Hence, the eigenvalues of $\bar{\mathscr{K}}$ are close to zero when $\hat{A}^{-1}$ and $\hat{G}^{-1}$ are close to A_s^{-1} and $(B^T\hat{A}^{-1}B)^{-1}$, respectively.

Theorem 11.1 *Let α and β be the rates of convergence of the inner iterations* (11.6) *and* (11.8), *respectively, defined by*

$$\alpha = \rho\left(I - \hat{A}^{-1}A_s\right) = \left\| I - \bar{A} \right\|_2 < 1,$$

and

$$\beta = \rho\big(I - \hat{G}^{-1}(B^T \hat{A}^{-1} B)\big) = \| I - \bar{G} \|_2 < 1,$$

then, in general, the iterative scheme (11.10) *is monotonically convergent if*

$$\max\{\alpha, \beta\} < \frac{\sqrt{5}-1}{2} \approx 0.6180.$$

Moreover, if $\beta \equiv 0$, *then a sufficient condition for convergence is* $\alpha < 1$, *and conversely, if* $\alpha \equiv 0$, *then it suffices to have* $\beta < 1$ *to guarantee convergence of Eq.* (11.10).

Proof We divide the proof into three cases.

Case 1: $\alpha \equiv 0$.
This special case corresponds to $(I - \bar{A}) \equiv 0$, i.e.,

$$\bar{\mathscr{K}} = \begin{bmatrix} 0 & -\bar{B}(I - \bar{G}) \\ 0 & (I - \bar{G}) \end{bmatrix} \implies \rho(\bar{\mathscr{K}}) = \| I - \bar{G} \|_2 = \beta.$$

Case 2: $\beta \equiv 0$.
This special case corresponds to $(I - \bar{G}) \equiv 0$, i.e.,

$$\bar{\mathscr{K}} = \begin{bmatrix} (I - \bar{B}\bar{B}^T)(I - \bar{A}) & 0 \\ \bar{B}^T(I - \bar{A}) & 0 \end{bmatrix} \implies \rho(\bar{\mathscr{K}}) = \rho\big[(I - \bar{B}\bar{B}^T)(I - \bar{A})\big].$$

Therefore since $\alpha = \| I - \bar{A} \|_2$, $\rho(\bar{\mathscr{K}}) \leq \alpha \| I - \bar{B}\bar{B}^T \|_2$.
Observing that in this case,

$$I - \bar{B}\bar{B}^T = I - \hat{A}^{-\frac{1}{2}} B \left(B^T \hat{A}^{-1} B\right)^{-1} B^T \hat{A}^{-\frac{1}{2}},$$

is an orthogonal projector, we have $\| I - \bar{B}\bar{B}^T \|_2 = 1$, and $\rho(\bar{\mathscr{K}}) \leq \alpha$.

Case 3: This is the general case in which $\alpha, \beta < 1$. From Eq. (11.16), it is clear that

$$\rho(\bar{\mathscr{K}}) \leq \left\| \begin{bmatrix} (I - \bar{A}) & 0 \\ 0 & -(I - \bar{G}) \end{bmatrix} \right\|_2 \times \left\| \begin{bmatrix} I - \bar{B}\bar{B}^T & \bar{B} \\ \bar{B}^T & -I \end{bmatrix} \right\|_2,$$

$$\leq \max\{\alpha, \beta\} \left\| \begin{bmatrix} I - \bar{B}\bar{B}^T & \bar{B} \\ \bar{B}^T & -I \end{bmatrix} \right\|_2.$$

Let the singular value decomposition of $\bar{B} = \hat{A}^{-\frac{1}{2}} B \hat{G}^{-\frac{1}{2}}$ be given by

$$\bar{B} = W \begin{bmatrix} \Omega \\ 0 \end{bmatrix} Y^T, \tag{11.20}$$

then

$$I - \bar{B}\bar{B}^T = W \begin{bmatrix} I_m - \Omega^2 & 0 \\ 0 & I_{n-m} \end{bmatrix} W^T,$$

where $W \in \mathbb{R}^{n\times n}$ and $Y \in \mathbb{R}^{m\times m}$ are orthogonal matrices and $\Omega \in \mathbb{R}^{m\times m}$ is a diagonal matrix containing the singular values ω_i of $\hat{A}^{-\frac{1}{2}} B \hat{G}^{-\frac{1}{2}}$ such that $1 > \omega_1 \geq \omega_2 \geq \cdots \geq \omega_m > 0$. Therefore,

$$\begin{bmatrix} I - \bar{B}\bar{B}^T & \bar{B} \\ \bar{B}^T & -I \end{bmatrix} = \begin{bmatrix} W & \\ & Y \end{bmatrix} \left[\begin{array}{cc|c} I_m - \Omega^2 & 0 & \Omega \\ 0 & I_{n-m} & 0 \\ \hline \Omega & 0 & -I_m \end{array}\right] \begin{bmatrix} W^T & \\ & Y^T \end{bmatrix},$$

and

$$\left\| \begin{bmatrix} I - \bar{B}\bar{B}^T & \bar{B} \\ \bar{B}^T & -I \end{bmatrix} \right\|_2 = \left\| \left[\begin{array}{cc|c} I_m - \Omega^2 & \Omega & \\ \Omega & -I_m & \\ \hline & & I_{n-m} \end{array}\right] \right\|_2 = \max\{1, \| T \|_2\},$$

in which T is the symmetric matrix

$$T = \begin{bmatrix} I_m - \Omega^2 & \Omega \\ \Omega & -I_m \end{bmatrix},$$

and where the eigenvalues of T are given by

$$\psi_i = -\frac{1}{2}\left[\omega_i^2 \pm \sqrt{\omega_i^4 + 4}\right], \quad i = 1, 2, \ldots, m.$$

Since $\omega_i^2 < 1$ due to the fact that $(I - \bar{G})$ is positive definite, then

$$\| T \|_2 = \frac{1}{2}\left[\omega_1^2 + \sqrt{\omega_1^4 + 4}\right] < \frac{1+\sqrt{5}}{2},$$

and

$$\rho(\bar{\mathscr{K}}) < \frac{1}{2}(1+\sqrt{5}) \max\{\alpha, \beta\}.$$

Thus, to guarantee that $\rho(\bar{\mathscr{K}}) < 1$, it is sufficient to have

$$\frac{1}{2}(\sqrt{5}+1) \max\{\alpha, \beta\} < 1,$$

or

$$\max\{\alpha, \beta\} < \frac{1}{2}(\sqrt{5}-1) \approx 0.6180,$$

which completes the proof of Theorem 11.1. □

Remark 11.2 The previous results can be summarized as follows:

- If $\alpha = 0$ then $\rho(\mathscr{K}) = \beta$, and all eigenvalues $\lambda(\mathscr{K})$ are real.
- If $\beta = 0$ then $\rho(\mathscr{K}) \leq \alpha$, and all eigenvalues $\lambda(\mathscr{K})$ are real, as will be seen later in Lemma 11.2.

- Otherwise the eigenvalues $\lambda(\mathscr{K})$ are complex, and with the appropriate α-β relationship, $\rho(\mathscr{K}) < 1$.

Lemma 11.1 *Let* $\max\{\alpha, \beta\} < (\sqrt{5}-1)/2$, *and* $(I - \bar{A})$ *be positive definite. Then the eigenvalues of the iteration matrix* $\mathscr{K} = \mathscr{I} - \hat{\mathscr{M}}^{-1}\mathscr{M}$, *of the preconditioned Richardson iteration* (11.10), *lie to the right of the imaginary axis, i.e.*, $\Re(\lambda(\mathscr{K})) > 0$, *in addition to the fact that* $|\lambda(\mathscr{K})| < 1$.

Proof Consider the eigenvalue problem $\mathscr{K}\,\mathbf{v} = \lambda\,\mathbf{v}$ and let

$$\mathscr{D} = \begin{bmatrix} \hat{A}^{\frac{1}{2}} & 0 \\ 0 & \hat{G}^{\frac{1}{2}} \end{bmatrix},$$

then

$$(\mathscr{D}\mathscr{K}\mathscr{D}^{-1})(\mathscr{D}\mathbf{v}) = \lambda\,(\mathscr{D}\mathbf{v}),$$

or

$$\begin{bmatrix} I - \bar{A} & 0 \\ 0 & I - \bar{G} \end{bmatrix} \begin{bmatrix} \mathbf{w}_1 \\ \mathbf{w}_2 \end{bmatrix} = \lambda \begin{bmatrix} I & \bar{B} \\ -\bar{B}^T & I - \bar{B}^T\bar{B} \end{bmatrix} \begin{bmatrix} \mathbf{w}_1 \\ \mathbf{w}_2 \end{bmatrix},$$

where $\bar{A}$, $\bar{B}$, and $\bar{G}$ are as given in Eqs. (11.17)–(11.19). Using the singular value decomposition of $\bar{B}$ in Eq. (11.20), we get the eigenvalue problem

$$\left[\begin{array}{cc|c} I_m & 0 & \Omega \\ 0 & I_{n-m} & 0 \\ \hline -\Omega & 0 & I_m - \Omega^2 \end{array}\right] \mathbf{z} = \frac{1}{\lambda} \begin{bmatrix} \widetilde{A} & 0 \\ 0 & I - \Omega^2 \end{bmatrix} \mathbf{z},$$

where $\widetilde{A} = W^T(I - \hat{A}^{-\frac{1}{2}}A_s\hat{A}^{-\frac{1}{2}})W$ is symmetric positive definite. Since $\omega_i < 1$, we have

$$\tau\,\|\,\mathbf{z}\,\|_2^2 = \mathbf{z}^\star \begin{bmatrix} \widetilde{A} & 0 \\ 0 & I - \Omega^2 \end{bmatrix} \mathbf{z} > 0,$$

where $\mathbf{z}^\star = [\mathbf{z}_1^\star\ \mathbf{z}_2^\star\ \mathbf{z}_3^\star]$, and

$$\Re\left(\frac{1}{\lambda}\right) = \frac{\|\,\mathbf{z}\,\|_2^2 - \mathbf{z}_3^\star\Omega^2\mathbf{z}_3}{\tau\,\|\,\mathbf{z}\,\|_2^2},$$

$$\geq \frac{1 - \omega_1^2}{\tau} > 0.$$

Hence, $\Re(\lambda) > 0$, i.e., all the eigenvalues of $\mathscr{K}$ lie to the right of the imaginary axis, an ideal situation for acceleration via GMRES. □

Lemma 11.2 *For special case 2, when $\hat{G} = (B^T \hat{A}^{-1} B)$, i.e., $\beta = 0$, the Richardson iteration matrix is given by*

$$\mathscr{K} = \begin{bmatrix} \hat{A}^{-\frac{1}{2}} & 0 \\ 0 & \hat{G}^{-\frac{1}{2}} \end{bmatrix} \begin{bmatrix} M & 0 \\ N & 0 \end{bmatrix} \begin{bmatrix} \hat{A}^{\frac{1}{2}} & 0 \\ 0 & \hat{G}^{\frac{1}{2}} \end{bmatrix},$$

where

$$\begin{aligned} M &= (I - P)(I - \bar{A}) & \in \mathbb{R}^{n \times n}, \\ N &= \bar{B}^T (I - \bar{A}) & \in \mathbb{R}^{m \times n}, \end{aligned}$$

in which $\bar{A}$ and $\bar{B}$ are as given in Eqs. (11.17) *and* (11.18), *and P is the orthogonal projector,*

$$P = \bar{B}\bar{B}^T = \hat{A}^{-\frac{1}{2}} B \big(B^T \hat{A}^{-1} B\big)^{-1} B^T \hat{A}^{-\frac{1}{2}}.$$

Then $\mathscr{K}$ has $2m$ zero eigenvalues, with $\rho(\mathscr{K}) \le \rho(I - \bar{A}) = \alpha < 1$, and the submatrix of interest,

$$\mathscr{K}_{11} = \hat{A}^{-\frac{1}{2}} (I - P)\big(I - \bar{A}\big)\hat{A}^{\frac{1}{2}},$$

has a complete set of eigenvectors X with

$$\kappa_2(X) \le (1 + \hat{\mu}),$$

where $\kappa_2(\cdot)$ denotes the spectral condition number, and

$$\hat{\mu} < \left(\frac{1 + \sqrt{5}}{2}\right)\left(\frac{1}{\lambda^2_{\min}(C)}\right).$$

Here,

$$C = (I - P)(I - \bar{A})(I - P), \tag{11.21}$$

and $0 < |\lambda_{\min}(C)| = \min_i \{|\lambda_i(C)| \ne 0\}$.

Proof The fact that $\mathscr{K}$ has $2m$ zero eigenvalues is obvious from its structure and the fact that $(I - P)$ is an orthogonal projector of rank $(n - m)$. Also, from Theorem 11.1, we have $\rho(\mathscr{K}) \le \|I - \bar{A}\|_2 = \alpha < 1$. Next, we consider the eigenvalue problem $\mathscr{K}_{11}\mathbf{z} = \lambda \mathbf{z}$, or

$$(I - P)\big(I - \bar{A}\big)\mathbf{w} = \lambda \mathbf{w}, \tag{11.22}$$

where $\mathbf{w} = \hat{A}^{\frac{1}{2}}\mathbf{z}$. Observing that the symmetric matrix C given by Eq. (11.21) has the same eigenvalues as Eq. (11.22) with an orthogonal set of eigenvectors $V = [V_1, V_2]$ such that

$$(I - P)\big(I - \bar{A}\big)(I - P)V_1 = V_1 \Lambda,$$

and

$$(I-P)\big(I-\bar{A}\big)(I-P)V_2=0,$$

in which $\Lambda=\operatorname{diag}(\lambda_i)$, with $\lambda_i\neq 0$, $i=1,2,\dots,n-2m$. Hence

$$(I-P)V_1=V_1,$$

and

$$(I-P)V_2=0.$$

Since

$$V^T(I-P)(I-\bar{A})V=\begin{bmatrix} V_1^T(I-\bar{A})V_1 & V_1^T(I-\bar{A})V_2 \\ 0 & 0 \end{bmatrix},$$

$$=\begin{bmatrix} \Lambda & \hat{E} \\ 0 & 0 \end{bmatrix},$$

we can construct the nonsingular matrix

$$X=V\begin{bmatrix} I_{n-2m} & -\Lambda^{-1}\hat{E} \\ 0 & I_{2m} \end{bmatrix},$$

so that

$$X^{-1}(I-P)\big(I-\bar{A}\big)X=\begin{bmatrix} \Lambda & 0 \\ 0 & 0 \end{bmatrix}.$$

Consider the matrix

$$\hat{H}=\begin{bmatrix} I_{n-2m} & -H \\ 0 & I_{2m} \end{bmatrix},$$

where $H=\Lambda^{-1}\hat{E}$, then

$$\|\hat{H}\|_2^2=1+\hat{\mu},$$

in which

$$\hat{\mu}=\Big[\mu^2+\sqrt{\mu^4+4\mu^2}\Big]\Big/2,$$

with μ being the largest singular value of H. Consequently,

$$\|X\|_2^2\le(1+\hat{\mu}).$$

This upper bound can be simplified further by observing that

$$\|H\|_2^2\le\frac{\alpha^2}{\lambda_{\min}^2(C)}<\frac{1}{\lambda_{\min}^2(C)},$$

in which $\alpha = \|I - \bar{A}\|_2$ and $\lambda_{\min}(C)$ is the smallest nonzero eigenvalue of C. Hence

$$\hat{\mu} < \left(\frac{1+\sqrt{5}}{2}\right)\left(\frac{1}{\lambda_{\min}^2(C)}\right),$$

and

$$\kappa_2(X) < 1 + \frac{1.618}{\lambda_{\min}^2(C)},$$

which completes the proof. □

11.6 Construction of $\hat{A}^{-1}$ and $\hat{G}^{-1}$

$\hat{A}^{-1}$ and $\hat{G}^{-1}$ are approximations of A_s^{-1} and $(B^T\hat{A}^{-1}B)^{-1}$, respectively. They are assumed to be symmetric and positive definite and are chosen such that $\alpha < 1$ and $\beta < 1$.

There are many ways to construct $\hat{A}^{-1}$ and $\hat{G}^{-1}$. For example, $\hat{A}$ can be taken as the incomplete Cholesky decomposition of A_s or other preconditioners of A_s. In this study, we always consider $\hat{A}$ and $\hat{G}$ corresponding to several iteration steps of a given iterative scheme for solving systems in Steps 5–7 of Algorithm 11.2. For example, suppose A_0 is a "simple" preconditioner for A_s, such that $\alpha_0 = \rho(I - A_0^{-1}A_s) < 1$, in which A_0 is obtained via a sparse approximate inverse scheme (SPAI), e.g. see [7, 33]. If we use the following "convergent" scheme, see Eq. (11.6), we have

$$\mathbf{x}_{k+1} = (I - A_0^{-1}A_s)\mathbf{x}_k + A_0^{-1}\underline{\mathbf{f}}, \quad k = 0, 1, 2, \tag{11.23}$$

for solving $A_s\,\mathbf{x} = \underline{\mathbf{f}}$. Choosing the initial iterate $\mathbf{x}_0 = \mathbf{0}$, we obtain

$$\mathbf{x}_3 = \left[(I - A_0^{-1}A_s)^2 + (I - A_0^{-1}A_s) + I\right]A_0^{-1}\underline{\mathbf{f}}.$$

Thus, we have *implicitly* generated $\hat{A}^{-1}$ as

$$\hat{A}^{-1} = \left[(I - A_0^{-1}A_s)^2 + (I - A_0^{-1}A_s) + I\right]A_0^{-1},$$

in which case it is easy to verify that

$$\rho(I - \hat{A}^{-1}A_s) = \rho^3(I - A_0^{-1}A_s) \ll 1.$$

Remark 11.3 One implicit matrix-vector multiplication with $\hat{A}^{-1}$ consists of 3 matrix-vector multiplications with A_0^{-1} and 2 matrix-vector multiplications with A_s. So the implicit acceleration via Eq. (11.23) results in more matrix-vector multiplications, but if A_0^{-1} is a diagonal matrix, for example, the additional cost is minimal, and the overall approach may be more economical than choosing a more accurate approximation of A_s^{-1}.

Remark 11.4 In solving the linear system in Step 6 of Algorithm 11.2, we use the conjugate gradient (CG) algorithm with $\hat{G} = B^T \hat{A}^{-1} B$, i.e., $\beta = 0$. Note the $B^T \hat{A}^{-1} B$ is never formed explicitly, and the major operation in each CG iteration is multiplying $\hat{G}$ by a vector. In our implementation in this study, however, we only solve the system in Step 6 approximately using a relaxed stopping criterion.

11.6.1 Construction of $\hat{A}^{-1}$

Recalling that each diagonal block of A_s is symmetric positive definite, and irreducibly diagonally dominant, one can construct a diagonal matrix A_0^{-1}, with positive elements such that $\rho(I - A_0^{-\frac{1}{2}} A_s A_0^{-\frac{1}{2}}) < 1$. Moreover, it can be easily verified that given such A_0^{-1}, then the action of the matrix $\hat{A}^{-1}$ can be *implicitly* generated via Eq. (11.23), such that $\rho(I - \hat{A}^{-\frac{1}{2}} A_s \hat{A}^{-\frac{1}{2}}) = \rho^3(I - A_0^{-\frac{1}{2}} A_s A_0^{-\frac{1}{2}}) = \alpha_0^3 \ll 1$.

Theorem 11.2 *Let the block diagonal matrix $A_s = [a_{ij}^{(s)}]$ be symmetric positive definite with each of its blocks irreducibly diagonally dominant, and let $A_0^{-1} = \mathrm{diag}(\delta_i)$ be the diagonal matrix that minimizes $\| I_n - A_s A_0^{-1} \|_F^2$. Then*

$$\delta_i = \frac{a_{ii}^{(s)}}{\| \mathbf{a}_i^{(s)} \|_2^2},$$

where $\mathbf{a}_i^{(s)}$ is the ith column of A_s, and the spectral radius

$$\rho\big(I - A_s A_0^{-1}\big) < 1.$$

Proof Let

$$\varphi = \| I_n - A_s A_0^{-1} \|_F^2 = \sum_{j=1}^{n} \| \mathbf{e}_j - \delta_j \mathbf{a}_j^{(s)} \|_2^2;$$

then φ is minimized when δ_j is chosen such that it minimizes $\| \mathbf{e}_j - \delta_j \mathbf{a}_j^{(s)} \|_2^2$, or

$$\delta_j = \frac{a_{jj}^{(s)}}{\| \mathbf{a}_j^{(s)} \|_2^2}.$$

The eigenvalue problem,

$$(I_n - A_s A_0^{-1})\mathbf{u} = \lambda\, \mathbf{u},$$

yields

$$(1-\lambda) = \frac{\mathbf{v}^T A_s \mathbf{v}}{\mathbf{v}^T A_0 \mathbf{v}},$$

where $\mathbf{v} = A_0^{-1}\mathbf{u}$. Since both A_s and A_0 are symmetric positive definite, we have

$$(1-\lambda) > 0,$$

i.e., either λ is negative or has a positive value less than 1.

Next, consider the symmetric matrix $S = 2A_0 - A_s$, and let

$$\begin{aligned} D &= \operatorname{diag}\big(a_{11}^{(s)}, a_{22}^{(s)}, \ldots, a_{nn}^{(s)}\big), \\ A_s &= D - E, \end{aligned}$$

thus $D^{-\frac{1}{2}} A_s D^{-\frac{1}{2}} = I - D^{-\frac{1}{2}} E D^{-\frac{1}{2}}$, with $\rho(D^{-\frac{1}{2}} E D^{-\frac{1}{2}}) < 1$, e.g., see [53, Theorem 7.1.5, page 120].

Consequently,

$$D^{-\frac{1}{2}} S D^{-\frac{1}{2}} = \Theta + D^{-\frac{1}{2}} E D^{-\frac{1}{2}},$$

where

$$\Theta = \operatorname{diag}\left(2\frac{\|\mathbf{a}_j^{(s)}\|_2^2}{a_{jj}^{(s)2}} - 1\right) = \operatorname{diag}(1+\gamma_j) = I + \Gamma,$$

in which $\Gamma = \operatorname{diag}(\gamma_j)$, and $\gamma_j = \frac{1}{a_{jj}^{(s)2}} \sum_{j\neq i} a_{ij}^{(s)2} > 0$.

Thus,

$$D^{-\frac{1}{2}} S D^{-\frac{1}{2}} = \Gamma + \big(I + D^{-\frac{1}{2}} E D^{-\frac{1}{2}}\big),$$

and since $\rho(D^{-\frac{1}{2}} E D^{-\frac{1}{2}}) < 1$, we see that S is symmetric positive definite. Moreover, it is easy to verify that

$$\begin{aligned} \mathbf{v}^T S \mathbf{v} &= 2\mathbf{v}^T A_0 \mathbf{v} - \mathbf{v}^T A_s \mathbf{v}, \\ &= \frac{(1+\lambda)}{(1-\lambda)} \mathbf{v}^T A_s \mathbf{v}, \end{aligned}$$

and as a result we have $-1 < \lambda < 1$, or $\rho(I - A_s A_0^{-1}) < 1$. □

In our numerical experiments, we consider also the more expensive approximate Cholesky factorization for obtaining $\hat{A} = R^T R \approx A_s$. In this case, we obtain the approximate factorization using a numerical drop tolerance as well as a prescribed maximum fill-in per row.

Table 11.2 shows a few problem instances in the particulate flow simulation. Results in Table 11.3 show that the reduction in the number of inner iterations

Table 11.2 Description of the set of problems

Time	$\Delta t = 0.01$, $Re = 100.0$, Newton Iteration 5, 20 Particles				
	$\frac{1}{2}\|A_{II} + A_{II}^T\|_F$	$\frac{1}{2}\|A_{II} - A_{II}^T\|_F$	size(A_{II})	size($\mathscr{A}$)	cond($\mathscr{A}$)
$5\Delta t$	4×10^3	13	3994	4733	10^8
$10\Delta t$	4×10^3	8	4336	5213	10^8
$20\Delta t$	4×10^3	15	4179	4920	10^8

Table 11.3 Results with SPAI(A_s, diag) vs. IC(A_s, 15, 1.0e-3)

GMRES(20)							
Problem	SPAI(A_s,diag)				IC(A_s,15,1.0e-3)		
Instance	α_0	$\alpha = \alpha_0^3$	inner	outer	α	inner	outer
$5\,\Delta t$	0.8748	0.6694	12	1	0.4912	8	1
$10\,\Delta t$	0.8842	0.6913	11	1	0.5217	8	1
$20\,\Delta t$	0.8617	0.6398	12	1	0.5021	8	1

(Richardson iterations in Algorithm 11.2) realized by using the expensive explicit generation of $\hat{A}$ via the approximate Cholesky factorization, is not sufficient to justify its use. Note that the number of inner iterations listed in Table 11.3 represents the number of iterations needed for a single call of Algorithm 11.2. Thus, for example, for the problem arising at time $5\Delta t$, with 1 outer iteration of GMRES(20), the total number of inner iterations is 240, which is still much more economical than solving systems of the form Eq. (11.3) directly within GMRES; see also Sect. 11.7.1.

11.6.2 Implicit Generation of Variable $\hat{G}_k^{-1}$

In Step 6 of Algorithm 11.2, we need to solve linear systems of the form

$$\hat{G}\mathbf{d}_k = \mathbf{h}_k,$$

to determine an approximate solution $\hat{\mathbf{d}}$ via the conjugate gradient (CG) scheme where we replace $\hat{G}$ by $(B^T \hat{A}^{-1} B)$. Thus, the approximation of the action of $(B^T \hat{A}^{-1} B)^{-1}$ on $\mathbf{h}_k$ varies in each CG iteration.

The following theorem gives an explanation as to why there is no need to solve the inner system (Step 6) accurately, i.e., at each CG iteration j, there is $\hat{G}_j$ such that $\hat{G}_j\,\hat{\mathbf{d}} = (B^T \hat{A}^{-1} B)\,\mathbf{d}$, where $\hat{\mathbf{d}}$ is close to $\mathbf{d}$ in the $(B^T \hat{A}^{-1} B)$-norm defined by

$$\|\mathbf{y}\|^2_{(B^T \hat{A}^{-1} B)} = \langle \mathbf{y}, \left(B^T \hat{A}^{-1} B\right)\mathbf{y}\rangle, \quad \forall \mathbf{y} \in \mathbb{R}^m,$$

where $\langle \cdot, \cdot \rangle$ is the usual Euclidean inner-product.

Theorem 11.3 (Bank, Welfert and Yserentant [6]) *Let $(B^T \hat{A}^{-1} B)$ be a symmetric and positive definite $m \times m$ matrix, and let $\mathbf{d}, \hat{\mathbf{d}} \in \mathbb{R}^m$ satisfy*

$$\| \mathbf{d} - \hat{\mathbf{d}} \|_{(B^T \hat{A}^{-1} B)} \leq \beta \| \mathbf{d} \|_{(B^T \hat{A}^{-1} B)},$$

with $0 \leq \beta < 1$. Then there exists a symmetric positive definite matrix $\hat{G}$ with

$$\hat{G} \hat{\mathbf{d}} = \left(B^T \hat{A}^{-1} B\right) \mathbf{d},$$

and

$$\| I - \hat{G}^{-\frac{1}{2}} \left(B^T \hat{A}^{-1} B\right) \hat{G}^{-\frac{1}{2}} \|_2 \leq \beta.$$

Proof The proof is by construction and can be found in [6]. □

Let each CG iteration j yield an approximate solution $\mathbf{d}_{k,j}$ with residual $\mathbf{r}_{k,j}$ given by

$$\begin{aligned} \mathbf{r}_{k,j} &= \mathbf{h}_k - \left(B^T \hat{A}^{-1} B\right) \mathbf{d}_{k,j}, \\ &= \mathbf{h}_k - \left(B^T \hat{A}^{-1} B\right) \hat{G}_j^{-1} \mathbf{h}_k, \\ &= \left[I - \left(B^T \hat{A}^{-1} B\right) \hat{G}_j^{-1} \right] \mathbf{h}_k. \end{aligned}$$

Therefore,

$$\left\| I - \hat{G}_j^{-\frac{1}{2}} \left(B^T \hat{A}^{-1} B\right) \hat{G}_j^{-\frac{1}{2}} \right\|_2 \geq \frac{\| \mathbf{r}_{k,j} \|_2}{\| \mathbf{h}_k \|}.$$

In general, there exists a $\widehat{\gamma} \approx 1$, e.g. see [68, p. 194], such that

$$\left\| I - \hat{G}_j^{-\frac{1}{2}} \left(B^T \hat{A}^{-1} B\right) \hat{G}_j^{-\frac{1}{2}} \right\|_2 \approx \widehat{\gamma} \frac{\| \mathbf{r}_{k,j} \|_2}{\| \mathbf{h}_{k,j} \|_2}.$$

Consequently, choosing the stopping criterion $\| \mathbf{r}_{k,j} \|_2 / \| \mathbf{h}_k \|_2 \leq 10^{-2}$ will almost guarantee a value of $\beta = \mathcal{O}(10^{-2})$.

In order to verify this last observation, we conducted a set of numerical experiments in which we solve Eq. (11.3) using the preconditioned Richardson iteration (11.10) using Algorithm 11.2 with a relative residual stopping criterion of 10^{-6}. Here, the system in Step 6 is solved using the conjugate gradient algorithm (without preconditioning) with $\hat{G}$ replaced by $(B^T \hat{A}^{-1} B)$. Table 11.4 shows the results for varying levels of the CG relative residuals stopping criterion (`tol_CG`), from 10^{-6} to 10^{-1}, for a sample problem.

The vectors $\mathbf{b} = [\mathbf{f}^T \, \mathbf{g}^T]^T$, $\widetilde{\mathbf{r}}_k = [\mathbf{r}_k^T \, \mathbf{s}_k^T]^T$, and $\mathbf{w}_k = [\mathbf{x}_k^T \, \mathbf{y}_k^T]^T$ are as given in Algorithm 11.2, and $\delta \mathbf{w}_k = \mathbf{w}_\star - \mathbf{w}_k$, in which $\mathbf{w}_\star$ is the exact solution of (11.3). It is clear that using a `tol_CG` of 10^{-2} produces just as satisfactory a result had we used a `tol_CG` of 10^{-6}. This result confirms Theorem 11.3.

Table 11.4 Inner-inner iteration: CG method in Step 6 of Algorithm 11.2

Problem Instance: $t = 20\,\Delta t$			
tol_CG	Richardson iters	$\|\tilde{\mathbf{r}}_k\|_2/\|\tilde{\mathbf{r}}_0\|_2$	$\|\delta\mathbf{w}_k\|_2/\|\mathbf{b}\|_2$
1.0e-6	12	8.4×10^{-7}	10^{-4}
1.0e-5	12	8.4×10^{-7}	10^{-4}
1.0e-4	12	8.4×10^{-7}	10^{-4}
1.0e-3	12	8.4×10^{-7}	10^{-4}
1.0e-2	12	8.4×10^{-7}	10^{-4}
1.0e-1	8	4.7×10^{-7}	10^{-3}

Table 11.5 Values of α

Problem	SPAI-0		IC
Instance	α_0	$\alpha = \alpha_0^3$	α
$5\,\Delta t$	0.8748	0.6694	0.4912
$10\,\Delta t$	0.8842	0.6913	0.5217
$20\,\Delta t$	0.8617	0.6398	0.5021

Table 11.6 Inner-outer iterations

$\hat{G} = (B^T \hat{A}^{-1} B) \implies \beta \approx 0$				
Problem	GMRES(20)			
Instance	SPAI(A_s, diag)		IC(A_s, 15, 1.0e−4)	
	Richardson iters	outer	Richardson iters	outer
$5\Delta t$	12	1	8	1
$10\Delta t$	11	1	8	1
$20\Delta t$	12	1	8	1

In what follows, we generate A_0^{-1} via SPAI-0, and then obtain *implicitly* the action of $\hat{A}^{-1}$ on a vector via Eq. (11.23). For a sample problem at different time Steps, Table 11.5 shows the spectral radii $\alpha = \rho(I - \hat{A}^{-1} A_s)$ as well as those when $\hat{A} = R^T R$ in which R^T is the approximate Cholesky factor of A_s. Also, for solving systems in Step 6 of Algorithm 11.2, $\hat{G}\mathbf{d} = \mathbf{h}$, we use the conjugate gradient scheme with a relaxed stopping criterion, in which $\hat{G} = (B^T \hat{A}^{-1} B)$.

Now, using our complete nested iterative scheme, illustrated in Fig. 11.1, on the same set of sample problems, with GMRES(20), yields a solution satisfying the outer iteration stopping criterion of a relative residual less than or equal to 10^{-6}, only after one outer GMRES iteration, see Table 11.6.

Again, we show that using an approximate Cholesky factorization to generate $\hat{A}$ does not reduce the number of inner (Richardson) iterations sufficiently to justify the additional cost in each time step.

Table 11.7 Robustness of the nested iterative scheme

$\hat{G} = (B^T \hat{A}^{-1} B) \implies \beta \approx 0$: CG method with $\| \tilde{\mathbf{r}}_k \|_2 / \| \tilde{\mathbf{r}}_0 \|_2 \leq 10^{-3}$

t	n_p	$(n+m)$	$\hat{A}$	$\alpha = \alpha_0^3$	GMRES(k)			
					inner	outer	k	ARMS
$20\,\Delta t$	20	8777	SPAI(A_s, diag)	0.6913	13	1	20	10
$100\,\Delta t$	240	95749	IC(A_s, 15, 10^{-4})	0.6514	10	2	50	†
			SPAI(A_s, diag)	0.7216	14	3	50	†
$200\,\Delta t$	240	111326	IC(A_s, 15, 10^{-4})	0.6911	12	3	50	†
			SPAI(A_s, diag)	0.7502	15	4	50	†

Table 11.8 Parameters for ARMS

bsize	nlev	fill_I	fill_last	fill_ILUT	droptol_I	$\text{droptol}_\text{last}$
500–1000	2–5	60	50	50	0.0001	0.001

11.7 Numerical Experiments

We show the robustness of our nested iterative scheme illustrated in Fig. 11.1 by solving linear systems (11.1) for varying sizes, as the number of particles increases and the mesh size decreases, and at different time steps from $10\Delta t$ to $200\Delta t$, as the dominance of A_s (vs. A_{ss}) decreases.

Adopting a stopping criterion of a 10^{-6} relative residual for the outer GMRES iterations, our results are shown in Tables 11.7, 11.8 and 11.10. In Table 11.7 we give the number of inner (Richardson) and outer (GMRES) iterations. We also note that for $\hat{A}^{-1}$ generated via SPAI-0, with the implicit acceleration (11.23), and the systems $(B^T \hat{A}^{-1} B)\,\mathbf{d} = \mathbf{h}$ solved via the CG scheme with a relaxed stopping criterion, the scheme is remarkably robust succeeding in solving all the linear systems arising in the particulate flow simulations of Newtonian fluids.

In Table 11.7, we compare our nested iterative scheme, and GMRES with Saad's "black-box" preconditioner, "Algebraic Recursive Multilevel Solver", see [60], applied to Eq. (11.1). In using ARMS(*nlev*), we employ $nlev = 2$ levels and in case of failure we increase *nlev* to 5. Table 11.8 shows the parameters that need to be set up for ARMS. These parameters can be fine-tuned for a particular system to assure success. In Table 11.9, we compare our scheme with GMRES preconditioned via the ILUT factorization of $\mathscr{A}$. Note that both general purpose preconditioners, ILUT and ARMS, could fail for our saddle-point problems. Finally, in Table 11.10 we illustrate "scalability" of our nested iterative scheme in the sense that the number of inner iterations in any given pass of Algorithm 11.2 remains *almost* constant with only a modest increase in the number of outer (GMRES) iterations.

Table 11.9 Comparison with ILUT

Size($\mathscr{A}$)	GMRES(k)								
	nested scheme					ILUT(p,τ)			
	$\hat{A}$	$\alpha = \alpha_0^3$	k	inner	outer	p	τ	iters	k
29816	IC(A_s, 15, 10^{-3})	0.5198	20	8	2	15	0.0001	89	20
65471	IC(A_s, 15, 10^{-3})	10^{-4}	20	2	2	15	0.0001	42	20
80945	IC(A_s, 15, 10^{-3})	10^{-4}	50	2	2	15	0.0001	–	100
95749	SPAI(A_s, diag)	0.7216	50	14	3	15	0.0001	–	100
111326	SPAI(A_s, diag)	0.7502	50	15	4	15	0.0001	–	100

Table 11.10 "Scalability" of nested iterative scheme

$n_p = 20$, $t = 70\,\Delta t$, $\hat{A}^{-1} = \mathtt{SPAI}(A_s, \mathtt{diag})$				
$(n+m)$	$\alpha = \alpha_0^3$	GMRES(k)		
		inner	outer	k
3872	0.6782	12	1	20
6157	0.6973	12	1	20
10217	0.7012	13	2	20
31786	0.7314	14	2	20
56739	0.7196	14	3	20
81206	0.7512	15	3	40
105213	0.7419	15	3	40

11.7.1 Comparison with Other Preconditioners

We compare our scheme with algorithms of the form displayed in Fig. 11.4, i.e., GMRES inner–outer iterations. Here we consider two preconditioners

$$\mathscr{M}_1 = \begin{bmatrix} A & 0 \\ 0 & B^T A^{-1} B \end{bmatrix}, \quad \text{and} \quad \mathscr{M}_2 = \begin{bmatrix} A_s & B \\ B^T & 0 \end{bmatrix}.$$

In the block-diagonal preconditioner case, the preconditioned matrix $\mathscr{P}_1 = \mathscr{M}_1^{-1}\mathscr{A}$ has at most four distinct eigenvalues [52], namely 0, 1, and $(1 \pm \sqrt{5})/2$. Thus, it directly follows that for any vector, the associated Krylov subspace is of dimension at most three if $\mathscr{P}_1$ is nonsingular (or four if $\mathscr{P}_1$ is singular). Thus, any Krylov subspace iterative method with an optimality property, such as GMRES, will terminate in at most three iterations in *exact arithmetic*. As for the indefinite preconditioner $\mathscr{M}_2$, more favorable distribution of the eigenvalues of the (left-)preconditioned linear system $\mathscr{P}_2 = \mathscr{M}_2^{-1}\mathscr{A}$, is expected. Since solving with the Schur complement $(B^T A^{-1} B)$ is too expensive, the block-diagonal precondi-

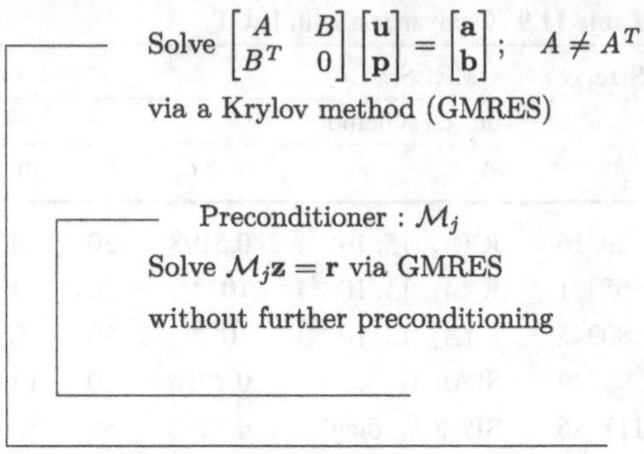

Solve $\begin{bmatrix} A & B \\ B^T & 0 \end{bmatrix} \begin{bmatrix} \mathbf{u} \\ \mathbf{p} \end{bmatrix} = \begin{bmatrix} \mathbf{a} \\ \mathbf{b} \end{bmatrix}$; $A \neq A^T$
via a Krylov method (GMRES)
Preconditioner : $\mathcal{M}_j$
Solve $\mathcal{M}_j \mathbf{z} = \mathbf{r}$ via GMRES
without further preconditioning

Fig. 11.4 A preconditioned Krylov method

tioner $\mathscr{M}_1$ is approximated by

$$\widetilde{\mathscr{M}}_1 = \begin{bmatrix} \text{diag}(A) & 0 \\ 0 & B^T(\text{diag}(A))^{-1}B \end{bmatrix},$$

and

$$\widetilde{\mathscr{M}}_2 = \begin{bmatrix} A & 0 \\ 0 & B^T(\text{diag}(A))^{-1}B \end{bmatrix}.$$

For the sake of completeness, we have added another popular symmetric indefinite preconditioner considered by Perugia and Simoncini in [55] for the solution of the stabilized symmetric saddle point problem that arises in mixed finite element approximations of magnetostatic problems,

$$\mathscr{M}_3 = \begin{bmatrix} I & B \\ B^T & 0 \end{bmatrix},$$

which is essentially a special case of Eq. (11.9) with $\hat{A} = I$ and $\hat{G} = B^T B$.

To evaluate the performance of the different preconditioners used in conjunction with inner–outer GMRES, e.g. see [58, 66, 67], numerical experiments have been performed on the set of problems displayed in Table 11.2. The results presented in Table 11.11 show that preconditioners $\widetilde{\mathscr{M}}_1$, $\widetilde{\mathscr{M}}_2$ and $\mathscr{M}_3$ require more than one outer iteration, whereas $\mathscr{M}_2$ seems to be the most effective competitor to our nested iterative scheme. Timing experiments on a uniprocessor, however, show that our nested scheme is at least 8 times faster than the other preconditioners shown in Table 11.11, used in the inner–outer GMRES setting of Fig. 11.4.

11.7.2 The Driven-Cavity Steady-State Case

Finally, we have used our nested scheme for obtaining the steady state solution of the Navier–Stokes equations modeling the incompressible fluid flow within a "leaky" two-dimensional lid-driven cavity problem in a square domain $-1 \leq x, y \leq$

Table 11.11 Performance of GMRES(20) with the different preconditioners

Time instance	$\Delta t = 0.01$, $Re = 100.0$, Newton Iteration 5				
	inner–outer iterations	Block-diagonal		Indefinite	
		$\widetilde{\mathscr{M}_1}$	$\widetilde{\mathscr{M}_2}$	$\mathscr{M}_2$	$\mathscr{M}_3$
$5\Delta t$	inner iters	10	6	7	14
$n_{\mathscr{A}} = 4733$	outer iters	8	2	1	11
$10\Delta t$	inner iters	14	8	15	20
$n_{\mathscr{A}} = 5213$	outer iters	8	2	1	12
$20\Delta t$	inner iters	12	7	10	16
$n_{\mathscr{A}} = 4920$	outer iters	8	2	1	11

Table 11.12 outer iteration: GMRES(k); a direct method for solving (11.3)

GMRES(10)			GMRES(20)		
mesh	outer	$\Vert$ residual $\Vert_2$	mesh	outer	$\Vert$ residual $\Vert_2$
8×8	3	1×10^{-6}	8×8	2	5×10^{-10}
16×16	4	2×10^{-7}	16×16	2	2×10^{-8}
32×32	4	1×10^{-7}	32×32	2	2×10^{-8}

1 with fluid viscosity of 0.01. The boundary condition for this model problem is $\mathbf{u}_x = \mathbf{u}_y = 0$ on the three walls ($x, y = -1$; $x = 1$), and $\mathbf{u}_x = 1, \mathbf{u}_y = 0$ on the moving wall ($y = 1$). Using Picard's iteration and mixed finite element ($Q2/Q1$) approximation of the resulting linearized equations (Oseen problems), we obtain systems of the form (11.1), derived from Picard's ninth iteration. Moreover, we see that even though the (1,1) block in Eq. (11.1) no longer has the advantage of the term $[(1/\Delta t) \times \text{mass matrix}]$, all the properties outlined above of its symmetric part still hold. Using a uniform mesh, the tables below show the effectiveness of our nested iterative scheme and its independence of the mesh size.

In Tables 11.12–11.13 we give the number of outer iterations of GMRES(10) and GMRES(20) needed to reach a residual of 2-norm less than or equal to 10^{-6} for solving Eq. (11.1). In Tables 11.12, similar to the Golub–Wathen study [31], we solve systems involving the preconditioner (11.2) using a direct scheme.

In Tables 11.13, we present similar results, except that we use our nested scheme shown in Fig. 11.1, with Algorithm 11.2 limited to only four iterations. The results shown in Tables 11.13 are the same whether the linear system in Step 6 of Algorithm 11.2 is solved directly, or solved using the conjugate gradient scheme with a relative residual stopping criterion of 10^{-2}. For much larger problems, however, the cost of direct solvers will be much higher than the CG scheme with a relaxed stopping criterion. Furthermore, the 2-norm of the residuals in Tables 11.12 and 11.13 are essentially the same. This demonstrates the effectiveness of our nested

Table 11.13 outer iteration: GMRES(k); the nested iterative scheme for solving (11.3)

GMRES(10)				GMRES(20)			
mesh	outer	$\|\text{residual}\|_2$	inner	mesh	outer	$\|\text{residual}\|_2$	inner
8×8	3	9×10^{-7}	4	8×8	2	5×10^{-10}	4
16×16	4	4×10^{-7}	4	16×16	2	7×10^{-8}	4
32×32	8	2×10^{-7}	4	32×32	4	5×10^{-8}	4

iterative scheme, not only for obtaining time-accurate solutions of the particulate flow problems, but also for the steady-state for driven cavity problem outlined here.

Finally, we would like to state that using flexible-type GMRES, to allow for changes in the preconditioner from one outer iteration to another, has resulted in inferior performance compared to that reported in Table 11.13. We would like also to mention that GMRES(20), without preconditioning, requires 592 iterations for the 8×8 mesh, and failed to achieve a residual of 2-norm $\leq 10^{-5}$ after 2000 iterations, for the 16×16 and 32×32 meshes.

11.8 Conclusion

We have presented a "nested iterative scheme" for solving saddle-point problems which can be regarded as a preconditioned inexact Uzawa algorithm with an additional correction step. The algorithm is essentially a preconditioned Krylov subspace method in which the preconditioner is itself a saddle-point problem. We propose a preconditioned Richardson iteration, with monotone convergence, for handling those inner iterations, i.e. for solving those systems involving the preconditioner. It should be noted that this Richardson scheme can be very effective in solving symmetric saddle-point problems in which the (1,1) block is symmetric positive definite.

We have used our nested iterative scheme for solving those nonsymmetric saddle-point problems that arise from the mixed finite element discretization of particulate flows, in which the fluid is incompressible. We have shown that an "inexpensive" preconditioner can be easily constructed, i.e. by constructing $\hat{A}^{-1}$ and $\hat{G}^{-1}$. In particular, we have shown that it is sufficient to have $\hat{A}^{-1} = \text{SPAI}(A_s, \text{diag})$, accelerated implicitly by three iterations, and to have the action of $\hat{G}^{-1}$ a close approximation of the action of $(B^T \hat{A}^{-1} B)^{-1}$. This latter implicit construction of $\hat{G}^{-1}$ is accomplished by solving systems involving $(B^T \hat{A}^{-1} B)$ via the Conjugate Gradient method with a relaxed stopping criterion.

We have compared our solution strategy of systems involving the adopted preconditioner with other preconditioners available in the literature. The resulting nested scheme proved to be more robust and more economical than others for handling those particulate flow simulations. Moreover, our scheme proved to be "scalable", and insensitive to changes in the fluid–particle system.

Finally, we should point out that all basic operations of our nested iterative scheme are amenable to efficient implementation on parallel computers.

Acknowledgements This work has been done in collaboration with Prof. Ahmed Sameh, and the author would like to acknowledge him for his continuous support.

References

1. Arrow, K., Hurwicz, L., Uzawa, H.: Studies in Nonlinear Programming. Stanford University Press, Stanford (1958)
2. Babuska, I.: The finite element method with Lagrangian multipliers. Numer. Math. **20**, 179–192 (1973)
3. Baggag, A.: Linear system solvers in particulate flows. Ph.D. thesis, Department of Computer Science, University of Minnesota (2003)
4. Baggag, A., Sameh, A.: A nested iterative scheme for indefinite linear systems in particulate flows. Comput. Methods Appl. Mech. Eng. **193**, 1923–1957 (2004)
5. Bank, R.E., Dupont, T., Yserentant, H.: The hierarchical basis multigrid method. Numer. Math. **52**, 427–458 (1988)
6. Bank, R.E., Welfert, B.D., Yserentant, H.: A class of iterative methods for solving saddle point problems. Numer. Math. **55**, 645–666 (1990)
7. Barnard, S., Grote, M.: A block version of the SPAI preconditioner. In: Hendrickson, B., Yelick, K., Bishof, C. (eds.) Proceedings of the Ninth SIAM Conference on Parallel Processing for Scientific Computing, March 22–24. SIAM, Philadelphia (1999). CD-ROM
8. Benzi, M., Golub, G.H.: An iterative method for generalized saddle point problems. SIAM J. Matrix Anal. (2012, to appear)
9. Braess, D., Sarazin, R.: An efficient smoother for the stokes problem. Appl. Numer. Math. **23**, 3–19 (1997)
10. Bramble, J.H., Leyk, Z., Pasciak, J.E.: Iterative schemes for non-symmetric and indefinite elliptic boundary value problems. Math. Comput. **60**, 1–22 (1993)
11. Bramble, J.H., Pasciak, J.E.: A preconditioning technique for indefinite systems resulting from mixed approximations of elliptic problems. Math. Comput. **50**, 1–18 (1988)
12. Bramble, J.H., Pasciak, J.E.: Iterative techniques for time dependent Stokes problems. Comput. Math. Appl. **33**, 13–30 (1997)
13. Bramble, J.H., Pasciak, J.E., Vassilev, A.T.: Analysis of the inexact Uzawa algorithm for saddle point problems. SIAM J. Numer. Anal. **34**, 1072–1092 (1997)
14. Bramble, J.H., Pasciak, J.E., Vassilev, A.T.: Uzawa type algorithms for nonsymmetric saddle point problems. Math. Comput. **69**, 667–689 (2000)
15. Brezzi, F., Fortin, M.: Mixed and Hybrid Finite Element Methods. Springer, New York (1991). ISBN 3-540-97582-9
16. Dyn, N., Ferguson, W.: The numerical solution of equality-constrained quadratic programming problems. Math. Comput. **41**, 165–170 (1983)
17. Elman, H., Silvester, D.: Fast nonsymmetric iterations and preconditioning for Navier–Stokes equations. SIAM J. Sci. Comput. **17**, 33–46 (1996)
18. Elman, H.C.: Multigrid and Krylov subspace methods for the discrete Stokes equations. Tech. Rep. 3302, Institute for Advanced Computer Studies (1994)
19. Elman, H.C.: Perturbation of eigenvalues of preconditioned Navier–Stokes operators. SIAM J. Matrix Anal. Appl. **18**, 733–751 (1997)
20. Elman, H.C.: Preconditioning for the steady-state Navier–Stokes equations with low viscosity. SIAM J. Sci. Comput. **20**, 1299–1316 (1999)
21. Elman, H.C.: Preconditioners for saddle point problems arising in computational fluid dynamics. Appl. Numer. Math. **43**, 75–89 (2002)

22. Elman, H.C., Golub, G.H.: Inexact and preconditioned Uzawa algorithms for saddle point problems. SIAM J. Numer. Anal. **31**, 1645–1661 (1994)
23. Elman, H.C., Silvester, D.J., Wathen, A.J.: Iterative methods for problems in computational fluid dynamics. In: Chan, R., Chan, T., Golub, G. (eds.) Iterative Methods in Scientific Computing. Springer, Singapore (1997)
24. Elman, H.C., Silvester, D.J., Wathen, A.J.: Performance and analysis of saddle point preconditioners for the discrete steady-state Navier–Stokes equations. Numer. Math. **90**, 641–664 (2002)
25. Falk, R.: An analysis of the finite element method using Lagrange multipliers for the stationary Stokes equations. Math. Comput. **30**, 241–269 (1976)
26. Falk, R., Osborn, J.: Error estimates for mixed methods. RAIRO. Anal. Numér. **14**, 249–277 (1980)
27. Fischer, B., Ramage, A., Silvester, D., Wathen, A.: Minimum residual methods for augmented systems. BIT Numer. Math. **38**, 527–543 (1998)
28. Gatica, G.N., Heuer, N.: Conjugate gradient method for dual-dual mixed formulation. Math. Comput. **71**, 1455–1472 (2001)
29. Girault, V., Raviart, P.: Finite Element Approximation of the Navier–Stokes Equations. Lecture Notes in Math., vol. 749. Springer, New York (1981)
30. Glowinski, R., Pan, T.W., Périaux, J.: Distributed Lagrange multiplier methods for incompressible viscous flow around moving rigid bodies. Comput. Methods Appl. Mech. Eng. **151**, 181–194 (1998)
31. Golub, G., Wathen, A.: An iteration for indefinite systems and its application to the Navier–Stokes equations. SIAM J. Sci. Comput. **19**, 530–539 (1998)
32. Golub, G., Wu, X., Yuan, J.Y.: SOR-like methods for augmented systems. BIT Numer. Math. **41**, 71–85 (2001)
33. Grote, M., Huckle, T.: Parallel preconditioning with sparse approximate inverses. SIAM J. Sci. Comput. **18**, 838–853 (1997)
34. Hu, H.: Direct simulation of flows of solid-liquid mixtures. Int. J. Multiph. Flow **22**, 335–352 (1996)
35. Johnson, A.A., Tezduyar, T.E.: Simulation of multiple spheres falling in a liquid-filled tube. Comput. Methods Appl. Mech. Eng. **134**, 351–373 (1996)
36. Johnson, A.A., Tezduyar, T.E.: 3D simulation of fluid–particle interactions with the number of particles reaching 100. Comput. Methods Appl. Mech. Eng. **145**, 301–321 (1997)
37. Johnson, A.A., Tezduyar, T.E.: Advanced mesh generation and update methods for 3D flow simulations. Comput. Mech. **23**, 130–143 (1999)
38. Johnson, A.A., Tezduyar, T.E.: Methods for 3D computation of fluid-object interactions in spatially-periodic flows. Comput. Methods Appl. Mech. Eng. **190**, 3201–3221 (2001)
39. Keller, C., Gould, N.I.M., Wathen, A.J.: Constraint preconditioning for indefinite linear systems. SIAM J. Matrix Anal. Appl. **21**, 1300–1317 (2000)
40. Klawonn, A.: An optimal preconditioner for a class of saddle point problems with a penalty term. SIAM J. Sci. Comput. **19**, 540–552 (1998)
41. Klawonn, A., Starke, G.: Block triangular preconditioners for nonsymmetric saddle point problems: Field-of-values analysis. Numer. Math. **81**, 577–594 (1999)
42. Knepley, M.: Parallel simulation of the particulate flow problem. Ph.D. thesis, Department of Computer Science, Purdue University (2000)
43. Knepley, M., Sarin, V., Sameh, A.: Parallel simulation of particulate flows. Appeared in Fifth Intl. Symp. on Solving Irregular Structured Problems in Parallel, IRREGULAR 98, LNCS, No. 1457, pp. 226–237, Springer (1998)
44. Krzyzanowski, P.: On block preconditioners for nonsymmetric saddle point problems. SIAM J. Sci. Comput. **23**, 157–169 (2001)
45. Little, L., Saad, Y.: Block LU preconditioners for symmetric and nonsymmetric saddle point problems. Tech. Rep. 1999-104, Minnesota Supercomputer Institute, University of Minnesota (1999)

46. Lou, G.: Some new results for solving linear systems arising from computational fluid dynamics problems. Ph.D. thesis, Department of Computer Science, University of Illinois U-C (1992)
47. Lukšan, L., Vlček, J.: Indefinitely preconditioned inexact Newton method for large sparse equality constrained nonlinear programming problems. Numer. Linear Algebra Appl. **5**, 219–247 (1998)
48. Maury, B.: Characteristics ALE method for the unsteady 3D Navier–Stokes equations with a free surface. Comput. Fluid Dyn. J. **6**, 175–188 (1996)
49. Maury, B.: A many-body lubrication model. C. R. Acad. Sci. Paris **325**, 1053–1058 (1997)
50. Maury, B.: Direct simulations of 2D fluid–particle flows in biperiodic domains. J. Comput. Phys. **156**, 325–351 (1999)
51. Maury, B., Glowinski, R.: Fluid–particle flow: a symmetric formulation. C. R. Acad. Sci. Paris **324**, 1079–1084 (1997)
52. Murphy, M.F., Golub, G.H., Wathen, A.J.: A note on preconditioning for indefinite linear systems. SIAM J. Sci. Comput. **21**, 1969–1972 (2000)
53. Ortega, J.M.: Numerical Analysis: a Second Course. Computer Science and Applied Mathematics Series. Academic Press, San Diego (1972)
54. Perugia, I., Simoncini, V.: An optimal indefinite preconditioner for mixed finite element method. Tech. Rep. 1098, Department of Mathematics, Università de Bologna, Italy (1998)
55. Perugia, I., Simoncini, V.: Block-diagonal and indefinite symmetric preconditioners for mixed finite element formulations. Numer. Linear Algebra Appl. **7**, 585–616 (2000)
56. Queck, W.: The convergence factor of preconditioned algorithms of the Arrow-Hurwicz type. SIAM J. Numer. Anal. **26**, 1016–1030 (1989)
57. Rusten, T., Winther, R.: A preconditioned iterative method for saddle point problem. SIAM J. Matrix Anal. Appl. **13**, 887–904 (1992)
58. Saad, Y.: A flexible inner–outer preconditioned GMRES algorithm. SIAM J. Sci. Comput. **14**, 461–469 (1993)
59. Saad, Y., Schultz, M.: GMRES: A generalized minimal residual algorithm for solving nonsymmetric linear systems. SIAM J. Sci. Stat. Comput. **7**, 856–869 (1986)
60. Saad, Y., Suchomel, B.: ARMS: an algebraic recursive multilevel solver for general sparse linear systems. Numer. Linear Algebra Appl. **9**, 359–378 (2002)
61. Sameh, A., Baggag, A.: Parallelism in iterative linear system solvers. In: Proceedings of the Sixth Japan-US Conference on Flow Simulation and Modeling, April 1, 2002
62. Sameh, A., Baggag, A., Wang, X.: Parallel nested iterative schemes for indefinite linear systems. In: Mang, H.A., Rammerstorfer, F.G., Eberhardsteiner, J. (eds.) Proceedings of the Fifth World Congress on Computational Mechanics, (WCCM V). Vienna University of Technology, Austria, July 7–12, 2002. ISBN 3-9501554-0-6
63. Silvester, D., Elman, H., Kay, D., Wathen, A.: Efficient preconditioning of the linearized Navier–Stokes equations for incompressible flow. J. Comput. Appl. Math. **128**, 261–279 (2001)
64. Silvester, D., Wathen, A.: Fast iterative solution of stabilized Stokes systems. part II: Using general block preconditioners. SIAM J. Numer. Anal. **31**, 1352–1367 (1994)
65. Silvester, D., Wathen, A.: Fast and robust solvers for time-discretized incompressible Navier–Stokes equations. Tech. Rep. 27, Department of Mathematics, University of Manchester (1995)
66. Simoncini, V., Szyld, D.: Flexible inner–outer Krylov subspace methods. SIAM J. Numer. Anal. **40**, 2219–2239 (2003)
67. Simoncini, V., Szyld, D.: Theory of inexact Krylov subspace methods and applications to scientific computing. SIAM J. Sci. Comput. **25**, 454–477 (2003)
68. Stewart, G.W.: Introduction to Matrix Computations. Academic Press, San Diego (1973)
69. Tezduyar, T.E.: Stabilized finite element formulations for incompressible flow computations. Adv. Appl. Mech. **28**, 1–44 (1991)
70. Vanderstraeten, D., Knepley, M.: Parallel building blocks for finite element simulations: Application to solid-liquid mixture flows. In: Emerson, D., Ecer, A., Periaux, J., Satofuka, N.

(eds.) Proceedings of Parallel CFD'99 Conf.: Recent Developments and Advances Using Parallel Computers, pp. 133–139. Academic Press, Manchester (1997)
71. Verfürth, R.: A combined conjugate gradient-multigrid algorithm for the numerical solution of the Stokes problem. IMA J. Numer. Anal. **4**, 441–455 (1984)
72. Verfürth, R.: A posteriori error estimators for the Stokes equations. Numer. Math. **55**, 309–325 (1989)
73. Wathen, A., Silvester, D.: Fast iterative solution of stabilized Stokes systems. part I: Using simple diagonal preconditioners. SIAM J. Numer. Anal. **30**, 630–649 (1993)
74. Zulehner, W.: A class of smoothers for saddle point problems. Computer **65**, 227–246 (2000)
75. Zulehner, W.: Analysis of iterative methods for saddle point problems: a unified approach. Math. Comput. **71**, 479–505 (2001)

Chapter 12
Effect of Ordering for Iterative Solvers in Structural Mechanics Problems

Sami A. Kilic

Abstract Direct solvers are commonly used in implicit finite element codes for structural mechanics problems. This study explores an alternative approach to solving the resulting linear systems by using the Conjugate Gradient algorithm. Preconditioning is applied by using the incomplete Cholesky factorization. The effect of ordering is investigated for the Reverse Cuthill–McKee scheme and the Approximate Minimum Degree Method. The solution time and the required storage space are reported for two test problems involving thin shell finite elements and hexahedral solid elements.

12.1 Introduction

Iterative solvers are gaining momentum in the area of structural mechanics. Problem sizes are getting larger as users access more powerful computer hardware. Direct solvers require large amounts of memory, and the full factorization procedure elongates the solution process.

In this study the effects of two ordering schemes on the iterative solver performance are demonstrated for two test problems obtained from the automotive industry. The Pre-conditioned Conjugate Gradient algorithm in combination with the incomplete Cholesky factorization is utilized to obtain the solution vectors [2]. The first problem consists of four node shell elements. The second problem contains eight node hexahedral volume elements. The performance of the iterative solver is demonstrated for the two problems with different topology of finite elements.

12.2 Description of the Test Problems

Table 12.1 shows the relevant properties of the test matrices provided by Dr. Roger Grimes of the Livermore Software Technology Company that develops the LS-Dyna

S.A. Kilic (✉)
Department of Civil Engineering, Faculty of Engineering, Bogazici University, 34342 Bebek, Istanbul, Turkey
e-mail: skilic@boun.edu.tr

M.W. Berry et al. (eds.), *High-Performance Scientific Computing*,
DOI 10.1007/978-1-4471-2437-5_12, © Springer-Verlag London Limited 2012

Table 12.1 Properties of test matrices

Problem	N	nzA	Condest(A)	Topology of finite elements
CAR_HOOD	235,962	12,860,848	$7.13E+12$	All shell elements
S_BLOCK	183,768	13,411,340	$7.54E+08$	All hexahedral elements

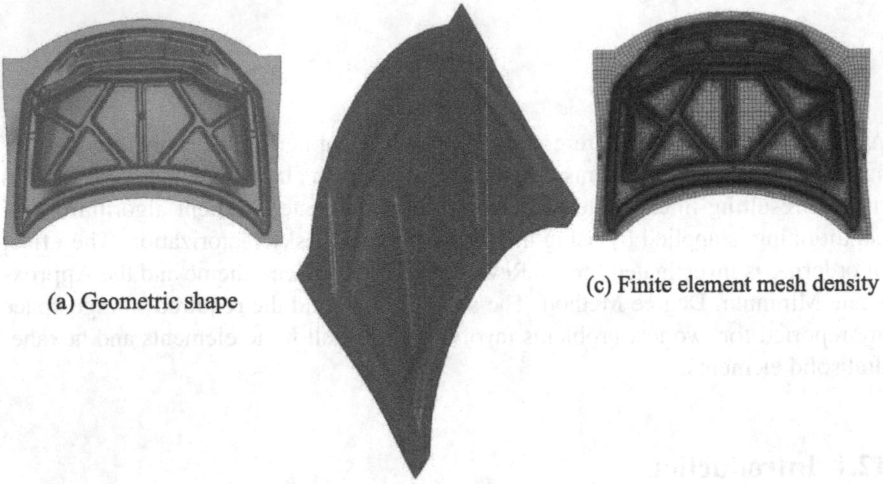

(a) Geometric shape

(b) Oblique view

(c) Finite element mesh density

Fig. 12.1 Illustrations of the CAR_HOOD test problem

finite element analysis software. The matrices are provided in order to test the performance of iterative solvers. The parameter N denotes the size of the matrix. The parameter nzA gives the number of non-zero entries in the upper triangular part of A. The test matrices represent the stiffness components of the non-linear static problems using the implicit solver of the commercial finite element code LS-Dyna [4].

Condition numbers (cond(A)) of the diagonally scaled coefficient matrix, $A = D^{-1/2}AD^{1/2}$ and the description of the finite element topology is also given in Table 12.1. The "CAR_HOOD" matrix was hand-picked from a set of problems coming from the commercial automobile industry users of LS-Dyna. It is a difficult problem to solve as it contains thin shell finite elements in a mesh with poor element aspect ratios and large variations in element sizes. The finite element mesh consists of four node shell elements. The type of the finite element analysis is sheet metal forming process during which the flat sheet of metal is stamped with a rigid surface in order to form the curved shape of the automobile engine hood. Implicit solver is used to simulate the spring-back of the metal sheet when it is released from the stamping process. It contains 40,241 four node shell elements and 41,425 nodes, and very few constraints. The shell element type is chosen as the Belytschko–Tsay shell element [1]. Figures 12.1a and 12.1b show the geometric shape of the

Fig. 12.2 Views of the S_BLOCK test problem

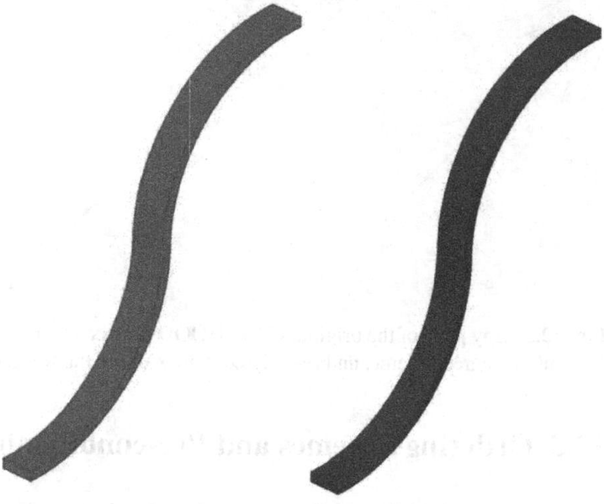

(a) Geometric shape (b) Finite element mesh density

"CAR_HOOD" problem. Figure 12.1c shows the finite element mesh density for the shell elements used in the analysis.

The second problem "S_BLOCK" is a generic S-shaped solid with eight node hexahedral elements as shown in Fig. 12.2. There are 10 elements through the thickness, 27 elements across the width, and 200 elements along the length. Figure 12.2b shows the finite element mesh with 54,000 hexahedral elements and 61,721 nodes. The S_BLOCK provides an opportunity for iterative solvers. Due to the 3-D topology of the hexahedral elements there is a large fill-in in the full factorization process [3]. Problems involving hexahedral elements generally yield better conditioned stiffness matrices compared to problems involving thin shell elements with 2-D topology. This is illustrated in Table 12.1 by the condition estimate (Condest(A) value) of the coefficient matrices. The CAR_HOOD problem has a condition estimate of $7.13E+12$, whereas the corresponding value for the S_BLOCK problem is $7.54E+08$. The shell elements have membrane and bending stiffness components [6]. The membrane stiffness is proportional to the thickness of the shell element. The bending stiffness is a function of the cube of the thickness. For thin shell elements the two types of stiffness yield values that are orders of magnitude apart. This physical fact causes a large condition number for the coefficient matrix that contains both types of stiffness component.

The coefficient matrices that represent the stiffness of the test problems are symmetric and positive-definite. For the non-linear static analyses of the test problems the stiffness matrix is updated at certain stages of the simulation process. However, the matrices remain symmetric and positive-definite during the entire simulation.

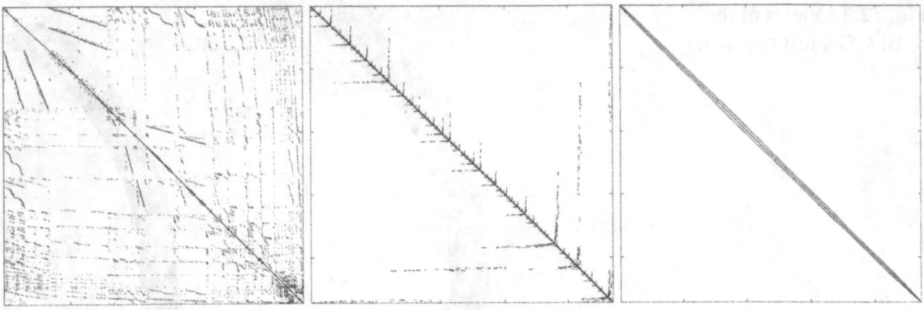

Fig. 12.3 Spy plots of the original CAR_HOOD matrix, re-ordered matrix using the Approximate Minimum Degree scheme, and re-ordered matrix using the Reverse Cuthill–McKee scheme

12.3 Ordering Schemes and Pre-conditioning

The Reverse Cuthill–McKee algorithm is designed to reduce the bandwidth of sparse symmetric matrices [3]. It is an application of the graph theory by the use of level sets. A level set is defined recursively as the set of all unmarked neighbors of all the nodes of a previous level set. As soon as a level set is traced, its nodes are marked and numbered. In the Cuthill–McKee ordering scheme the elements of a level set are numbered from the nodes of lowest degree to those of the highest degree. The Reverse Cuthill–McKee algorithm is obtained by reversing the index numbers of the Cuthill–McKee algorithm, which generally results in a better solution.

The minimum degree algorithm is used to permute the rows and columns of a symmetric sparse matrix before applying the Cholesky decomposition in order to reduce the number of non-zeros in the factorization process [3]. This results in reduced storage requirements and also fewer arithmetic operations in the iterative solution process.

At each step of the Gaussian elimination process row and column permutations are performed so as to minimize the number of off-diagonal non-zeros in the pivot row and column. The main goal is to find a permutation matrix P so that the Cholesky factorization of $PAP^\top = LL^\top$ has fewer nonzero entries than the Cholesky factorization of A. The most costly part of the minimum degree algorithm is the recomputation of the degrees of nodes adjacent to the current pivot element. Rather than keeping track of the exact degree, the approximate minimum degree algorithm finds an upper bound on the degree that is easier to compute. For nodes of least degree, this bound tends to be tight. Using the approximate degree instead of the exact degree leads to substantial savings in the solution time, particularly for irregularly structured matrices.

Figures 12.3 and 12.4 show the spy plots of the two test matrices obtained from the MATLAB software [7]. The first spy plot shows the original matrix, the second spy plot shows the effect of the Approximate Minimum Degree ordering, and the third spy plot shows the effect of the Reverse Cuthill–McKee ordering. The elements of the matrix are clustered around the main diagonal for the Reverse Cuthill–McKee

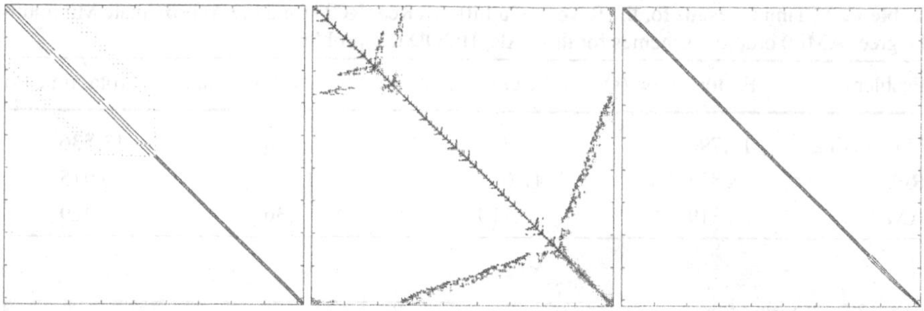

Fig. 12.4 Spy plots of the original S_BLOCK matrix, re-ordered matrix using the Approximate Minimum Degree scheme, and re-ordered matrix using the Reverse Cuthill–McKee scheme

ordering since the goal is to reduce the bandwidth. This effect is clearly illustrated in Figures 12.3 and 12.4.

The Pre-conditioned Conjugate Gradient method was used in order to obtain the solution vectors. All calculations were done using the MATLAB software [7]. The incomplete Cholesky factorization of the coefficient matrix A constructs the pre-conditioner in the Conjugate Gradient iterations. In this study the incomplete Cholesky function `cholinc` of the MATLAB software with a user-specified drop tolerance was utilized for the factorization process. The drop tolerance of the "cholinc" function was chosen as $1.E-06$. An initial parametric study was done in order to assess the effect of the drop tolerance on the number of iterations and the solution time for the values of $1.E-04$, $1.E-05$, $1.E-06$, and $1.E-07$. The value of $1.E-06$ provided the optimal results for the set considered. The larger values of $1.E-04$ and $1.E-05$ for the drop tolerance tend to produce a pre-conditioner that poorly represents the original coefficient matrix. The smallest value of $1.E-07$ causes excessive run times for the Incomplete Cholesky factorization of the pre-conditioner.

12.4 Iterative Solver Performance Results

The timing results for the CAR_HOOD problem are given in Table 12.2. The Approximate Minimum Degree ordering has a clear advantage over the Reverse Cuthill–McKee problem. The CAR_HOOD problem represents the typical case in automotive engineering and aerospace engineering applications. The body of an automobile is made of sheet metal, which is modeled with thin shell elements. Similarly, the fuselage skin of airplanes is also modeled with thin shell elements. These problems usually have large condition numbers due to their reduced shell thickness values. The large condition number is attributed to the gap between the bending and membrane stiffness coefficients of the thin shell elements [6]. As the thickness value gets smaller, the gap between the bending and membrane stiffness coefficients widens further, resulting in large condition numbers. The thin shell problems represent the most difficult set of solid mechanics domain for the application of iterative solvers.

Table 12.2 Timing results for the Reverse Cuthill–McKee (RCM) and the Approximate Minimum Degree (AMD) ordering schemes for the CAR_HOOD test problem

Problem	Factorization (s)	PCG iterations (s)	Iteration count	Total time (s)
No ordering	12,790	746	90	13,536
RCM	3,832	47,083	3,748	50,915
AMD	319	110	36	429

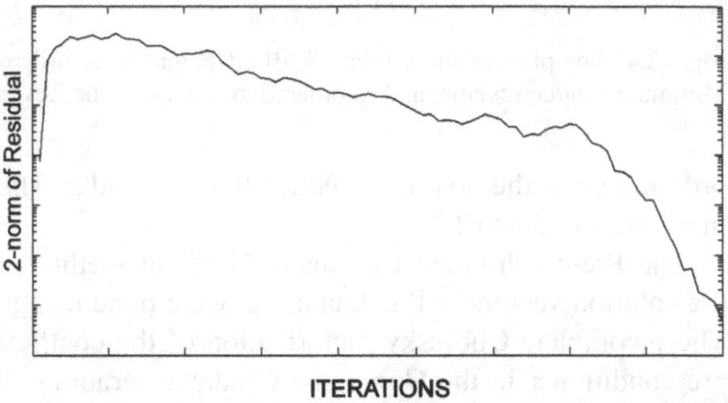

Fig. 12.5 2-norm of the residual vector versus PCG iterations for the CAR_HOOD problem without ordering

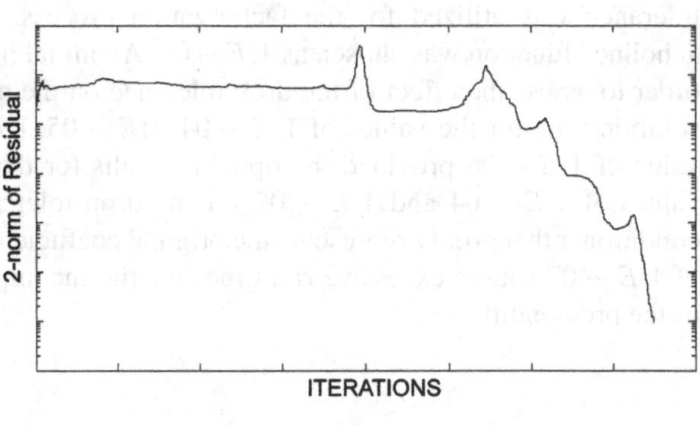

Fig. 12.6 2-norm of the residual vector versus PCG iterations for the CAR_HOOD problem with RCM ordering

Figure 12.5 shows the 2-norm of the residual versus the PCG iterations for the CAR_HOOD matrix without ordering. The total number of iterations is 90 and the total solution time is 13,536 seconds.

Figure 12.6 shows the 2-norm of the residual versus the PCG iterations for the CAR_HOOD matrix with the RCM ordering. The total number of iterations is 3,748 and the total solution time is 50,915 seconds. The RCM ordering makes the PCG method perform with a poor performance for the thin shell finite elements.

Figure 12.7 shows the 2-norm of the residual versus the PCG iterations for the CAR_HOOD matrix for the AMD ordering. The total number of iterations is 36 and the total solution time is 429 seconds. The AMD ordering provides the least number of iterations and the fastest timing results for the PCG solver. The AMD ordering provides a clear advantage over the original ordering and the RCM ordering cases for thin shell problems.

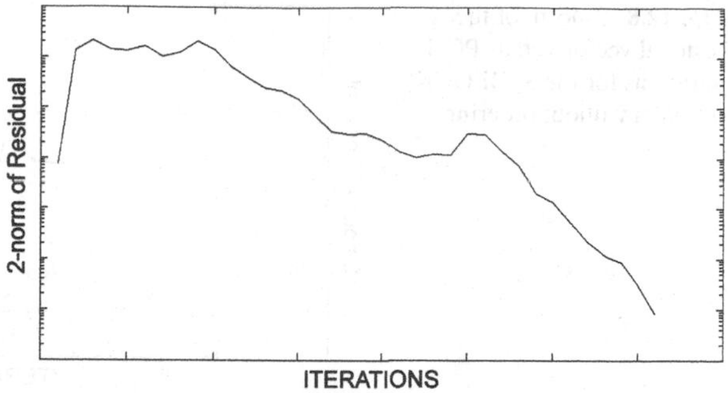

Fig. 12.7 2-norm of the residual vector versus PCG iterations for the CAR_HOOD problem with AMD ordering

Table 12.3 Timing results for the S_BLOCK test problem consisting of hexahedral elements

Problem	Factorization (s)	PCG iterations (s)	Iteration count	Total time (s)
No ordering	5,797	16,313	106	22,110
RCM	1,755	295	17	2,050
AMD	4,335	1,039	77	5,374

Table 12.4 Storage required for the factorization process for the ordering schemes used in the study

Problem	Nnz of factorized matrix for no ordering	Nnz of factorized matrix for RCM	Nnz of factorized matrix for AMD
CAR_HOOD	122,386,086	219,958,874	50,427,528
S_BLOCK	258,263,545	188,533,104	175,503,315

The timing results for the S_BLOCK problem are given in Table 12.3. The fastest time to solution is obtained by the RCM ordering of the coefficient matrix. The S_BLOCK problem contains eight node hexahedral elements. The relative advantage of the AMD ordering scheme is lost when the topology of the problem consists of volumetric finite elements.

Table 12.4 gives the number of non-zeros in the factorized coefficient matrix for the CAR_HOOD and the S_BLOCK problems. For the CAR_HOOD problem the AMD ordering significantly reduces the amount of fill-in. The superior performance of the Pre-conditioned Conjugate Gradient solver can be attributed to the reduction of the fill-in. On the other hand, the RCM ordering produces a large fill-in, which leads to the large number of iterations as shown in Table 12.2. The original ordering of the coefficient matrix yields the medium level in the fill-in results, and provides the medium level of performance in terms of the timing values presented in Table 12.2. The amount of fill-in for the S_BLOCK problem is also presented in Table 12.4. The AMD ordering scheme again provides the least amount of fill-in for the incomplete Cholesky factorization. However, the RCM ordering scheme yields fewer number of PCG iterations and faster time to solution as given in Table 12.3. The S_BLOCK problem consists of hexahedral elements with 3-D topology. The

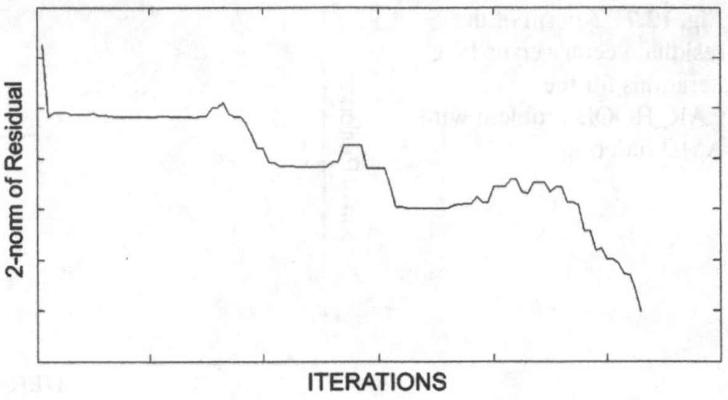

Fig. 12.8 2-norm of the residual vector versus PCG iterations for the S_BLOCK problem without ordering

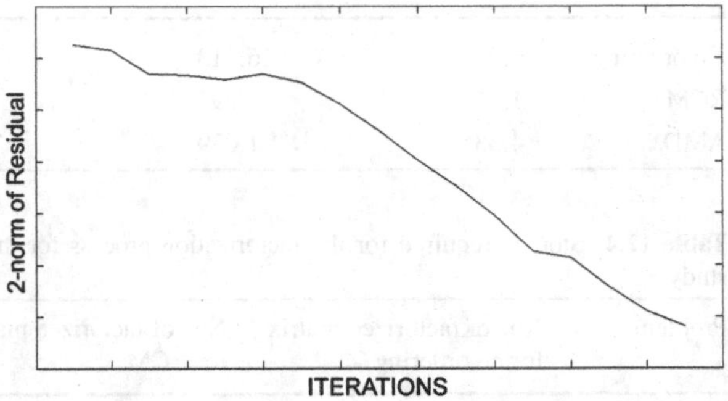

Fig. 12.9 2-norm of the residual vector versus PCG iterations for the S_ BLOCK problem with RCM ordering

RCM scheme provides an opportunity for iterative solvers to perform faster. For problems with 3-D finite element topology direct solvers produce even a higher level of fill-in in the full factorization process. Therefore, iterative solvers have the upper hand for such finite element mesh topologies [5].

Figure 12.8 shows the 2-norm of the residual versus the PCG iterations for the S_BLOCK matrix without ordering. The total number of iterations is 106 and the total solution time is 22,110 seconds.

Figure 12.9 shows the 2-norm of the residual versus the PCG iterations for the S_BLOCK matrix with the RCM ordering. The total number of iterations is 17 and the total solution time is 2,050 seconds. The RCM ordering provides the fastest solution time for the PCG iterations.

Figure 12.10 shows the 2-norm of the residual versus the PCG iterations for the S_BLOCK matrix for the AMD ordering. The total number of iterations is 77 and the total solution time is 5,374 seconds. Although the AMD ordering results in the least amount of fill-in, it does not provide the fastest time to solution and causes a higher number of iterations for the PCG solver for problems involving 3-D topology of finite elements.

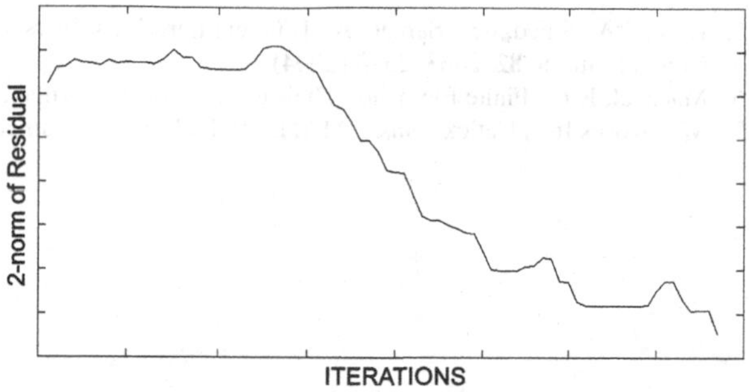

Fig. 12.10 2-norm of the residual vector versus PCG iterations for the S_ BLOCK problem with AMD ordering

12.5 Conclusions and Recommendations

The solution of linear systems of equations for the structural mechanics problems involving thin shell finite elements is a challenging task. A large number of applications in the automotive and the aeronautical industry contain thin shell problems. The body panels of cars and the fuselage of aircraft are examples of the thin shell problems. Thin shell elements yield ill-conditioned coefficient matrices with large condition numbers. The type of ordering in fact affects the performance of the Preconditioned Conjugate Gradient method. For the thin shell problem investigated in this study the Approximate Minimum Degree ordering showed an advantage over the Reverse Cuthill–McKee method due to the reduced amount of fill-in. On the other hand, the Reverse Cuthill–McKee method provided a better solution time for the hexahedral topology problem presented in this study. The advantage of the Approximate Minimum Degree ordering requires further investigation for problems involving a mixture of thin shell and hexahedral finite elements. Therefore, further work is suggested in order to extend the conclusions obtained from the test problems investigated in this study.

Acknowledgements The author would like to express his gratitude to Professor Ahmed Sameh and Dr. Faisal Saied of Purdue University for the insight they provided into the fascinating world of iterative solvers. Concrete foundations of this study were laid out at the Computing Research Institute under the provision of Professor Ahmed Sameh. The test matrices were provided by Dr. Roger Grimes of the Livermore Software Technology Company of Livermore, California. The generous support of the Bogazici University Research Fund through contract number 07HT102 is gratefully acknowledged.

References

1. Belytschko, T., Liu, W.K., Moran, B.: Nonlinear Finite Elements for Continua and Structures. Wiley, New York (2000)
2. Braess, D.: Finite Elements: Theory, Fast Solvers, and Applications in Solid Mechanics. Cambridge University Press, Cambridge (2001)
3. George, A., Liu, J.W.H.: Computer solution of large sparse positive definite systems. Prentice-Hall, New York (1981)
4. Hallquist, J.O.: LS-Dyna Theoretical Manual (2006)

5. Kilic, S.A., Saied, F.S., Sameh, A.: Efficient iterative solvers for structural dynamics problems. Comput. Struct. **82**, 2363–2375 (2004)
6. Macneal, R.H.: Finite Elements: Their Design and Performance. Dekker, New York (1994)
7. Mathworks Inc., Natick, Mass.: MATLAB Technical Documentation (2010)

Chapter 13
Scaling Hypre's Multigrid Solvers to 100,000 Cores

Allison H. Baker, Robert D. Falgout, Tzanio V. Kolev, and Ulrike Meier Yang

Abstract The *hypre* software library (http://www.llnl.gov/CASC/hypre/) is a collection of high performance preconditioners and solvers for large sparse linear systems of equations on massively parallel machines. This paper investigates the scaling properties of several of the popular multigrid solvers and system building interfaces in *hypre* on two modern parallel platforms. We present scaling results on over 100,000 cores and even solve a problem with over a trillion unknowns.

13.1 Introduction

The need to solve increasingly large, sparse linear systems of equations on parallel computers is ubiquitous in scientific computing. Such systems arise in the numerical simulation codes of a diverse range of phenomena, including stellar evolution, groundwater flow, fusion plasmas, explosions, fluid pressures in the human eye, and many more. Generally these systems are solved with iterative linear solvers, such as the conjugate gradient method, combined with suitable preconditioners, see e.g. [17, 19].

A particular challenge for parallel linear solver algorithms is scalability. An application code is scalable if it can use additional computational resources effectively. In particular, in this paper we focus on *weak scalability*, which requires that if the size of a problem and the number of cores are increased proportionally, the computing time should remain approximately the same. Unfortunately, in practice, as simulations grow to be more realistic and detailed, computing time may increase dramatically even when more cores are added to solve the problem. Recent machines

A.H. Baker (✉) · R.D. Falgout · T.V. Kolev · U.M. Yang
Lawrence Livermore National Laboratory, Center for Applied Scientific Computing, Livermore, CA 94551-0808, USA
e-mail: abaker@llnl.gov

R.D. Falgout
e-mail: rfalgout@llnl.gov

T.V. Kolev
e-mail: tzanio@llnl.gov

U.M. Yang
e-mail: umyang@llnl.gov

M.W. Berry et al. (eds.), *High-Performance Scientific Computing*,
DOI 10.1007/978-1-4471-2437-5_13, © Springer-Verlag London Limited 2012

with tens or even hundreds of thousands of cores offer both enormous computing possibilities and unprecedented challenges for achieving scalability.

The *hypre* library was developed with the specific goal of providing users with advanced parallel linear solvers and preconditioners that are scalable on massively parallel architectures. Scalable algorithms are essential for combating growing computing times. The library features parallel multigrid solvers for both structured and unstructured problems. Multigrid solvers are attractive for parallel computing because of their scalable convergence properties. In particular, if they are well-designed, then the computational cost depends linearly on the problem size, and increasingly larger problems can be solved on (proportionally) increasingly larger numbers of cores with approximately the same number of iterations to solution. This natural algorithmic scalability of the multigrid methods combined with the robust and efficient parallel algorithm implementations in *hypre* result in preconditioners that are well-suited for large numbers of cores.

The *hypre* library is a vital component of a broad array of application codes both at and outside of Lawrence Livermore National Laboratory (LLNL). For example, the library was downloaded more than 1800 times from 42 countries in 2010 alone, approaching nearly 10,000 total downloads from 70 countries since its first open source release in September of 2000. The scalability of its multigrid solvers has a large impact on many applications, particularly because simulation codes often spend the majority of their runtime in the linear solve.

The objective of this paper is to demonstrate the scalability of the most popular multigrid solvers in *hypre* on current supercomputers. We present scaling studies for conjugate gradient, preconditioned with the structured-grid solvers PFMG, SMG, SysPFMG, as well as the algebraic solver BoomerAMG, and the unstructured Maxwell solver AMS. Note that previous investigations beyond 100,000 cores focused only on the scalability of BoomerAMG on various architectures and can be found in [2, 5].

The paper is organized as follows. First we provide more details about the overall *hypre* library in Sect. 13.2 and the considered multigrid linear solvers in Sect. 13.3. Next, in Sect. 13.4, we specify the machines and the test problems we used in our experimental setup. We present and discuss the corresponding scalability results in Sect. 13.5, and we conclude by summarizing our findings in Sect. 13.6.

13.2 The *hypre* Library

In this section we give a general overview of the *hypre* library. More detailed information can be found in the User's Manual available on the *hypre* web page [15].

13.2.1 Conceptual Interfaces

We first discuss three of the so-called *conceptual interfaces* in *hypre*, which provide different mechanisms for describing a linear system on a parallel machine. These interfaces not only facilitate the use of the library, but they make it possible to provide linear solvers that take advantage of additional information about the application.

The *Structured Grid Interface* (`Struct`) is a stencil-based interface that is most appropriate for scalar finite-difference applications whose grids consist of unions of logically rectangular (sub)grids. The user defines the matrix and right-hand side in terms of the stencil and the grid coordinates. This geometric description, for example, allows the use of the PFMG solver, a parallel algebraic multigrid solver with geometric coarsening, described in more detail in the next section.

The *Semi-Structured Grid Interface* (`SStruct`) is essentially an extension of the Structured Grid Interface that can accommodate problems that are mostly structured, but have some unstructured features (e.g., block-structured, composite or overset grids). It can also accommodate multiple variables and variable types (e.g., cell-centered, edge-centered, etc.), which allows for the solution of more general problems. This interface requires the user to describe the problem in terms of structured grid *parts*, and then describe the relationship between the data in each part using either stencils or finite element stiffness matrices.

The *Linear-algebraic Interface* (`IJ`) is a standard linear-algebraic interface that requires that the users map the discretization of their equations into row-column entries in a matrix structure. Matrices are assumed to be distributed across P MPI tasks in contiguous blocks of rows. In each task, the matrix block is split into two components which are each stored in compressed sparse row (CSR) format. One component contains the coefficients that are local to the task, and the second, which is generally much smaller than the local one, contains the coefficients whose column indices point to rows located in other tasks. More details of the parallel matrix structure, called ParCSR, can be found in [10].

13.2.2 Solvers

The *hypre* library contains highly efficient and scalable specialized solvers as well as more general-purpose solvers that are well-suited for a variety of applications.

The specialized multigrid solvers use more than just the matrix to solve certain classes of problems, a distinct advantage provided by the conceptual interfaces. For example, the structured multigrid solvers SMG, PFMG, and SysPFMG all take advantage of the structure of the problem. As a result, these solves are typically more efficient and scalable than a general-purpose solver alternative. The SMG and PFMG solvers require the use of the `Struct` interface, and SysPFMG requires the `SStruct` interface.

For electromagnetic problems, *hypre* provides the unstructured Maxwell solver, AMS, which is the first provably scalable solver for definite electromagnetic problems on general unstructured meshes. The AMS solver requires matrix coefficients plus the discrete gradient matrix and the vertex coordinates which can be described with the `IJ` or `SStruct` interface.

For problems on arbitrary unstructured grids, *hypre* provides a robust parallel implementation of algebraic multigrid (AMG), called BoomerAMG. BoomerAMG can be used with any interface (currently not supported through `Struct`), as it only requires the matrix coefficient information.

The *hypre* library also provides common general-purpose iterative solvers, such as the GMRES and Conjugate Gradient (CG) methods. While these algorithms are not scalable as stand-alone solvers, they are particularly effective (and scalable) when used in combination with a scalable multigrid preconditioner.

13.2.3 Considerations for Large-Scale Computing

Several features of the *hypre* library are required to efficiently solve very large problems on current supercomputers. Here we describe three of these features: support for 64-bit integers, scalable interface support for large numbers of MPI tasks, and the use of a hybrid programming model.

64-bit integer support has recently been added in *hypre*. This support is needed to solve problems in ParCSR format with more than 2 billion unknowns (previously a limitation due to 32-bit integers). To enable the 64-bit integer support, *hypre* must be configured with the `-enable-bigint` option. When this feature is turned on, the user must pass *hypre* integers of type `HYPRE_Int`, which is the 64-bit integer (usually a 'long long int' type in C). Note that this 64-bit integer option converts all integers to 64-bit, which does affect performance and increases memory use.

Scalable interfaces as well as solver algorithms are required for a code utilizing *hypre* to be scalable. When using one of *hypre*'s interfaces, the problem data is passed to *hypre* in its distributed form. However, to obtain a solution via a multigrid method or any other linear solver algorithm, MPI tasks need to obtain nearby data from other tasks. For a task to determine which tasks own the data that it needs, i.e. their communication partners or neighbors, some information regarding the global distribution of the data is required. Storing and querying a global description of the data, which is the information detailing which MPI task owns what data or the *global partition*, is either too costly or not possible when dealing with tens of thousands or more tasks. Therefore, to determine inter-task communication in a scalable manner, we developed new algorithms that employ an *assumed partition* to answer queries through a type of rendezvous algorithm, instead of storing global data descriptions. This strategy significantly reduces storage, communication, and computational costs for the solvers in *hypre* and improves scalability as shown in [4]. Note that this optimization requires configuring *hypre* with the `-no-global-partition` option and is most beneficial when using tens of thousands of tasks.

A *hybrid programming model* is used in *hypre*. While we have obtained good scaling results in the past using an MPI-only programming model, see e.g. [11], with increasing numbers of cores per node on multicore architectures, the MPI-only model is expected to be increasingly insufficient due to the limited off-node bandwidth and decreasing amounts of memory per core. Therefore, in *hypre* we also employ a mixed or hybrid programming model which combines both MPI and the shared memory programming model OpenMP. The OpenMP code in *hypre* is largely used in loops and divides a loop among k threads into roughly k equal-sized portions. Therefore, basic matrix and vector operations, such as the matrix-vector multiply or dot product, are straightforward, but the use of OpenMP within other more complex parts of the multigrid solver algorithm, such as in parts of the setup phase (described in Sect. 13.3), may be non-trivial. The right choice for hybrid MPI/OpenMP partitioning in terms of obtaining optimal performance is dependent on the specific target machine's node architecture, interconnect, and operating system capabilities [5]. See [5] or [2] for more discussion on the performance of BoomerAMG with a hybrid programming model.

13.3 The Multigrid Solvers

As mentioned in Sect. 13.1, multigrid solvers are algorithmically scalable, meaning that they require O(N) computations to solve a linear system with N variables. This desirable property is obtained by cleverly utilizing a sequence of smaller (or coarser) grids, which are computationally cheaper to compute on than the original (finest) grid. A multigrid method works as follows. At each grid level, a smoother is applied to the system, which serves to resolve the high-frequency error on that level. The improved guess is then transferred to a smaller, or coarser, grid, the smoother is applied again, and the process continues. The coarsest level is generally chosen to be a size that is reasonable to solve directly, and the goal is to eliminate a significant part of the error by the time this coarsest level is reached. The solution to the coarse grid solve is then interpolated, level by level, back up to the finest grid level, applying the smoother again at each level. A simple cycle down and up the grid is referred to as a V-cycle. To obtain good convergence, the smoother and the coarse-grid correction process must complement each other to remove all components of the error.

A multigrid method has two phases: the setup phase and the solve phase. The setup phase consists of defining the coarse grids, interpolation operators, and coarse-grid operators for each of the coarse-grid levels. The solve phase consists of performing the multilevel cycles (i.e., iterations) until the desired convergence is obtained. In the scaling studies, we often time the setup phase and solve phase separately. Note that while multigrid methods may be used as linear solvers, they are more typically used as preconditioners for Krylov methods such as GMRES or conjugate gradient.

The challenge for multigrid methods on supercomputers is turning an efficient serial algorithm into a robust and scalable parallel algorithm. Good numerical properties need to be preserved when making algorithmic changes needed for parallelism. This non-trivial task affects all aspects of a multigrid algorithm, including coarsening, interpolation, and smoothing.

In the remainder of this section, we provide more details for the most commonly used solvers in *hypre* for which we perform our scaling study.

13.3.1 PFMG, SMG, and SysPFMG

PFMG [1, 12] is a semicoarsening multigrid method for solving scalar diffusion equations on logically rectangular grids discretized with up to 9-point stencils in 2D and up to 27-point stencils in 3D. It is effective for problems with variable coefficients and anisotropies that are uniform and grid-aligned throughout the domain. The solver automatically determines the "best" direction of semicoarsening, but the user may also control this manually. Interpolation is determined algebraically. The coarse-grid operators are also formed algebraically, either by Galerkin or by the non-Galerkin process described in [1]. The latter is available only for 5-point (2D) and 7-point (3D) problems, but maintains these stencil patterns on all coarse grids, reducing cost and improving performance. Relaxation options are either weighted Jacobi or red/black Gauss–Seidel. The solver can also be run in a mode that skips relaxation on certain grid levels when the problem is (or becomes) isotropic, further reducing cost and increasing performance.

PFMG also has two constant-coefficient modes, one where the entire stencil is constant throughout the domain, and another where the diagonal is allowed to vary. Both modes require significantly less storage and can also be somewhat faster than the full variable-coefficient solver, depending on the machine. The variable diagonal case is the most effective and flexible of the two modes. The non-Galerkin options here are similar to the variable case, but maintain the constant-coefficient format on all grid levels.

SMG [7, 9, 12, 20] is also a semicoarsening multigrid method for solving scalar diffusion equations on logically rectangular grids. It is more robust than PFMG, especially when the equations exhibit anisotropies that vary in either strength or direction across the domain, but it is also much more expensive per iteration. SMG coarsens in the z direction and uses a plane smoother. The xy plane solves in the smoother are approximated by employing one cycle of a 2D SMG method, which in turn coarsens in y and uses x-line smoothing. The plane and line solves are also used to define interpolation, and the solver uses Galerkin coarse-grid operators.

SysPFMG is a generalization of PFMG for solving systems of elliptic PDEs. Interpolation is defined only within the same variable using the same approach as PFMG, and the coarse-grid operators are Galerkin. The smoother is of nodal type and solves all variables at a given point simultaneously.

13.3.2 BoomerAMG

BoomerAMG [14] is the unstructured algebraic multigrid (AMG) solver in *hypre*. AMG is a particular type of multigrid method that is unique because it does not require an explicit grid geometry. This attribute greatly increases the types of problem that can be solved with multigrid because often the actual grid information may not be known or the grid may be highly unstructured. Therefore, in AMG the "grid" is simply the set of variables, and the coarsening and interpolation processes are determined entirely based on the entries of the matrix. For this reason, AMG is a rather complex algorithm, and it is challenging to design parallel coarsening and interpolation algorithms that combine good convergence, low computational complexity as well as low memory requirements. See [25], for example, for an overview of parallel AMG. Note that the AMG setup phase can be costly, compared to that of a geometric multigrid method. Classical coarsening schemes [8, 18] have led to slow coarsening, especially for 3D problems, resulting in large computational complexities per V-cycle, increased memory requirements and decreased scalability. In order to achieve scalable performance, one needs to use reduced complexity coarsening methods, such as HMIS and PMIS [23], which require distance-two interpolation operators, such as extended+i interpolation [22], or even more aggressive coarsening schemes, which need interpolation with an even longer range [24, 26]. The parallel implementation of long range interpolation schemes generally involves much more complicated and costly communication patterns than nearest neighbor interpolation. Additionally, communication requirements on coarser grid levels can become more costly as the stencil size typically increases with coarsening, which results in MPI tasks having many more neighbors (see, e.g., [13] for a discussion). The AMG solve phase consists largely of matrix-vector multiplies and the application of a (typically) inexpensive smoother, such as hybrid (symmetric) Gauss–Seidel, which applies sequential (symmetric) Gauss–Seidel locally on each core and uses delayed updates across cores. Note that hybrid smoothers depend on the number of cores as well as the distribution of data across tasks, and therefore one cannot expect to achieve exactly the same results or the same number of iterations when using different configurations. However, the number of iterations required to converge to the desired tolerance should be fairly close.

13.3.3 AMS

The Auxiliary-space Maxwell Solver (AMS) is an algebraic solver for electromagnetic diffusion problems discretized with Nedelec (edge) finite elements. AMS can be viewed as an AMG-type method with multiple coarse spaces, in each of which a BoomerAMG V-cycle is applied to a variationally constructed scalar/vector nodal problems. Unlike BoomerAMG, AMS requires some fine-grid information besides the matrix: the coordinates of the vertices and the list of edges in terms of their vertices (the so-called discrete gradient matrix), which allows it to be scalable and robust with respect to jumps in material coefficients. More details about the AMS algorithm and its performance can be found in [16].

13.4 Experimental Setup

For the results in this paper, we used version 2.7.1a of the *hypre* software library. In this section, we describe the machines and test runs used in our scaling studies.

13.4.1 Machine Descriptions

The *Dawn* machine is a Blue Gene/P system at LLNL. This system consists of 36,864 compute nodes, and each node contains a quad-core 850 MHz PowerPC 450 processor, bringing the total number of cores to 147,456. All four cores on a node have a common and shared access to the complete main memory of 4.0 GB. This guarantees uniform memory access (UMA) characteristics. All nodes are connected by a 3D torus network. We compile our code using IBM's C and OpenMP/C compilers and use IBM's MPICH2-based MPI implementation.

The *Hera* machine is a multicore/multi-socket Linux cluster at LLNL with 864 compute nodes connected by Infiniband network. Each compute node has four sockets, each with an AMD Quadcore (8356) 2.3 GHz processor. Each processor has its own memory controller and is attached to a quarter of the node's 32 GB memory. While a core can access any memory location, the non-uniform memory access (NUMA) times depend on the location of the memory. Each node runs CHAOS 4, a high-performance computing Linux variant based on Redhat Enterprise Linux. Our code is compiled using Intel's C and OpenMP/C compiler and uses MVAPICH for the MPI implementation.

13.4.2 Test Runs

Because we are presenting a scaling study, and not a convergence study, we chose relatively simple problems from a mathematical point of view. However, these problems are sufficient for revealing issues with scaling performance.

3D Laplace: A 3D Laplace equation with Dirichlet boundary conditions, discretized with seven-point finite differences on a uniform Cartesian grid.

3D Laplace System: A system of two 3D Laplace equations as above, with weak inter-variable coupling at each grid point. Each Laplacian stencil had a coefficient of 6 on the diagonal and -1 on the off-diagonals, and the inter-variable coupling coefficient was 10^{-5}.

3D Electromagnetic Diffusion: A simple 3D electromagnetic diffusion problem posed on a structured grid of the unit cube. The problem has unit conductivity and homogeneous Dirichlet boundary conditions and corresponds to the example code `ex15` from the *hypre* distribution.

We use the notation PFMG-*n*, CPFMG-*n*, SMG-*n*, SysPFMG, AMG-*n*, and AMS-*n* to represent conjugate gradient solvers preconditioned respectively by PFMG, constant-coefficient PFMG, SMG, SysPFMG, BoomerAMG, and AMS, where n signifies a local grid of dimension $n \times n \times n$ on each core. We use two different parameter choices for PFMG (and CPFMG), and denote them by appending a '-1' or '-2' to the name as follows:

- PFMG-*n*-1—Weighted Jacobi smoother and Galerkin coarse-grid operators;
- PFMG-*n*-2—Red/black GS smoother, non-Galerkin coarse-grid operators, and relaxation skipping.

For AMG, we use aggressive coarsening with multipass interpolation on the finest level, and HMIS coarsening with extended+i interpolation truncated to at most 4 elements per row on the coarser levels. The coarsest system is solved using Gaussian elimination. The smoother is one iteration of symmetric hybrid Gauss–Seidel. For AMS, we use the default parameters of `ex15` plus the ℓ_1^*-GS smoother from [3] (option `-rlx 4`).

For PFMG, CPFMG, SMG, and AMG, we solve the 3D Laplace problem with global grid size $N = np \times np \times np$, where $P = p^3$ is the total number of cores. For SysPFMG, we solve the 3D Laplace System problem with a local grid size of 40^3. For AMS, we solve the 3D Electromagnetic Diffusion problem, where here n^3 specifies the local number of finite elements on each core.

In the scaling studies, P ranged from 64 to 125,000 on Dawn and 64 to 4096 on Hera, with specific values for p given as follows:

- $p = 4, 8, 12, 16, 20, 24, 28, 32, 36, 40, 44, 48, 50$ on Dawn; and
- $p = 4, 8, 12, 16$ on Hera.

For the hybrid MPI/OpenMP runs, the same global problems were run, but they were configured as in the following table.

Machine	Threads	Problem size per MPI task	Number of MPI tasks
Dawn (smp)	4	$2n \times 2n \times n$	$p/2 \times p/2 \times p$
Hera (4 × 4)	4	$n \times 2n \times 2n$	$p \times p/2 \times p/2$
Hera (1 × 16)	16	$2n \times 2n \times 4n$	$p/2 \times p/2 \times p/4$

13.5 Scaling Studies

In this section, we present the scaling results. We first comment on the solver performance and conclude the section with comments on the times spent to set up the problems using *hypre*'s interfaces. For all solvers, iterations were stopped when the L^2 norm of the relative residual was smaller than 10^{-6}. The number of iterations are listed in Table 13.1.

In Fig. 13.1, we give results for PFMG using MPI only. Since communication latency speeds are several orders of magnitude slower than MFLOP speeds on todays architectures (Dawn included), communication costs tend to dominate for the smaller PFMG-16 problems. Because of this, the setup phase is slower than the solve phase, due primarily to the assumed partition and global partition components of the code. The global partition requires $O(P \log P)$ communications in the setup phase, while the current implementation of the assumed partition requires $O((\log P)^2)$ communications (the latter is easier to see in Fig. 13.2). It should be possible to reduce the communication overhead of the assumed partition algorithm in the setup phase to $O(\log P)$ by implementing a feature for coarsening the *box manager* in *hypre* (the box manager serves the role of the *distributed directory* in

Table 13.1 Iteration counts on Dawn (top two tables) and Hera (bottom table). For all solvers, iterations were stopped when the relative residual was smaller than 10^{-6}

P	PFMG-				CPFMG-				SMG-40	SysPFMG
	16-1	16-2	40-1	40-2	16-1	16-2	40-1	40-2		
64	9	9	10	10	9	9	9	9	5	9
512	9	10	10	11	9	9	10	10	5	9
1728	10	10	10	11	9	9	10	10	5	9
4096	10	10	10	11	10	9	10	10	5	9
8000	10	11	10	11	10	10	10	10	6	9
13824	10	11	11	11	10	10	10	10	6	9
21952	10	11	11	12	10	10	10	11	6	9
32768	10	11	11	12	10	10	10	10	6	9
46656	10	11	11	12	10	10	10	11	6	9
64000	10	11	11	12	10	10	10	10	6	9
85184	10	11	11	13	10	10	10	10	6	9
110592	10	11	11	13	10	10	10	11	6	
125000	10	11	11	13	10	10	10	12	6	

P	AMG-				AMS-	
	16	16 (smp)	40	40 (smp)	16	32
64	12		13			
512	14	14	15	16	9	10
1728	15		17		10	10
4096	15	15	18	18	10	11
8000	16		19		11	11
13824	17	16	21	20	11	11
21952	17		22		11	12
32768	18	17	22	22	11	13
46656	18		23		11	14
64000	19	18	24	23	12	13
85184	19		24		12	14
110592	19	19	24	24	12	14
125000	19		27		12	14

P	AMG-16			AMG-40		
	(16x1)	(4x4)	(1x16)	(16x1)	(4x4)	(1x16)
64	12	12	12	13	13	14
512	14	13	14	15	16	16
1728	15	14	14	17	17	18
4096	15	15	15	18	19	20

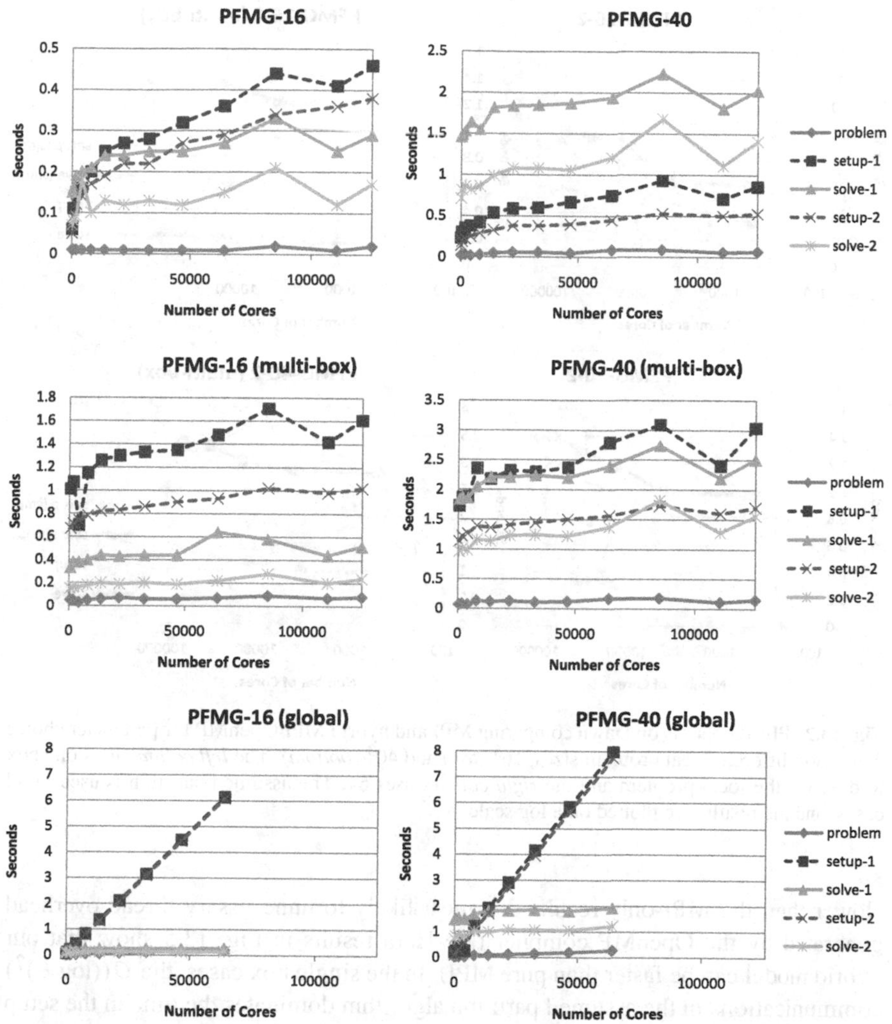

Fig. 13.1 PFMG results on Dawn for two different local problem sizes, 16^3 (*left*) and 40^3 (*right*). The *top* and *bottom rows* use one box to describe the local problem, while the *second row* splits the local problem into $64 = 4 \times 4 \times 4$ boxes. The *top two rows* use the assumed partition algorithm and the *bottom row* uses the global partition. Setup and solve phase times are given for two different parameter choices

[4]). For PFMG-40 with the assumed partition, the setup phase is cheaper than the solve phase for the single box case, and roughly the same cost for the multi-box case. For both multi-box cases, describing the data with 64 boxes leads to additional overhead in all phases, but the effect is more pronounced for the setup phase. The problem setup uses the `Struct` interface and is the least expensive component.

In Fig. 13.2, we give results for PFMG, comparing MPI to hybrid MPI/OpenMP. With the exception of the setup phase in the single box cases, the hybrid results are

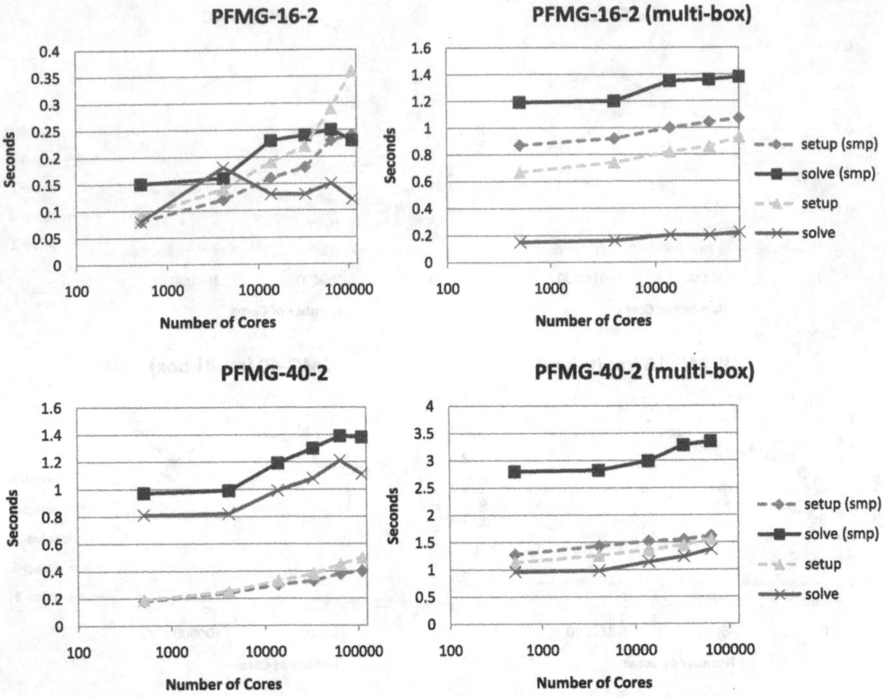

Fig. 13.2 PFMG results on Dawn comparing MPI and hybrid MPI/OpenMP for parameter choice 2 and two different local problem sizes, 16^3 (*top*) and 40^3 (*bottom*). The *left column* uses one box to describe the local problem and the *right column* uses 64. The assumed partition is used in all cases, and the results are plotted on a log scale

slower than the MPI-only results due most likely to unnecessary thread overhead generated by the OpenMP compiler (the Hera results in Fig. 13.5 show that our hybrid model can be faster than pure MPI). In the single box cases, the $O((\log P)^2)$ communications in the assumed partition algorithm dominates the time in the setup phase, so the pure MPI runs are slower than the hybrid runs due to the larger number of boxes to manage. This scaling trend is especially apparent in the PFMG-16-2 plot. For the multi-box cases, thread overhead is multiplied by a factor of 64 (each threaded loop becomes 64 threaded loops) and becomes the dominant cost.

The constant-coefficient CPFMG solver saves significantly on memory, but it was only slightly faster than the variable-coefficient PFMG case with nearly identical scaling results. The memory savings allowed us to run CPFMG-200-2 to solve a 1.049 trillion unknown problem on 131,072 cores ($64 \times 64 \times 32$) in 11 iterations and 83.03 seconds. Problem setup took 0.84 seconds and solver setup took 1.10 seconds.

In Fig. 13.3, we show results for SMG-40 and SysPFMG. We see that SMG is a much more expensive solver than PFMG, albeit more robust. We also see that the setup phase scales quite poorly, likely due to the $O((\log P)^3)$ number of coarse

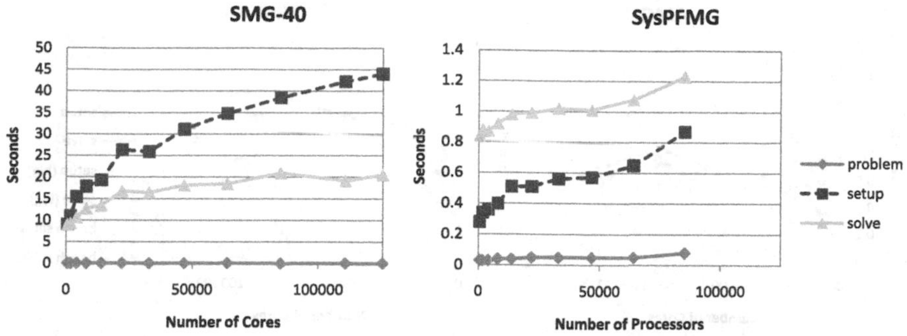

Fig. 13.3 SMG-40 and SysPFMG results on Dawn using the assumed partition algorithm

grids and our current approach for building a box manager on each grid level. See [12] for more discussion on expected scaling behavior of SMG (and PFMG).

For SysPFMG, we see that the scaling of both the setup and solve phases are not flat, even though the iteration counts in Table 13.1 are constant. However, this increase is the same increase seen for PFMG-40 at 85,184 cores, so we expect the times to similarly decrease at larger core counts. The problem setup uses the `SStruct` interface and is the least expensive component.

Figure 13.4 presents the performance of BoomerAMG-CG on Dawn. The two top figures show the performance obtained when using MPI only and the effect of using 32-bit integers versus 64-bit integers. The 32-bit version could only be used to up to 32,768 cores for the 40^3 Laplace problem. We also include results that were obtained using the Sequoia benchmark AMG [21], which is a stand-alone version of BoomerAMG, in which only those integers that are required to have a longer format were converted, an approach that was unfortunately too work extensive to be applied to all of *hypre*. The benchmark results are very close to the 32-bit integer results.

The two bottom figures show the effect of using a global partition when setting up the communication pattern and its linear dependence on the number of MPI tasks. The problem setup uses the IJ interface and takes very little time compared to solve and setup.

The middle figures show timings obtained using a hybrid OpenMP/MPI programming model with one MPI task per node using 4 OpenMP threads on Dawn. On Dawn, the use of OpenMP for AMG leads to larger run times than using MPI only and is therefore not recommended. While the solve phase of BoomerAMG is completely threaded, portions of it, like the multiplication of the transpose of the interpolation operator with a vector cannot be implemented as efficiently using OpenMP as MPI with our current data structure. Also, portions of the setup phase, like the coarsening and part of the interpolation, are currently not threaded, which also negatively affects the performance when using OpenMP.

On Hera, the use of a hybrid MPI/OpenMP versus an MPI only programming model yields different results, see Fig. 13.5. Since at most 4096 cores were available to us, we used the global partition. We compare the MPI only implementations with

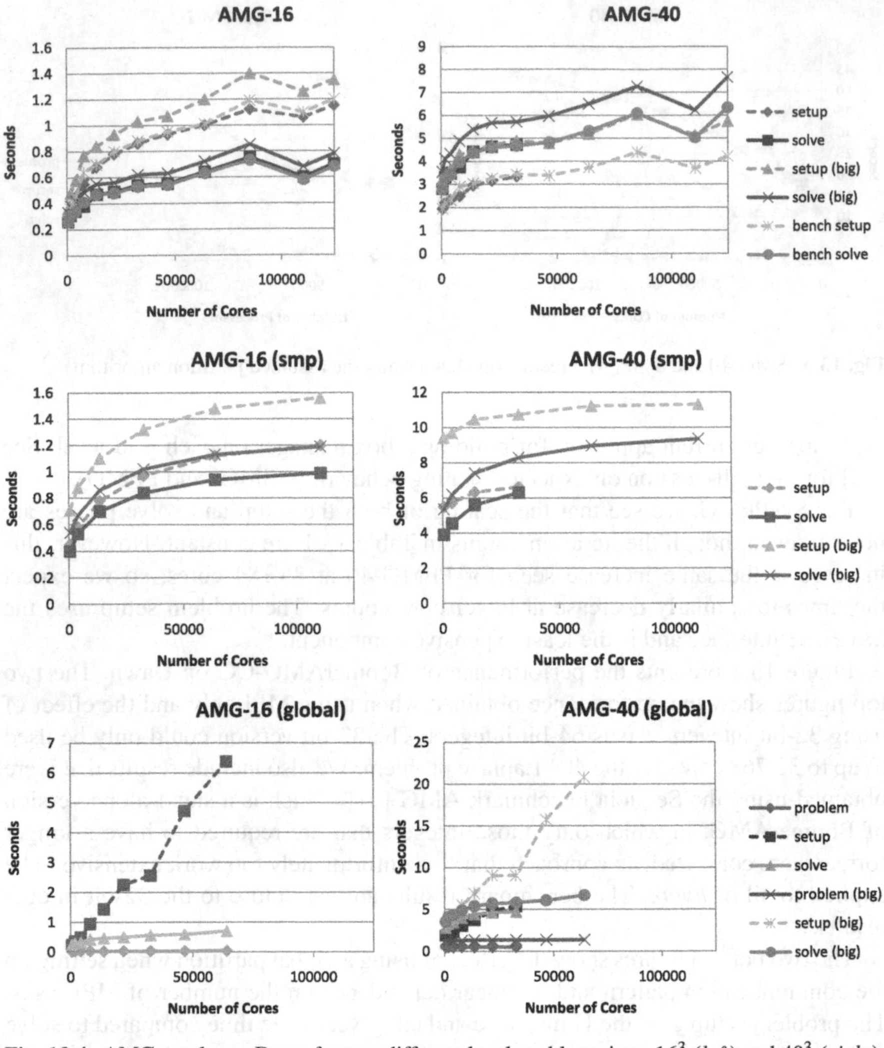

Fig. 13.4 AMG results on Dawn for two different local problem sizes, 16^3 (*left*) and 40^3 (*right*). The *top figures* show setup and solve times for the 32-bit version of *hypre*, the 64-bit version (big), and the AMG Sequoia benchmark (bench), using MPI only and assumed partitioning. The *middle figures* show times using the hybrid programming model MPI/OpenMP and assumed partitioning and include the problem setup times. The *bottom figures* use MPI only and global partitioning

16 MPI tasks per node (16x1) to a hybrid model using 4 MPI tasks with 4 OpenMP threads each (4x4), which is best adapted to the architecture, and a hybrid model using 1 MPI task with 16 OpenMP threads per node (1x16). For the smaller problem with 16^3 unknowns per core, we obtain the worst times using MPI only, followed by the 4x4 hybrid model. The best times are achieved using the 1x16 hybrid model, which requires the least amount of communication, since communication is very expensive compared to computation on Hera and communication dominates over

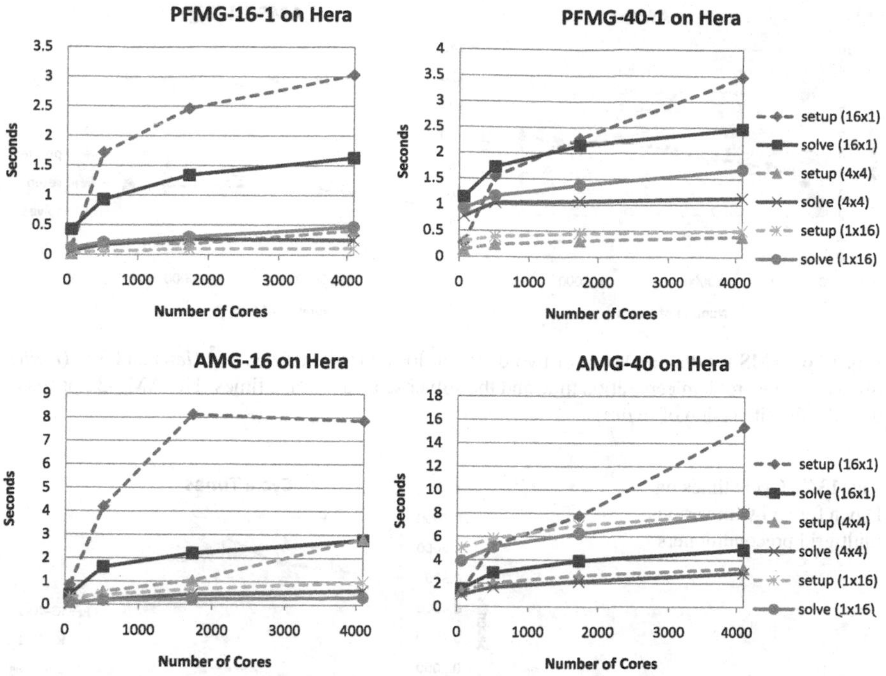

Fig. 13.5 Setup and solve times on Hera for PFMG and AMG, using an MPI only programming model (16x1), and two hybrid MPI/OpenMP programming models using 4 MPI tasks with 4 OpenMP threads each (4x4) and using 1 MPI task with 16 OpenMP threads (1x16) per node

computation for smaller problem sizes. The picture changes for the larger problem with 40^3 unknowns per core. Now the overall best times are achieved using the 4x4 hybrid model, which has less communication overhead than the MPI only model, but is not plagued by NUMA effects like the 1x16 hybrid model, where all memory is located in the first memory module, causing large access times and contention for the remaining 12 cores, see also [6].

In Fig. 13.6 we show the scalability results for CG with AMS preconditioner applied to the constant coefficient electromagnetic diffusion problem described in the previous section. We consider a coarse (AMS-16) and a 64-bit fine problem (AMS-32) with parallel setups corresponding to 16^3 and 32^3 elements per core. The respective largest problem sizes were around 1.5 and 12 billion on 125, 000 cores. Both cases were run with the assumed partition version of *hypre*.

Though not perfect, both the AMS setup and solve times in Fig. 13.6 show good parallel scalability, especially in the case of AMS-32 where the larger amount of local computations offsets better the communications cost. The slight growth in the solve times can be partially explained by the fact that the number of AMS-CG iterations varies between 9 to 12, for AMS-16, and 10 to 14, for AMS-32.

In Fig. 13.7, we show timings for one iteration (cycle) of a few solvers considered earlier. This removes the effect of the iteration counts given in Table 13.1 on the overall solve times. We see that the cycle time for AMG-40 is only slightly larger

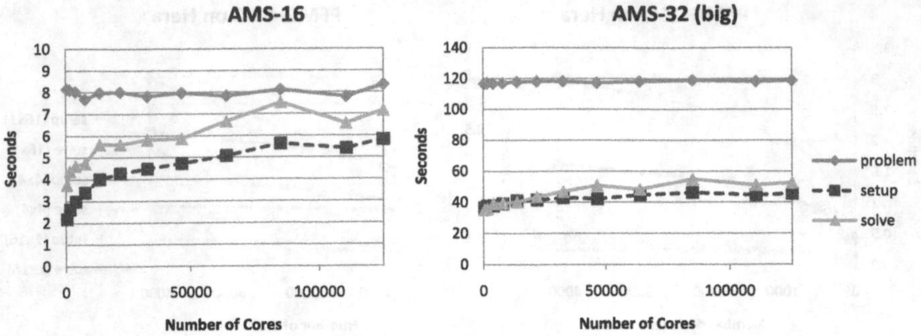

Fig. 13.6 AMS results on Dawn for two different local problem sizes, 16^3 (*left*) and 32^3 (*right*). Shown are the problem generation time and the solver setup and solve times. The AMS-32 problem uses the 64-bit version of *hypre*

Fig. 13.7 Cycle times on Dawn for CG with various multigrid preconditioners

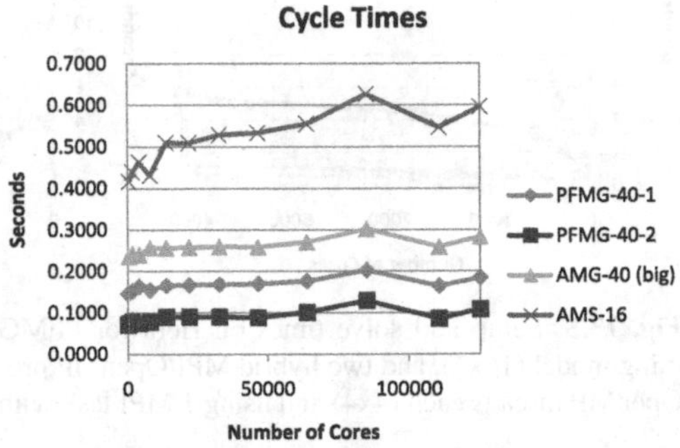

than PFMG-40-1, even though it is a fully unstructured code. This is mainly due to the dominant cost of communication, but it also suggests that improvements may be possible in the PFMG computational kernels where we should be able to take better advantage of structure. PFMG-40-2 is faster than PFMG-40-1 because it maintains 7-pt operators on all grid levels, reducing both communication and computation costs. AMS-16 is the slowest even though the grid is smaller, because it is solving a much harder problem that involves essentially four AMG V-cycle plus additional smoothing. It is interesting to note that all of the curves have almost the same shape, with a slight increase at 85,184 cores. This is most likely due to a poor mapping of the problem data to the hardware, which resulted in more costly long-distance communication.

We now comment on the performance improvements that can be achieved when using the struct interface and PFMG or CPFMG over the IJ interface and Boomer-AMG for suitable structured problems. For the smaller Laplace problem, PFMG-16-1 is about 2.5 times faster than the 32-version of AMG-16 and the AMG benchmark, and 3 times faster than the 64-bit version. The non-Galerkin version, PFMG-16-2, is about 3 to 4 times faster than AMG. For the larger problem, PFMG-40-2 is about 7 times faster than the 64-bit version of AMG and about 5 times faster than the benchmark.

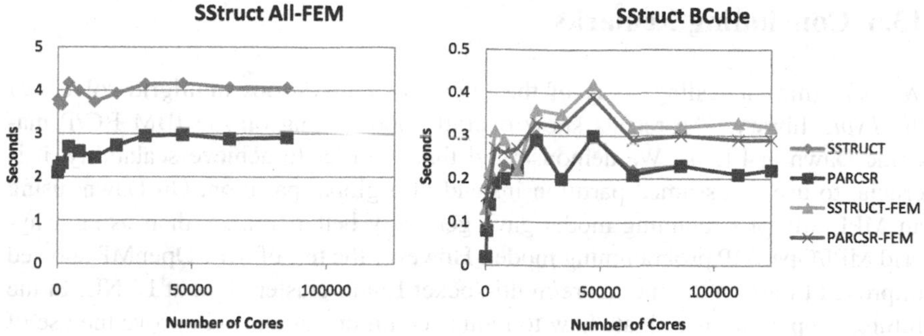

Fig. 13.8 Scaling results on Dawn for the `SStruct` interface, comparing the cost of setting up the `SSTRUCT` and `PARCSR` object types, and the cost of using a stencil-based or FEM-based approach

Finally, we comment on the times it takes to set up the problems via *hypre*'s interfaces. Figures 13.1, 13.3, and 13.4 showed that for many problems, the setup takes very little time compared to setup and solve times of the solvers. Note that the times for the problem setup via the IJ interface in Fig. 13.4 includes setting up the problem directly in the ParCSR data structure and then using the information to set up the matrix using the IJ data interface. Using the IJ interface directly for the problem setup takes only about half as much time.

The problem generation time in Fig. 13.6 corresponds to the assembly of the edge element Maxwell stiffness matrix and load vector, as well as the computation of the rectangular discrete gradient matrix and the nodal vertex coordinates needed for AMS. These are all done with the `sstruct` interface, using its finite element functionality for the stiffness matrix and load vector. Overall, the problem generation time scales very well. Though its magnitude may appear somewhat large, one should account that it also includes the (redundant) computation of all local stiffness matrices, and the penalty from the use of the 64-bit version of *hypre* in AMS-32.

In Fig. 13.8, we give results for the `sstruct` interface. The BCube stencil-based results involve one cell-centered variable and two parts connected by the `GridSetNeighborPart()` routine, while the BCube FEM-based results use a node-centered variable. The All-FEM results involve 7 different variable types (all of the supported types except for cell-centered) and three parts.

For the BCube example, we see that building the `SSTRUCT` and `PARCSR` objects takes about the same time for the stencil-based case, with `SSTRUCT` taking slightly longer in the FEM-based case. This is probably due to the extra communication required to assemble the structured part of the matrix for non-cell-centered variable types along part boundaries. This difference is even more pronounced in the All-FEM example. We also see from the BCube example that the FEM-based approach is more expensive than the stencil-based approach. This is due to the additional cost of assembling the matrix and also partly due to the fact that the FEM interface does not currently support assembling a box of stiffness matrices in one call to reduce overhead.

13.6 Concluding Remarks

We performed a scaling study of the interfaces and various multigrid solvers of the *hypre* library. The results show overall good scaling on the IBM BG/P machine Dawn at LLNL. We demonstrated that in order to achieve scalability, it is crucial to use an assumed partition instead of a global partition. On Dawn, using an MPI only programming model gave generally better timings than using a hybrid MPI/OpenMP programming model. However the use of MPI/OpenMP showed improved times on the multicore/multi-socket Linux cluster Hera at LLNL. In the future, we plan to investigate how to reduce communication, to improve the use of threads, as well as to employ more suitable data structures. Our goal is to achieve good performance on exascale computers, which are expected to have millions of cores with memories that are orders of magnitudes smaller than on current machines.

Acknowledgements This work performed under the auspices of the U.S. Department of Energy by Lawrence Livermore National Laboratory under Contract DE-AC52-07NA27344 (LLNL-JRNL-479591).

References

1. Ashby, S.F., Falgout, R.D.: A parallel multigrid preconditioned conjugate gradient algorithm for groundwater flow simulations. Nucl. Sci. Eng. **124**(1), 145–159 (1996) UCRL-JC-122359
2. Baker, A., Falgout, R., Gamblin, T., Kolev, T., Schulz, M., Yang, U.M.: Scaling algebraic multigrid solvers: On the road to exascale. In: Wittum, G. (ed.) Competence in High Performance Computing 2010. Springer, Berlin (2012). LLNL-PROC-463941
3. Baker, A.H., Falgout, R.D., Kolev, T.V., Yang, U.M.: Multigrid smoothers for ultra-parallel computing. SIAM J. Sci. Comput. **33**, 2864–2887 (2011). LLNL-JRNL-473191
4. Baker, A.H., Falgout, R.D., Yang, U.M.: An assumed partition algorithm for determining processor inter-communication. Parallel Comput. **32**(5–6), 394–414 (2006). UCRL-JRNL-215757
5. Baker, A.H., Gamblin, T., Schulz, M., Yang, U.M.: Challenges of scaling algebraic multigrid across modern multicore architectures. In: 25th IEEE International Symposium on Parallel and Distributed Processing, IPDPS 2011, Anchorge, AK, USA, 16–20 May, 2011 – Conference Proceedings, pp. 275–286, IEEE, 2011. LLNL-CONF-458074
6. Baker, A.H., Schulz, M., Yang, U.M.: On the performance of an algebraic multigrid solver on multicore clusters. In: Palma, J.M.L.M., et al. (ed.) VECPAR 2010. Lecture Notes in Computer Science, vol. 6449, pp. 102–115. Springer, Berlin (2011). http://vecpar.fe.up.pt/2010/papers/24.php
7. Baldwin, C., Brown, P.N., Falgout, R.D., Jones, J., Graziani, F.: Iterative linear solvers in a 2D radiation-hydrodynamics code: methods and performance. J. Comput. Phys. **154**, 1–40 (1999) UCRL-JC-130933
8. Brandt, A., McCormick, S.F., Ruge, J.W.: Algebraic multigrid (AMG) for sparse matrix equations. In: Evans, D.J. (ed.) Sparsity and Its Applications. Cambridge University Press, Cambridge (1984)
9. Brown, P.N., Falgout, R.D., Jones, J.E.: Semicoarsening multigrid on distributed memory machines. SIAM J. Sci. Comput. **21**(5), 1823–1834 (2000). Special issue on the Fifth Copper Mountain Conference on Iterative Methods. UCRL-JC-130720

10. Falgout, R., Jones, J., Yang, U.M.: Pursuing scalability for hypre's conceptual interfaces. ACM Trans. Math. Softw. **31**, 326–350 (2005)
11. Falgout, R.D.: An introduction to algebraic multigrid. Comput. Sci. Eng. **8**(6), 24–33 (2006). Special issue on Multigrid Computing. UCRL-JRNL-220851
12. Falgout, R.D., Jones, J.E.: Multigrid on massively parallel architectures. In: Dick, E., Riemslagh, K., Vierendeels, J. (eds.) Multigrid Methods VI. Lecture Notes in Computational Science and Engineering, vol. 14, pp. 101–107. Springer, Berlin (2000). Proc. of the Sixth European Multigrid Conference held in Gent, Belgium, September 27-30, 1999. UCRL-JC-133948
13. Gahvari, H., Baker, A.H., Schulz, M., Yang, U.M., Jordan, K.E., Gropp, W.: Modeling the performance of an algebraic multigrid cycle on HPC platforms. In: Proceedings of the International Conference on Supercomputing (ICS 2011), pp. 172–181. ACM, New York (2011)
14. Henson, V., Yang, U.: BoomerAMG: A parallel algebraic multigrid solver and preconditioner. Appl. Numer. Math. **41**, 155–177 (2002)
15. hypre: High performance preconditioners. http://www.llnl.gov/CASC/hypre/
16. Kolev, T., Vassilevski, P.: Parallel auxiliary space AMG for H(curl) problems. J. Comput. Math. **27**, 604–623 (2009). Special issue on Adaptive and Multilevel Methods in Electromagnetics. Also available as a Lawrence Livermore National Laboratory technical report UCRL-JRNL-237306
17. Meier, U., Sameh, A.: The behavior of conjugate gradient algorithms on a multivector processor with a hierarchical memory. J. Comput. Appl. Math. **24**, 13–32 (1988)
18. Ruge, J.W., Stüben, K.: Algebraic multigrid (AMG). In: McCormick, S.F. (ed.) Multigrid Methods. Frontiers in Applied Mathematics, vol. 3, pp. 73–130. Philadelphia, SIAM (1987)
19. Saad, Y.: Iterative Methods for Sparse Linear Systems, 2nd edn. Society for Industrial and Applied Mathematics, Philadelphia (2003)
20. Schaffer, S.: A semi-coarsening multigrid method for elliptic partial differential equations with highly discontinuous and anisotropic coefficients. SIAM J. Sci. Comput. **20**(1), 228–242 (1998)
21. Sequoia: ASC sequoia benchmark codes. https://asc.llnl.gov/sequoia/benchmarks/
22. Sterck, H.D., Falgout, R.D., Nolting, J.W., Yang, U.M.: Distance-two interpolation for parallel algebraic multigrid. Numer. Linear Algebra Appl. **15**(2–3), 115–139 (2008). Special issue on Multigrid Methods. UCRL-JRNL-230844
23. Sterck, H.D., Yang, U.M., Heys, J.J.: Reducing complexity in parallel algebraic multigrid preconditioners. SIAM J. Matrix Anal. Appl. **27**(4), 1019–1039 (2006)
24. Stüben, K.: Algebraic multigrid (AMG): an introduction with applications. In: Hackbusch, U., Oosterlee, C., Schueller, A. (eds.) Multigrid. Academic Press, San Diego (2000)
25. Yang, U.: Parallel algebraic multigrid methods – high performance preconditioners. In: Bruaset, A., Tveito, A. (eds.) Numerical Solution of Partial Differential Equations on Parallel Computers, vol. 51, pp. 209–236. Springer, Berlin (2006)
26. Yang, U.M.: On long range interpolation operators for aggressive coarsening. Numer. Linear Algebra Appl. **17**, 453–472 (2010). LLNL-JRNL-417371

Chapter 14
A Riemannian Dennis-Moré Condition

Kyle A. Gallivan, Chunhong Qi, and P.-A. Absil

Abstract In this paper, we generalize from Euclidean spaces to Riemannian manifolds an important result in optimization that guarantees Riemannian quasi-Newton algorithms converge superlinearly.

14.1 Introduction

Roughly speaking, a manifold is a generalization of the Euclidean space $\mathbb{R}^n$ on which the notion of a differentiable scalar field still exists. One can think of a nonlinear manifold as a smooth, curved surface, even though this simple picture does not fully do justice to the generality of the concept. Retaining the notion of differentiability opens the way for preserving concepts such as gradient vector fields and derivatives of vector fields, which are instrumental in many well-known optimization methods in $\mathbb{R}^n$, such as steepest descent, Newton, trust regions or conjugate gradients.

Optimization on manifolds can be intuitively thought of as unconstrained optimization over a constrained search space. As such, optimization algorithms on manifolds are not fundamentally different from classical algorithms for unconstrained optimization in $\mathbb{R}^n$. Indeed, new optimization algorithms on manifolds are often obtained by starting from an algorithm for unconstrained optimization in $\mathbb{R}^n$, extracting the underlying concepts, and rewriting them in such a way that they are well-defined on abstract manifolds. Generally speaking, applying the techniques of optimization on manifolds to a given computational problem involves the following steps. First, one needs to rephrase the problem as an optimization problem on

K.A. Gallivan (✉) · C. Qi
Department of Mathematics, Florida State University, Tallahassee, FL, 32312, USA
e-mail: gallivan@math.fsu.edu

C. Qi
e-mail: cqi@math.fsu.edu

P.-A. Absil
Department of Mathematical Engineering, ICTEAM Institute, Université catholique de Louvain, 1348 Louvain-la-Neuve, Belgium

M.W. Berry et al. (eds.), *High-Performance Scientific Computing*,
DOI 10.1007/978-1-4471-2437-5_14, © Springer-Verlag London Limited 2012

a manifold. Clearly, this is not possible for all problems, but examples abound [2, 7, 9]. Second, one needs to pick an optimization method, typically from the several classical optimization methods that have been formulated and analyzed for manifold search spaces; these include line-search methods [12], conjugate gradients [11], BFGS [8, 10], various direct-search methods [6], and trust-region methods [1, 4].

The final step is to turn the generic optimization method into a practical numerical algorithm. This entails choosing a representation of the manifold, e.g., encoding via a particular quotient manifold or embedded submanifold, and providing numerical expressions for a handful of differential-geometric objects, such as a Riemannian metric and a retraction. The compartmentalization of the representation of the elements of the manifold, the differential-geometric objects and the algorithm that uses them lends itself to the development of very general and quite powerful software. Generic prototype implementation of algorithms on a Riemannian manifold as well as more specific implementations exploiting the structure of particular problems and manifolds can be obtained, for example, from http://www.math.fsu.edu/~cbaker/GenRTR.

Some basic intuition behind the adaptation of algorithms for unconstrained optimization in $\mathbb{R}^n$ can be seen by considering the fact that many have a basic step of $x_{k+1} = x_k + \alpha_k p_k$ where the direction vector p_k may be determined first followed by a one-dimensional search to set the step α_k as in a line search method, or $\alpha_k p_k$ may be set by considering a local constrained optimization of a simplified model of the cost function as in a trust-region method. In either case, the main concern is the ability to generalize the notion of motion for some distance on a line given by a direction vector. Hence, much of the initial manifold work centered around the evaluation of geodesics with the associated concern over excessive computational cost.

As an example, consider Newton's method for finding a stationary point of a differentiable function f. In $\mathbb{R}^n$, the method reads

$$x_+ = x - (\operatorname{Hess} f(x))^{-1} \operatorname{grad} f(x),$$

where x is the current iterate, x_+ is the new iterate, $\operatorname{grad} f(x) = \left[\partial_1 f(x) \dots \partial_n f(x)\right]^T$ is the gradient of f at x and $\operatorname{Hess} f(x)$ is the Hessian matrix of f at x defined by $(\operatorname{Hess} f(x))_{ij} = \partial_i \partial_j f(x)$. When f is a function on a nonlinear Riemannian manifold $\mathcal{M}$, most of these operations become undefined. However, the notion of a gradient still exists on an abstract Riemannian manifold. Given a smooth scalar field f on a Riemannian manifold $\mathcal{M}$ with Riemannian metric g defined at each $x \in \mathcal{M}$, denoted g_x, the gradient of f at x, denoted by $\operatorname{grad} f(x)$, is defined as the unique element of $T_x\mathcal{M}$, the tangent space of x, that satisfies

$$g_x(\operatorname{grad} f(x), \xi) = \mathrm{D} f(x)[\xi], \quad \forall \xi \in T_x\mathcal{M}. \tag{14.1}$$

where $\mathrm{D} f(x)[\xi]$ is the Riemannian directional derivative. Therefore, if one sees the Newton method as iteration that defines x_+ as $x + \eta_x$ where η_x is the vector along which the derivative of the gradient is equal to the negative of the gradient, one is led to the following Riemannian Newton equation:

$$\nabla_{\eta_x} \operatorname{grad} f = -\operatorname{grad} f(x)$$

where ∇ is the Levi-Civita connection. When taking a step on a manifold $\mathcal{M}$ from a point $x \in \mathcal{M}$ along a vector $\eta_x \in T_x\mathcal{M}$, it is natural to think about following the geodesic curve γ with initial velocity η_x and define the new point as $\gamma(1)$. This yields the update $x_+ = \operatorname{Exp}_x(\eta_x)$, where Exp denotes the Riemannian exponential. In the 1990s, this was considered "the" Newton iteration on manifolds; see, e.g., [11].

However, whereas geodesics admit closed-form expressions for some specific manifolds, in general, they are the solution of an ordinary differential equation, and are thus costly to compute accurately. Fortunately, in most optimization algorithms one is content with first-order approximations of the geodesic. This prompted Adler et al. [3] to introduce the concept of retraction.

Quite similarly, when one has to subtract two tangent vectors ξ_x and ξ_y at two different points x and $y = \operatorname{Exp}_x(\eta_x)$, it is natural to think about parallel translating one tangent vector to the foot of the other along the curve $t \mapsto \operatorname{Exp}_x(t\eta_x)$. Here again, apart from some specific manifolds where parallel translation admits a closed-form expression, in general, parallel translation requires solving an ordinary differential equation. This prompted the relaxation of the idea and the introduction of the concept of *vector transport*, of which parallel translation is a particular instance [2]. The definition below invokes the Whitney sum $T\mathcal{M} \oplus T\mathcal{M}$, which is defined as the set of all ordered pairs of tangent vectors with same foot.

Definition 14.1 A *vector transport* on a manifold $\mathcal{M}$ is a smooth mapping

$$T\mathcal{M} \oplus T\mathcal{M} \to T\mathcal{M} : (\eta_x, \xi_x) \mapsto \mathcal{T}_{\eta_x}(\xi_x) \in T\mathcal{M}$$

satisfying the following properties for all $x \in \mathcal{M}$.

- (Associated retraction) There exists a retraction R, called the *retraction associated with* $\mathcal{T}$, such that the following diagram commutes:

$$\begin{array}{ccc} (\eta_x, \xi_x) & \xrightarrow{\;\mathcal{T}\;} & \mathcal{T}_{\eta_x}(\xi_x) \\ \big\downarrow & & \big\downarrow \pi \\ \eta_x & \xrightarrow[R]{} & \pi(\mathcal{T}_{\eta_x}(\xi_x)) \end{array}$$

  where $\pi(\mathcal{T}_{\eta_x}(\xi_x))$ denotes the foot of the tangent vector $\mathcal{T}_{\eta_x}(\xi_x)$.
- (Consistency) $\mathcal{T}_{0_x}\xi_x = \xi_x$ for all $\xi_x \in T_x\mathcal{M}$;
- (Linearity) $\mathcal{T}_{\eta_x}(a\xi_x + b\zeta_x) = a\mathcal{T}_{\eta_x}(\xi_x) + b\mathcal{T}_{\eta_x}(\zeta_x)$.

The first point in Definition 14.1 means that $\mathcal{T}_{\eta_x}\xi_x$ is a tangent vector in $T_{R_x(\eta_x)}\mathcal{M}$, where R is the retraction associated with $\mathcal{T}$. When it exists, $(\mathcal{T}_{\eta_x})^{-1}(\zeta_{R_x(\eta_x)})$ belongs to $T_x\mathcal{M}$. If η and ξ are two vector fields on $\mathcal{M}$, then $(\mathcal{T}_\eta)^{-1}\xi$ is naturally defined as the vector field satisfying

$$\left((\mathcal{T}_\eta)^{-1}\xi\right)_x = (\mathcal{T}_{\eta_x})^{-1}(\xi_{R_x(\eta_x)}).$$

It was shown in [2, Sect. 8.2.1] that when any vector transport is used in an approximate Newton method to find zeros of functions defined on a manifold where the Jacobian (or Hessian if in an optimization context) is approximated by finite differences, the resulting algorithm enjoys convergence properties akin to those of approximate Newton method in $\mathbb{R}^n$. As with the introduction of retraction to replace the exponential map, replacing parallel translation by the more general, and sometimes more efficient, concept of vector transport is a key part of developing efficient Riemannian optimization algorithms.

Recently, a Riemannian BFGS method (RBFGS) based on vector transport was developed and a convergence theory derived in the Ph.D. thesis of C. Qi [10]. A key element of that theory is proving that RBFGS has superlinear convergence. The proof relies on a Riemannian version of the Dennis-Moré Condition on Euclidean spaces [5, Theorem 8.2.4].

In this paper, we generalize the Dennis-Moré Condition to Riemannian manifolds to give conditions when it is guaranteed the basic Riemannian quasi-Newton algorithm $x_{k+1} = R_{x_k}(\eta_k)$, where $\eta_k = -B_k^{-1}F(x_k)$ and B_k is a linear operator on $T_{x_k}\mathcal{M}$, converges superlinearly.

14.2 The Riemannian Dennis-Moré Condition

In the discussions that follow, coordinate expressions in a neighborhood and in tangent spaces are used. For elements of the manifold, $v \in \mathcal{M}$, $\hat{v} \in \mathbb{R}^d$ will denote the coordinates defined by a chart ϕ over a neighborhood $\mathcal{U}$, i.e., $\hat{v} = \phi(v)$ for $v \in \mathcal{U}$. Coordinate expressions, $\hat{F}(x)$, for elements, $F(x)$, of a vector field F on $\mathcal{M}$ are written in terms of the canonical basis of the associated tangent space, $T_x\mathcal{M}$ via the coordinate vector fields defined by the chart ϕ (see, e.g., [2, Sect. 3.5]). The main result is stated as Theorem 14.1, but several preparatory lemmas are needed, some of which are interesting for their own sake.

Lemma 14.1 *Let $\mathcal{M}$ be a Riemannian manifold and $\mathcal{U}$ be a compact coordinate neighborhood in $\mathcal{M}$, and let the hat denote coordinate expressions. Then there are $c_2 > c_1 > 0$ such that, for all $x, y \in \mathcal{U}$, we have*

$$c_1\|\hat{x} - \hat{y}\| \le \operatorname{dist}(x, y) \le c_2\|\hat{x} - \hat{y}\|,$$

where $\|\cdot\|$ denotes the Euclidean norm.

Proof Proof of the first inequality:

Let $\Gamma_{\hat{x},\hat{y}}$ be the set of all smooth curves $\hat{\gamma}$ with $\hat{\gamma}(0) = \hat{x}$ and $\hat{\gamma}(1) = \hat{y}$. We have

$$\begin{aligned}\operatorname{dist}(x, y) &= \inf_{\hat{\gamma}\in\Gamma_{\hat{x},\hat{y}}} \int_0^1 \sqrt{\dot{\hat{\gamma}}(t)^T \hat{G}(\hat{\gamma}(t))\dot{\hat{\gamma}}(t)}\,\mathrm{d}t \\ &\geq \sqrt{\lambda_{\min}} \inf_{\hat{\gamma}\in\Gamma_{\hat{x},\hat{y}}} \int_0^1 \sqrt{\dot{\hat{\gamma}}(t)^T \dot{\hat{\gamma}}(t)}\,\mathrm{d}t \\ &\geq \sqrt{\lambda_{\min}} \|\hat{y} - \hat{x}\|,\end{aligned}$$

where $\hat{G}(\hat{v})$ is the matrix expression of the Riemannian metric on $T_v\mathcal{M}$ (see, e.g., [2, (3.29)]) and where

$$\lambda_{\min} = \min_{\hat{x}\in\hat{\mathscr{U}}} \lambda_{\min}\big(\hat{G}(\hat{x})\big).$$

Proof of the second inequality:
Taking $\hat{\gamma}(t) = \hat{x} + t(\hat{y} - \hat{x})$, we have

$$\begin{aligned}\operatorname{dist}(x, y) &\leq \int_0^1 \sqrt{\dot{\hat{\gamma}}(t)^T \hat{G}_{\hat{\gamma}(t)}\dot{\hat{\gamma}}(t)}\,\mathrm{d}t \leq \sqrt{\lambda_{\max}} \int_0^1 \sqrt{\dot{\hat{\gamma}}(t)^T \dot{\hat{\gamma}}(t)}\,\mathrm{d}t \\ &= \sqrt{\lambda_{\max}} \|\hat{x} - \hat{y}\|,\end{aligned}$$

where

$$\lambda_{\max} = \max_{\hat{x}\in\mathscr{U}} \lambda_{\max}\big(\hat{G}(\hat{x})\big).$$

We have the proof by taking $c_1 = \sqrt{\min_{\hat{x}\in\hat{\mathscr{U}}} \lambda_{\min}(\hat{G}(\hat{x}))}$ and $c_2 = \sqrt{\lambda_{\max}}$. □

Lemma 14.2 *Let $\mathcal{M}$ be a Riemannian manifold endowed with a vector transport $\mathscr{T}$ and an associated retraction R, and let $x_* \in \mathcal{M}$. Let F be a smooth vector field on $\mathcal{M}$. Then there is a neighborhood $\mathscr{U}$ of x_* and $L > 0$ s.t., $\forall x, y \in \mathscr{U}$:*

$$\big|\|\mathscr{T}^{-1}_{R_y^{-1}x} F(x)\|_y^2 - \|F(x)\|_x^2\big| \leq L\|F(x)\|_x^2 \operatorname{dist}(x, y),$$

where $\|F(v)\|_v$ denotes the norm in $T_v\mathcal{M}$ defined by the Riemannian metric.

Proof Let $L(y, x)$ denote $\mathscr{T}^{-1}_{R_y^{-1}x}$. We work in a coordinate chart and let the hat denote the coordinate expressions. We have

$$\big|\|\mathscr{T}^{-1}_{R_y^{-1}x} F(x)\|_y^2 - \|F(x)\|_x^2\big| = \big|\hat{F}(x)^T\big(\hat{L}(\hat{y},\hat{x})^T\hat{G}(\hat{y})\hat{L}(\hat{y},\hat{x}) - \hat{G}(\hat{x})\big)\hat{F}(x)\big| \tag{14.2}$$

$$\leq \|\hat{F}(x)\|^2\|H(\hat{y},\hat{x})\| \tag{14.3}$$

$$\leq c_1\|\hat{F}(x)\|^2\|\hat{y} - \hat{x}\| \tag{14.4}$$

$$\leq c_2\|F(x)\|_x^2 \operatorname{dist}(x, y), \tag{14.5}$$

where $H(\hat{y},\hat{x}) = \hat{L}(\hat{y},\hat{x})^T\hat{G}(\hat{y})\hat{L}(\hat{y},\hat{x}) - \hat{G}(\hat{x})$, $\|\hat{F}(x)\|$ denotes the classical Euclidean norm of $\hat{F}(x) \in \mathbb{R}^d$, where d is the dimension of $\mathcal{M}$, and $\|H(\hat{y},\hat{x})\|$ denotes the induced matrix norm (spectral norm). To get Eq. (14.4), take $\mathcal{U}$ bounded and observe that H is smooth and that $H(\hat{x},\hat{x}) = 0$ for all $\hat{x}$. To get Eq. (14.5), use Lemma 14.1. □

Lemma 14.3 *Under the assumptions of Lemma* 14.2, *there is a neighborhood* $\mathcal{U}$ *of* x_* *and* $L' > 0$ *s.t.*, $\forall x, y \in \mathcal{U}$:

$$\left| \|\mathcal{T}^{-1}_{R_y^{-1}x} F(x)\|_y - \|F(x)\|_x \right| \le L' \|F(x)\|_x \mathrm{dist}(x,y). \tag{14.6}$$

Proof If $\|\mathcal{T}^{-1}_{R_y^{-1}x} F(x)\|_y + \|F(x)\|_x = 0$, then both sides of Eq. (14.6) are zero and the claim holds. Otherwise,

$$\begin{aligned}
\left| \|\mathcal{T}^{-1}_{R_y^{-1}x} F(x)\|_y - \|F(x)\|_x \right| &= \frac{\left| \|\mathcal{T}^{-1}_{R_y^{-1}x} F(x)\|_y^2 - \|F(x)\|_x^2 \right|}{\|\mathcal{T}^{-1}_{R_y^{-1}x} F(x)\|_y + \|F(x)\|_x} \\
&\le \frac{L\|F(x)\|^2 \mathrm{dist}(x,y)}{c_3\|F(x)\|} \\
&\le L' \|F(x)\|_x \mathrm{dist}(x,y).
\end{aligned}$$

□

Definition 14.2 (Nondegenerate zero) Let F be a smooth vector field on a Riemannian manifold $\mathcal{M}$. A point $x_* \in \mathcal{M}$ is termed a *nondegenerate zero* of F if $F(x_*) = 0$ and $\nabla_{\xi_{x_*}} F \neq 0$, $\forall \xi_{x_*} \neq 0 \in T_{x_*}\mathcal{M}$ for some (and thus all, see [2, p. 96]) affine connection ∇ on $\mathcal{M}$.

Lemma 14.4 (Lemma 7.4.7, [2]) *Let* $\mathcal{M}$ *be a Riemannian manifold with Levi-Civita connection* ∇. *Let* $x \in \mathcal{M}$, *let* $\mathcal{U}$ *be a normal neighborhood of* x, *and let* ζ *be a* C^1 *tangent vector field on* $\mathcal{M}$, *then, for all* $y \in \mathcal{U}$,

$$P_\gamma^{0\leftarrow 1}\zeta_y = \zeta_x + \nabla_\xi \zeta + \int_0^1 \left(P_\gamma^{0\leftarrow\tau}\nabla_{\gamma'(\tau)}\zeta - \nabla_\xi\zeta\right) \mathrm{d}\tau,$$

where γ *is the unique geodesic in* $\mathcal{U}$ *satisfying* $\gamma(0) = x$ *and* $\gamma(1) = y$, $P_\gamma^{b\leftarrow a}$ *denotes parallel transport along* $\gamma(t)$ *from* a *to* b, *and* $\xi = \mathrm{Exp}_x^{-1} y = \gamma'(0)$.

Lemma 14.5 *Let* F *be a smooth vector field on a Riemannian manifold* $\mathcal{M}$. *Let* $x_* \in \mathcal{M}$ *be a nondegenerate zero of* F, *then there exists a neighborhood* $\mathcal{U}$ *of* x_* *and* $c_0, c_1 > 0$ *such that, for all* $x \in \mathcal{U}$,

$$c_0 \,\mathrm{dist}(x, x_*) \le \|F(x)\| \le c_1 \,\mathrm{dist}(x, x_*). \tag{14.7}$$

Proof Let $\mathbb{D}F(x)$ denote the linear transformation of $T_x\mathcal{M}$ defined by $\mathbb{D}F(x)[\xi_x] = \nabla_{\xi_x} F$, $\forall \xi_x \in T_x\mathcal{M}$, where ∇ denotes the Levi-Civita connection. Let $\mathcal{U}$ be a normal neighborhood of x_* and, for all $x \in \mathscr{U}$, let γ_x denote the unique geodesic in $\mathscr{U}$ satisfying $\gamma_x(0) = x_*$ and $\gamma_x(1) = x$.

From Taylor (Lemma 14.4), it follows that

$$P_{\gamma_x}^{0\leftarrow 1} F(x) = \mathbb{D}F(x_*)\big[\gamma_x'(0)\big] + \int_0^1 \big(P_{\gamma_x}^{0\leftarrow\tau} \mathbb{D}F(\gamma_x(\tau))\big[\gamma_x'(\tau)\big] - \mathbb{D}F(x_*)\big[\gamma_x'(0)\big]\big)\, d\tau. \quad (14.8)$$

Since F is smooth and since $\|\gamma_x'(\tau)\| = \text{dist}(x_*, x)$, $\forall\tau \in [0, 1]$, we have the following bound for the integral:

$$\begin{aligned} &\left\| \int_0^1 \big(P_{\gamma_x}^{0\leftarrow\tau} \mathbb{D}F(\gamma_x(\tau))\big[\gamma_x'(\tau)\big] - \mathbb{D}F(x_*)\big[\gamma_x'(0)\big]\big)\, d\tau \right\| \\ &= \left\| \int_0^1 \big(P_{\gamma_x}^{0\leftarrow\tau} \circ \mathbb{D}F(\gamma_x(\tau)) \circ P_{\gamma_x}^{\tau\leftarrow 0} - \mathbb{D}F(x_*)\big)\big[\gamma_x'(0)\big]\, d\tau \right\| \\ &\le \varepsilon(\text{dist}(x_*, x))\text{dist}(x_*, x), \end{aligned}$$

where $\lim_{t\to 0} \varepsilon(t) = 0$.

Since $\mathbb{D}F(x_*)$ is nonsingular, there exists c_0, c_1 such that

$$2c_0\|\xi_{x_*}\| \le \|\mathbb{D}F(x_*)[\xi_{x_*}]\| \le \frac{1}{2}c_1\|\xi_{x_*}\|, \quad \forall\xi_{x_*} \in T_{x_*}M. \quad (14.9)$$

Take $\mathscr{U}$ sufficiently small such that $\varepsilon(\text{dist}(x_*, x)) < c_0$ and $< \frac{1}{2}c_1$ for all $x \in \mathscr{U}$.

Applying Eq. (14.8) yields

$$\begin{aligned} \|F(x)\| = \big\|P_{\gamma_x}^{0\leftarrow 1} F(x)\big\| &\le \frac{1}{2}c_1\text{dist}(x_*, x) + \frac{1}{2}c_1\text{dist}(x_*, x) \\ &= c_1\text{dist}(x_*, x), \quad \text{for all } x \in \mathscr{U} \end{aligned}$$

and

$$\begin{aligned} \|F(x)\| = \big\|P_{\gamma_x}^{0\leftarrow 1} F(x)\big\| &\ge 2c_0\text{dist}(x_*, x) - c_0\text{dist}(x_*, x) \\ &= c_0\text{dist}(x_*, x), \quad \text{for all } x \in \mathscr{U}. \end{aligned}$$

□

Lemma 14.6 *Let F be a smooth vector field on a Riemannian manifold $\mathscr{M}$ endowed with a vector transport $\mathscr{T}$ and associated retraction R. Let $x_* \in \mathscr{M}$ be a nondegenerate zero of F. Then there exists a neighborhood $\mathscr{V}$ of $0_{x_*} \in T_{x_*}\mathscr{M}$ and $c_0, c_1 > 0$ such that, for all $\xi \in \mathscr{V}$,*

$$c_0\|\xi\| \le \|\mathscr{T}_\xi^{-1}\big(F(R_{x_*}(\xi))\big)\| \le c_1\|\xi\|. \quad (14.10)$$

Proof Let $G(\xi) = \mathscr{T}_\xi^{-1}(F(R_{x_*}(\xi)))$ and $E(\varepsilon) = G(\varepsilon\xi)$. Let $\widetilde{\mathbb{D}}F(x)$ denote the derivative at 0_x of the function $T_x\mathscr{M} \to T_x\mathscr{M} : \eta \mapsto \mathscr{T}_\eta^{-1}F(R_x(\eta))$. We have

$$\mathscr{T}_\xi^{-1}\big(F(R_{x_*}(\xi))\big) = E(1)$$

$$= E(0) + E'(0) + \int_0^1 E'(\tau) - E'(0)\, \mathrm{d}\tau \tag{14.11}$$

$$= E(0) + \mathrm{D}G(0)\xi + \int_0^1 [\mathrm{D}G(\tau\xi) - \mathrm{D}G(0)]\xi\, \mathrm{d}\tau \tag{14.12}$$

$$= 0 + \widetilde{\mathbb{D}}F(x_*)\xi + \int_0^1 [\mathrm{D}G(\tau\xi) - \mathrm{D}G(0)]\xi\, \mathrm{d}\tau. \tag{14.13}$$

The above Eq. (14.11) follows from the fundamental theorem $E(1) - E(0) = \int_0^1 E'(\tau)\,\mathrm{d}\tau$, and Eq. (14.12) comes by the chain rule. Observe that G is a function from $T_{x_*}\mathscr{M}$ to $T_{x_*}\mathscr{M}$, which are vector spaces, thus $\mathrm{D}G$ is the classical derivative of G. To get Eq. (14.13), observe that $E(0) = \mathscr{T}_{0_{x_*}}^{-1}(F(R_{x_*}(0_{x_*}))) = F(x_*) = 0$.

It can be checked that $\widetilde{\mathbb{D}}$ is an affine connection. Hence, since x_* is a nondegenerate zero of F, $\widetilde{\mathbb{D}}F(x_*)$ is invertible. We have

$$\|\xi\| = \left\|\widetilde{\mathbb{D}}F(x_*)^{-1}\widetilde{\mathbb{D}}F(x_*)\xi\right\| \le \left\|\widetilde{\mathbb{D}}F(x_*)^{-1}\right\|\left\|\widetilde{\mathbb{D}}F(x_*)\xi\right\|,$$

i.e.

$$\left\|\widetilde{\mathbb{D}}F(x_*)\xi\right\| \ge \frac{\|\xi\|}{\|\widetilde{\mathbb{D}}F(x_*)^{-1}\|}. \tag{14.14}$$

From Eq. (14.13), we have

$$\begin{aligned}
\left\|\mathscr{T}_\xi^{-1}\big(F(R_{x_*}(\xi))\big)\right\| &\ge \left\|\widetilde{\mathbb{D}}F(x_*)\xi\right\| - \left\|\int_0^1 [\mathrm{D}G(\tau\xi) - \mathrm{D}G(0)]\xi\, \mathrm{d}\tau\right\| \\
&\ge \frac{1}{\|\widetilde{\mathbb{D}}F(x_*)^{-1}\|}\|\xi\| - \int_0^1 \|\mathrm{D}G(\tau\xi) - \mathrm{D}G(0)\|\|\xi\|\, \mathrm{d}\tau \\
&\ge \frac{1}{\|\widetilde{\mathbb{D}}F(x_*)^{-1}\|}\|\xi\| - \int_0^1 \alpha\tau\|\xi\|\|\xi\|\, \mathrm{d}\tau,\ \forall\xi \in \mathscr{V}, \\
&\ge \frac{1}{\|\widetilde{\mathbb{D}}F(x_*)^{-1}\|}\|\xi\| - \frac{1}{2}\alpha\|\xi\|^2,\ \forall\xi \in \mathscr{V},
\end{aligned} \tag{14.15}$$

where Eq. (14.15) relies on Lipschitz continuity of $\mathrm{D}G$, which holds by taking $\mathscr{V}$ bounded since G is smooth. Taking $\mathscr{V}$ smaller if necessary, we have

$$\left\|\mathscr{T}_\xi^{-1}\big(F(R_{x_*}(\xi))\big)\right\| \ge \frac{1}{2\|(\widetilde{\mathbb{D}}F(x_*))^{-1}\|}\|\xi\|, \quad \forall\xi \in \mathscr{V}.$$

Let $c_0 = \frac{1}{2\|(\widetilde{\mathbb{D}}F(x_*))^{-1}\|}$, this concludes the first inequality in Eq. (14.10).
From Eq. (14.13), we have

$$\left\|\mathscr{T}_\xi^{-1}\big(F(R_{x_*}(\xi))\big)\right\| \le \left\|\widetilde{\mathbb{D}}F(x_*)\xi\right\| + \left\|\int_0^1 [\mathrm{D}G(\tau\xi) - \mathrm{D}G(0)]\xi\, \mathrm{d}\tau\right\|$$

$$
\begin{aligned}
&\le \|\widetilde{\mathbb{D}}F(x_*)\| \|\xi\| + \int_0^1 \|\mathrm{D}G(\tau\xi) - \mathrm{D}G(0)\| \|\xi\| \,\mathrm{d}\tau \\
&\le \|\widetilde{\mathbb{D}}F(x_*)\| \|\xi\| + \int_0^1 \alpha\tau \|\xi\| \|\xi\| \,\mathrm{d}\tau, \ \forall \xi \in \mathscr{V}, \\
&\le \|\widetilde{\mathbb{D}}F(x_*)\| \|\xi\| + \frac{1}{2}\alpha \|\xi\|^2, \ \forall \xi \in \mathscr{V} \\
&\le \|\widetilde{\mathbb{D}}F(x_*)\| \|\xi\| + \frac{1}{2}\alpha \|\xi\|, \ \forall \xi \in \mathscr{V}, (\|\xi\| \le 1).
\end{aligned}
$$

Let $c_1 = \|\widetilde{\mathbb{D}}F(x_*)\| + \frac{1}{2}\alpha$, this concludes the second inequality in Eq. (14.10). □

Finally we note that since $c_1 \|\widehat{R_x(\xi)} - \hat{x}\| \le \mathrm{dist}(x, R_x(\xi)) \le c_2 \|\widehat{R_x(\xi)} - \hat{x}\|$, by Lemma 14.1, and $\widehat{R_x(\xi)} = \hat{\xi} + O(\hat{\xi}^2)$, for the retraction R there exist $\mu > 0, \tilde{\mu} > 0$ and $\delta_{\mu,\tilde{\mu}} > 0$ such that $\forall x$ in a sufficiently small neighborhood of x^* and $\xi \in T_x\mathscr{M}$, $\|\xi\| \le \delta_{\mu,\tilde{\mu}}$

$$
\frac{1}{\tilde{\mu}} \|\xi\| \le \mathrm{dist}(x, R_x(\xi)) \le \frac{1}{\mu} \|\xi\|. \tag{14.16}
$$

We are now in a position to state and prove the main result of a necessary and sufficient condition for superlinear convergence of a Riemannian quasi-Newton algorithm.

Theorem 14.1 (Riemannian Dennis-Moré Condition) *Let $\mathscr{M}$ be a Riemannian manifold endowed with a C^2 vector transport $\mathscr{T}$ and an associated retraction R. Let F be a C^2 tangent vector field on $\mathscr{M}$. Also let $\mathscr{M}$ be endowed with an affine connection ∇. Let $\mathbb{D}F(x)$ denote the linear transformation of $T_x\mathscr{M}$ defined by $\mathbb{D}F(x)[\xi_x] = \nabla_{\xi_x} F$ for all tangent vectors ξ_x to $\mathscr{M}$ at x. Let $\{\mathscr{B}_k\}$ be a bounded sequence of nonsingular linear transformations of $T_{x_k}\mathscr{M}$, where $k = 0, 1, \dots, x_{k+1} = R_{x_k}(\eta_k)$, and $\eta_k = -\mathscr{B}_k^{-1} F(x_k)$. Assume that $\mathbb{D}F(x_*)$ is nonsingular, $x_k \ne x_*, \forall k$, and $\lim_{k\to\infty} x_k = x_*$. Then $\{x_k\}$ converges superlinearly to x_* and $F(x_*) = 0$ if and only if*

$$
\lim_{k\to\infty} \frac{\|[\mathscr{B}_k - \mathscr{T}_{\xi_k}\mathbb{D}F(x_*)\mathscr{T}_{\xi_k}^{-1}]\eta_k\|}{\|\eta_k\|} = 0, \tag{14.17}
$$

where $\xi_k \in T_{x_*}\mathscr{M}$ is defined by $\xi_k = R_{x_*}^{-1}(x_k)$, i.e. $R_{x_*}(\xi_k) = x_k$.

Proof Assume first that Eq. (14.17) holds. Since, for $\xi_k \in T_{x^*}\mathscr{M}$ and $\eta_k \in T_{x_k}\mathscr{M}$ we have

$$
\begin{aligned}
0 &= \mathscr{B}_k\eta_k + F(x_k) \\
&= \left(\mathscr{B}_k - \mathscr{T}_{\xi_k}\mathbb{D}F(x_*)\mathscr{T}_{\xi_k}^{-1}\right)\eta_k + F(x_k) + \mathscr{T}_{\xi_k}\mathbb{D}F(x_*)\mathscr{T}_{\xi_k}^{-1}\eta_k,
\end{aligned} \tag{14.18}
$$

and we have

$$\begin{aligned} -\mathscr{T}_{\eta_k}^{-1} F(x_{k+1}) &= \big(\mathscr{B}_k - \mathscr{T}_{\xi_k} \mathbb{D} F(x_*) \mathscr{T}_{\xi_k}^{-1}\big)\eta_k \\ &\quad + \big(-\mathscr{T}_{\eta_k}^{-1} F(x_{k+1}) + F(x_k) + \mathscr{T}_{\xi_k} \mathbb{D} F(x_*) \mathscr{T}_{\xi_k}^{-1}\eta_k\big) \\ &= \big(\mathscr{B}_k - \mathscr{T}_{\xi_k} \mathbb{D} F(x_*) \mathscr{T}_{\xi_k}^{-1}\big)\eta_k \\ &\quad + \big(-\mathscr{T}_{\eta_k}^{-1} F(x_{k+1}) + F(x_k) + \widetilde{\mathbb{D}} F(x_k)\eta_k\big) \\ &\quad + \big(\mathscr{T}_{\xi_k} \widetilde{\mathbb{D}} F(x_*) \mathscr{T}_{\xi_k}^{-1} - \widetilde{\mathbb{D}} F(x_k)\big)\eta_k \\ &\quad + \mathscr{T}_{\xi_k} \big(\mathbb{D} F(x_*) - \widetilde{\mathbb{D}} F(x_*)\big) \mathscr{T}_{\xi_k}^{-1} \eta_k. \end{aligned} \tag{14.19}$$

Recall that $\widetilde{\mathbb{D}} F(x)$ denotes the derivative at 0_x of the function $T_x\mathscr{M} \to T_x\mathscr{M}$: $\eta \mapsto \mathscr{T}_\eta^{-1} F(R_x(\eta))$, so we have

$$\lim_{k\to\infty} \frac{\|(-\mathscr{T}_{\eta_k}^{-1} F(x_{k+1}) + F(x_k) + \widetilde{\mathbb{D}} F(x_k)\eta_k)\|}{\|\eta_k\|} = 0. \tag{14.20}$$

Since F is C^2, we have

$$\lim_{k\to\infty} \frac{\|(\mathscr{T}_{\xi_k} \widetilde{\mathbb{D}} F(x_*) \mathscr{T}_{\xi_k}^{-1} - \widetilde{\mathbb{D}} F(x_k))\eta_k\|}{\|\eta_k\|} = 0. \tag{14.21}$$

Since $\lim_{k\to\infty} x_k = x_*$, we have $\lim_{k\to\infty} \|\eta_k\| = 0$ and $\lim_{k\to\infty} \|F(x_k)\| = 0$ if $\mathscr{B}_k$ is bounded. So $F(x_*) = 0$.

Since $F(x_*) = 0$, we have $\widetilde{\mathbb{D}} F(x_*) = \mathbb{D} F(x_*)$, [2, p. 96], hence

$$\frac{\|\mathscr{T}_{\xi_k} (\mathbb{D} F(x_*) - \widetilde{\mathbb{D}} F(x_*)) \mathscr{T}_{\xi_k}^{-1} \eta_k\|}{\|\eta_k\|} = 0. \tag{14.22}$$

Thus Eq. (14.19) yields

$$\lim_{k\to\infty} \frac{\|\mathscr{T}_{\eta_k}^{-1} F(x_{k+1})\|}{\|\eta_k\|} = 0. \tag{14.23}$$

From Lemma 14.6, we have

$$\big\|\mathscr{T}_{\xi_{k+1}}^{-1} F(x_{k+1})\big\| \geq \alpha \|\xi_{k+1}\|, \quad \forall k \geq k_0 \tag{14.24}$$

where $\xi_{k+1} \in T_{x^*}\mathscr{M}$ and $R_{x_*}(\xi_{k+1}) = x_{k+1}$. Therefore, we have

$$\begin{aligned} &\big\|\mathscr{T}_{\eta_k}^{-1} F(x_{k+1})\big\| \\ &\quad = \big\|\mathscr{T}_{\eta_k}^{-1} F(x_{k+1})\big\| - \|F(x_{k+1})\| + \|F(x_{k+1})\| \\ &\qquad - \big\|\mathscr{T}_{\xi_{k+1}}^{-1} F(x_{k+1})\big\| + \big\|\mathscr{T}_{\xi_{k+1}}^{-1} F(x_{k+1})\big\| \end{aligned} \tag{14.25}$$

$$
\begin{aligned}
&\geq \left\|\mathscr{T}_{\xi_{k+1}}^{-1} F(x_{k+1})\right\| \\
&\quad - \left| \left\|\mathscr{T}_{\eta_k}^{-1} F(x_{k+1})\right\| - \|F(x_{k+1})\| \right| - \left| \|F(x_{k+1})\| - \left\|\mathscr{T}_{\xi_{k+1}}^{-1} F(x_{k+1})\right\| \right| \\
&\geq \alpha\|\xi_{k+1}\| - L'\|F(x_{k+1})\|\mathrm{dist}(x_k, x_{k+1}) - L'\|F(x_{k+1})\|\mathrm{dist}(x_*, x_{k+1}) \quad (14.26)\\
&\geq \alpha\|\xi_{k+1}\| - c_4\|\xi_{k+1}\|(\mathrm{dist}(x_k, x_{k+1}) + \mathrm{dist}(x_*, x_{k+1})), \quad (14.27)
\end{aligned}
$$

where Eq. (14.26) follows from Eq. (14.24) with k_0 sufficiently large and Lemma 14.3, and Eq. (14.27) follows from Lemma 14.5 and Eq. (14.16).

We have also

$$1/\tilde{\mu}\|\eta_k\| \leq \mathrm{dist}(x_k, x_{k+1}) \leq \mathrm{dist}(x_k, x_*) + \mathrm{dist}(x_{k+1}, x_*) \leq 1/\mu\|\xi_k\| + 1/\mu\|\xi_{k+1}\|,$$

that is

$$\|\eta_k\| \leq \tilde{\mu}/\mu(\|\xi_k\| + \|\xi_{k+1}\|).$$

We also have

$$
\begin{aligned}
0 &= \lim_{k\to\infty} \frac{\|\mathscr{T}_{\eta_k}^{-1} F(x_{k+1})\|}{\|\eta_k\|} \\
&\geq \lim_{k\to\infty} \frac{\alpha\|\xi_{k+1}\|}{\|\eta_k\|}\left(1 - \frac{c_4}{\alpha}(\mathrm{dist}(x_k, x_{k+1}) + \mathrm{dist}(x_{k+1}, x_*))\right) \\
&= \lim_{k\to\infty} \frac{\alpha\|\xi_{k+1}\|}{\|\eta_k\|} \geq \lim_{k\to\infty} \frac{\alpha\|\xi_{k+1}\|}{\tilde{\mu}/\mu(\|\xi_k\| + \|\xi_{k+1}\|)} \\
&= \lim_{k\to\infty} \frac{\alpha\|\xi_{k+1}\|/\|\xi_k\|}{\tilde{\mu}/\mu(1 + \|\xi_{k+1}\|/\|\xi_k\|)}.
\end{aligned}
$$

Hence

$$\lim_{k\to\infty} \frac{\|\xi_{k+1}\|}{\|\xi_k\|} = 0.$$

This is superlinear convergence and this concludes the if portion of the proof.

Conversely, assume that $\{x_k\}$ converges superlinearly to x_* and $F(x_*) = 0$. It is sufficient to show that

$$\lim_{k\to\infty} \frac{\|\mathscr{T}_{\eta_k}^{-1} F(x_{k+1})\|}{\|\eta_k\|} = 0, \quad (14.28)$$

as this allows us to conclude using Eqs. (14.19), (14.20), (14.21), and (14.22). To show Eq. (14.28), observe that

$$
\begin{aligned}
\left\|\mathscr{T}_{\eta_k}^{-1} F(x_{k+1})\right\| &= \left\|\mathscr{T}_{\eta_k}^{-1} F(x_{k+1})\right\| - \|F(x_{k+1})\| + \|F(x_{k+1})\| \\
&\quad - \left\|\mathscr{T}_{\xi_{k+1}}^{-1} F(x_{k+1})\right\| + \left\|\mathscr{T}_{\xi_{k+1}}^{-1} F(x_{k+1})\right\| \\
&\leq \left\|\mathscr{T}_{\xi_{k+1}}^{-1} F(x_{k+1})\right\| + \left| \left\|\mathscr{T}_{\eta_k}^{-1} F(x_{k+1})\right\| - \|F(x_{k+1})\| \right|
\end{aligned}
$$

$$+ \left| \|F(x_{k+1})\| - \|\mathscr{T}_{\xi_{k+1}}^{-1} F(x_{k+1})\| \right|$$
$$\leq c_1 \|\xi_{k+1}\| + c_4 \|\xi_{k+1}\| \mathrm{dist}(x_k, x_{k+1}) + (\mathrm{dist}(x_*, x_{k+1}))$$

and that

$$\|\eta_k\| \geq \mu \mathrm{dist}(x_k, x_{k+1}) \geq \mu(\mathrm{dist}(x_k, x_*) - \mathrm{dist}(x_{k+1}, x_*)) \geq \frac{\mu}{\tilde{\mu}}(\|\xi_k\| - \|\xi_{k+1}\|).$$

Hence

$$\lim_{k\to\infty} \frac{\|\mathscr{T}_{\eta_k}^{-1} F(x_{k+1})\|}{\|\eta_k\|} \leq \lim_{k\to\infty} \frac{c_1 \|\xi_{k+1}\|}{\frac{\mu}{\tilde{\mu}}(\|\xi_k\| - \|\xi_{k+1}\|)}$$
$$\leq \lim_{k\to\infty} \frac{c_1 \tilde{\mu}}{\mu} \frac{\|\xi_{k+1}\|/\|\xi_k\|}{1 - \|\xi_{k+1}\|/\|\xi_k\|} = 0$$

since

$$\lim_{k\to\infty} \frac{\|\xi_{k+1}\|}{\|\xi_k\|} = 0$$

by the superlinear convergence assumption. □

Acknowledgements This paper presents research results of the Belgian Network DYSCO (Dynamical Systems, Control, and Optimization), funded by the Interuniversity Attraction Poles Programme, initiated by the Belgian State, Science Policy Office. The scientific responsibility rests with its authors.

This work was performed in part while the first author was a Visiting Professor at the Institut de mathématiques pures et appliquées (MAPA) at Université catholique de Louvain.

References

1. Absil, P.A., Baker, C.G., Gallivan, K.A.: Trust-region methods on Riemannian manifolds. Found. Comput. Math. **7**(3), 303–330 (2007) doi:10.1007/s10208-005-0179-9
2. Absil, P.-A., Mahony, R., Sepulchre, R.: Optimization Algorithms on Matrix Manifolds. Princeton University Press, New Jersey (2008)
3. Adler, R.L., Dedieu, J.P., Margulies, J.Y., Martens, M., Shub, M.: Newton's method on Riemannian manifolds and a geometric model for the human spine. IMA J. Numer. Anal. **22**(3), 359–390 (2002)
4. Baker, C.G.: Riemannian manifold trust-region methods with applications to eigenproblems. Ph.D. thesis, School of Computational Science, Florida State University (2008)
5. Dennis, J.E., Schnabel, R.B.: Numerical Methods for Unconstrained Optimization and Nonlinear Equations. Springer, New Jersey (1983)
6. Dreisigmeyer, D.W.: Direct search algorithms over Riemannian manifolds (2006). Optimization Online 2007-08-1742
7. Edelman, A., Arias, T.A., Smith, S.T.: The geometry of algorithms with orthogonality constrains. SIAM J. Matrix Anal. Appl. **20**(2), 303–353 (1998)
8. Gabay, D.: Minimizing a differentiable function over a differential manifold. J. Optim. Theory Appl. **37**(2), 177–219 (1982)
9. Helmke, U., Moore, J.: Optimization and Dynamical Systems. Springer, Berlin (1994)

10. Qi, C.: Numerical optimization on Riemannian manifolds. Ph.D. thesis, Florida State University, Tallahassee, FL, USA (2011)
11. Smith, S.T.: Optimization techniques on Riemannian manifolds. In: Bloch, A. (ed.) Hamiltonian and Gradient Flows, Algorithms and Control. Fields Inst. Commun., vol. 3, pp. 113–136. Amer. Math. Soc., Providence (1994)
12. Yang, Y.: Globally convergent optimization algorithms on Riemannian manifolds: Uniform framework for unconstrained and constrained optimization. J. Optim. Theory Appl. **132**(2), 245–265 (2007). doi:10.1007/s10957-006-9081-0

Chapter 15
A Jump-Start of Non-negative Least Squares Solvers

Mu Wang and Xiaoge Wang

Abstract Non-negative least squares fitting is a basic block in many applications. Following the classical active-set method by Lawson and Hanson (Solving least squares problems, Prentice-Hall, 1974), much research has been directed toward improving that algorithm. In this paper we present a new method that produces an initial setting for this classical algorithm. This initialization method exploits the relationship between projection based methods and the active set methods. Two quantitative measurements are introduced to evaluate the quality of initial settings for active set method. Experimental results indicate that the proposed initialization provides a good jump-start for the active set method leading to a reduction of the number of iterations of active set methods.

15.1 Introduction

The non-negative least squares (NNLS) problem has many applications [18, 20], for instance in image processing and data mining. In some scenarios, the non-negative constraint comes from the requirements of the background application or the demand to reduce dimensions. The non-negativity constraint makes the solution methods very different than the solution methods for the unconstrained least squares (ULS) problems The latter may be direct or iterative while the former are always iterative. In iterative methods, initialization can influence the number of iterations used to reach a satisfactory approximation to the solution. For example, as shown in [4], a scheme based on the singular value decomposition (SVD) can be used to "jump-start" algorithms for non-negative matrix factorization (NMF). In this paper, the initialization for active set methods is studied and a new initialization method is proposed.

The main contributions of this article are:

M. Wang (✉) · X. Wang
Tsinghua University, Beijing, P.R. China
e-mail: wangmu04@mails.tsinghua.edu.cn

X. Wang
e-mail: wangxg@tsinghua.edu.cn

M.W. Berry et al. (eds.), *High-Performance Scientific Computing*,
DOI 10.1007/978-1-4471-2437-5_15, © Springer-Verlag London Limited 2012

1. The relation between two classes of solvers, active set methods and projection based iterative methods, is studied. We show that a sequence of sets induced from the sequence of solution vectors generated by a projection based algorithm converges to the slack set of the active set methods.
2. Two quantitative measurements, `Precision` and `Recall`, are introduced to evaluate the quality of initialization of active set methods.
3. A projection based initialization method for active set methods is proposed. Experimental results indicate that the proposed initialization method could generate better initialization and achieve better performance than active set method due to the better initial setting.

The rest of the paper is organized as follows. In Sect. 15.2 we review the two main categories of existing methods and briefly discuss their characteristics. The description of the proposed initialization algorithm, its principles and design are given in Sect. 15.3. Numerical experiments programmed in MATLAB are described in Sect. 15.4. Section 15.5 contains our conclusion and comments on some possible directions of future work.

15.2 Background and Related Work

The NNLS problem to be solved is the following: Given a matrix $A \in R^{m\times n}$ and vector $b \in R^{m\times 1}$, find the vector $x \in R^{n\times 1}$ with the constraint $x \geq 0$ that minimizes the objective function $f(x) = \|Ax - b\|_2$, where $\|\cdot\|_2$ is the L_2-norm. It can be denoted as

$$\min \|Ax - b\|_2, \quad \text{subject to } x \geq 0 \tag{15.1}$$

A non-negative vector x is the solution to the above problem if and only if it satisfies the Karush–Kuhn–Tucker (KKT) conditions [13, 14], namely that if $w = A^T(b - Ax)$, there exists a partitioning of the integers from 1 to n into two sets S and E such that

$$\forall j \in S, \quad x_j > 0 \quad \text{and} \quad \forall j \in E, \quad x_j = 0$$
$$\forall j \in S, \quad w_j = 0 \quad \text{and} \quad \forall j \in E, \quad w_j \leq 0$$

In addition, if A is full column rank, then the solution is unique. Moreover, let A_S denote the $m \times n$ matrix by constrain A on set S such that the jth column of A_S equals to jth column of A for $\forall j \in S$, otherwise it equals to the vector of all zeros. x is the solution vector of the associated unconstrained least squares (ULS) problem defined by S (cf. [14]):

$$\min \|A_S x - b\|_2 \tag{15.2}$$

In the rest of the paper, the set S is called slack set, and the set E is called equality set. Solving NNLS problem is equivalent to determine the slack (equality) set $S(E)$ followed by solving the corresponding ULS problem shown in Eq. (15.2).

A comprehensive review of NNLS solvers is given in [6]. The algorithms for solving NNLS problems can be classified into two categories: active set methods which focus on determining the slack set by iteratively moving candidate elements from active set to passive set, and projection based methods which focus on computation of feasible solution vector. The work presented in this paper is related to both approaches. In the following subsections we briefly review the two types of method.

15.2.1 Active Set Methods

In 1974, Lawson and Hanson [14] proposed the first widely used active set algorithm. The main idea of the algorithm is to find the slack set iteratively. It defines an index set P called passive set and its complement set Z called active set. The algorithm starts with empty passive set P. It consists of two nested loops. In each step of the outer loop k, a candidate index is chose from active set to be added to the passive set P, and then computes the solution x of the associated ULS problem defined by P. If the solution contains zero or negative elements, the P cannot be a valid candidate of the slack set, so the inner loop is invoked. The function of the inner loop is to modify the passive set to become a valid candidate. In the inner loop, the indices whose corresponding position in the solution vector x is zero or negative are removed form the passive set. After the update of the passive set, the solution of the associated ULS problem defined by the updated passive set is recomputed and checked. This process is repeated until there is no further update. Then whether the passive set is the slack set is tested by checking whether the Karush–Kuhn–Tucker conditions is satisfied. If the condition is satisfied, the outer loop is terminated. Since there are only finite possible combinations of the passive set, and the objective function strictly decreases during the course of the iterations as proved in [14], this algorithm must stop in a finite iterations. The solution at the termination of the outer iteration is the solution to the NNLS problem. Later works were aimed at reducing redundant computations as well as taking advantage of the fact that many times, several NNLS problems with the same coefficient matrix A need to be solved; cf. [5] and [19]. Reference [9] deals with active set methods when the coefficient matrix is sparse and has more columns than rows. There has also been research on how to improve the implementation of NNLS algorithms on modern computer systems, e.g. taking advantage of multicore technology [16].

The main weakness of the standard active set method is that the algorithm starts from empty passive set and only one index is added in each outer iteration step. Although the QR up-dating and down-dating techniques can be applied to speed up computing the solutions of associated ULS problems, these computations consist mostly of level 1 or 2 BLAS, which may not fully take advantages of modern computer architecture in the implementation. So some modifications have been proposed to further improve its performance.

One way to improve the performance of active set methods is to add two or more indices at each outer iteration. but this should be done very carefully because it may cause the endless loop in the algorithm. A "backup exchange rule" introduced in [12, 17] is one way to deal with this problem. It consists of a method to select a block of indices and a rule of backup to ensure the correct function of the algorithm. The method is expected to work well when A is full column rank but may fail otherwise. Another way is to start the outer iteration with a non-empty initial passive set, which could be considered as a jump-start for active set methods compared with the original versions which start with an empty set. The work presented in [19] uses the initial passive set that comes from the solution to the LS problem. Such initialization may work well when there is only little difference in the solutions of two successive problems or when the non-negative constraint is enforced only to eliminate the noise caused by observation, measurement or rounding errors of floating point computation. In those cases, the solution of the ULS problem provides a reasonable guess of the solution to the NNLS problem.

15.2.2 Projection Based Methods

The methods in this category use gradient information to approach to the solution. Their advantage is that they can handle multiple active constraints at each iteration. A state-of-the-art method in this category is the Projected Quasi-Newton (PQN) method by Kim, Sra, and Dhillon [10, 11]. As its name indicates, the idea of PQN is like the Newton method. At iteration k, a "search direction" is determined based on current solution vector x_k and the gradient scaling matrix, along which the objective function decrease in a neighborhood. Along that direction an adjustment of solution vector is made so that the new solution vector would reduce the objective function and satisfy the non-negative constraint as well. In the course of the iterations, a sequence of solution vectors is generated whose corresponding objective function values are strictly decreasing. Theoretical termination is when the KKT condition is satisfied; in practice, some specific stopping criteria are set and tested.

As indicated in [10], a characteristic of the above projection based methods is that in the course of the iterations, the objective function decreases fast early on but after a few steps, this decrease slows down considerably. This phenomenon is known as "jamming". This phenomenon was also observed in other projection based methods, e.g. [2, 3, 15]. It is thus interesting to see if it is possible to design NNLS solvers in which the good performance of projection based methods is preserved, while the jamming in the later iterations is avoided. This is exactly the inspiration of the method proposed in this paper.

There are some other methods in this category. The approach presented in [8] is to optimize at a single coordinate with all other coordinates fixed. Because this method is sequential and most of the computation is in the form of vector-vector (Level 1 BLAS) computation, it may not suitable for optimizations designed to take advantage of modern computer systems. The random projection method [7] only gives a vector which is close to the final solution in a relative error approximation sense with high probability. These works are not as much related to the ideas presented in this paper, so we do not discuss them any further.

15.3 Jump-Start for Active Set Methods

In this section, two quantitative measurements of the quality of an initial passive set for active set algorithm are introduced. Then a study on the relationship between projection based methods and active set methods is presented. Based on the theoretical analysis and observations, a projection based initialization strategy for accelerating active set methods is introduced and discussed. Then, PiNNLS, the algorithm of solving NNLS with the initialization, is presented.

15.3.1 Performance Indicators

Intuitively, if we wish to provide a jump-start for active set methods, we would like for the initial passive set to be close to the slack set, that is, the initial guess should include more elements in the slack set and fewer elements in the equality set. Unfortunately, to the best of our knowledge, there is no quantitative measurement about this. So our investigation starts from the introduction of two quantities which measure the quality of the initial passive set in active set methods. We require the matrix A to be to be full column rank here for convenience, so the final solution and the slack set is unique. We will show later that this requirement could be removed without tampering the correctness of our method.

Let S be the slack set and P_0 be the initial passive set. Let $|\cdot|$ denote the number of elements in a set. To quantify the measurement on how close the initial guess P_0 is to S, we borrow the concepts of recall and precision from the field of information retrieval with S corresponding to the relevant documents and P_0 corresponding to the retrieved documents. Similarly, we define two quantities, `Recall` and `Precision`, as follows:

$$\texttt{Recall} = \frac{|P_0 \cap S|}{|S|} \tag{15.3}$$

$$\texttt{Precision} = \frac{|P_0 \cap S|}{|P_0|} \tag{15.4}$$

`Recall` is the fraction of the elements that are correctly guessed by initial passive set among all the elements of the slack set. It measures the probability of elements in the slack set to be correctly included by the initial guess.

`Precision` is the fraction of the elements of correct guess among all guessed elements by initial guess. It quantitatively measures the precision of the initial passive set.

$$1 - \texttt{Precision} = \frac{|P_0 \setminus S|}{|P_0|} \tag{15.5}$$

gives the information on how many incorrect guesses among the initial guesses.

Together, these two quantities give the evaluation on how close is the set P_0 to S. `Recall` indicates how much benefits we could get from the initialization and the `Precision` indicates how much extra cost we have to pay to remove the false elements from the initial guess. In the ideal case, when both quantities are 100%, it means P_0 is exactly the same as S.

15.3.2 Relation Between Projection Based Methods and Active Set Methods

As mentioned in previous section, Active set methods deal with the passive set and the solution vector of the associated ULS problem defined by the passive set, while projection based methods deal with only the solution vectors. We denote x_k, $k = 1, 2, \ldots$ the vector sequence generated by a projection based algorithm. As proved in [11], this sequence converges to the solution x. Does this sequence of vectors tell us something about the slack set? To answer this question, we prove the following.

Theorem 15.3.1 *Let x_k, $k = 1, 2, \ldots$ be the solution vector sequence generated by a projection based algorithm that converges to the final solution x. Then there exists $\delta > 0$ such that if we define $P_k = \{i | x_k^i > \delta\}$, $k = 1, 2, \ldots$, where the superscript i denotes the ith element, then there exists K, such that for all $k > K$, $P_k = S$, where $S = \{i | x^i > 0\}$ is the slack set.*

Proof From x_k convergent to x, we know that

$$\forall \varepsilon > 0, \quad \exists K, \quad \text{s.t.,} \quad \forall k > K, \quad i \in \{1, \ldots, n\}, \quad |x_k^i - x^i| < \varepsilon$$

Then, we have

$$\forall k > K, \quad i \in \{1, \ldots, n\}, \quad x_k^i > x^i - \varepsilon \quad \text{and} \quad x_k^i < x^i + \varepsilon$$

Notice that x_k and x are non-negative, so if we choose δ so that

$$0 < \delta < \min\{x^i | x^i > 0\} \quad \text{and} \quad \varepsilon = \min\left\{ \min_{i \in 1, \ldots, n} \{x^i | x^i > 0\} - \delta, \delta \right\}$$

Then, we have

$$x^i > 0 \quad \Rightarrow \quad x_k^i > \delta, \quad \text{and,} \quad x^i = 0 \quad \Rightarrow \quad x_k^i < \delta$$

Recall the definition of S and P_k,

$$S = \{i | x^i > 0\} \quad \text{and} \quad P_k = \{i | x_k^i > \delta\}$$

And this leads to

$$\forall k > K, \quad j \in \{1, \cdots, n\}, \quad j \in S \Leftrightarrow j \in P_k$$
$$\forall k > K, \quad P_k = S.$$

□

From the proof of the theorem, we can see that δ plays the role of threshold to distinguish small values from zero in order to rule out the inconsistency brought by the convergence to zero because a sequence could converge to zero with all its elements being positive. It would need fewer steps to reach the slack set if δ is properly chosen. Theoretically, any value satisfying the condition $0 < \delta < \min\{x^i | x^i > 0\}$ is enough from the proof. But in practice, such value is not known until we get the final solution. Therefore, we simply assign δ to be the numerical tolerance considering that, in a numerical implementation, the tolerance is used to differentiate small values from zero.

The theorem says that projection based methods can be used to determine the slack set. The theorem proves the existence of K by the convergence condition, though we do not know quantitatively how large K would be. If K is relatively small in practice, then the slack set can be determined in several projection iteration steps. If K is very large, which means P_k is equal to S only at convergence, will P_k with small k be useful? Recall the discussion in Sect. 15.2.2 where we said that after few iterations of a projection based method, the objective function decreases at a slow pace. We assume that this means that x_k is close to the final solution x but still has some way to go to be considered a satisfactory approximation. So it could be that after few steps, P_k is close, but not equal, to the slack set S. This inspires the basic idea of our algorithm, namely to use a projection based algorithms to generate an initial passive set for active set methods.

When finding an initial passive set for an active set algorithm, there are two requirements.

- Quality requirement: the initial passive set should be close to the slack set.
- Cost requirement: the computation of the initialization should not be costly.

These two requirements are important to make sure that the initialization is applicable. Otherwise, it may not be better than simply using an empty passive set. We next elaborate on these issues.

15.3.3 Observations

Let FNNLSi be the active set method with the initial passive set taken from the solution vector of ULS problem. Take FNNLSi as an example to show the applicability according to the two requirements. Suppose P_0 is the initial passive set and S is the slack set. The `Recall` and `Precision` are defined as in Eqs. (15.3) and (15.4). In FNNLSi, $P_0 = \{i | x^i > 0\}$, where x is the solution of $\min \|Ax - b\|_2$. Solving the ULS does not impose a heavy computational burden compared to the total cost of an active set algorithm. So the cost requirement is met. To evaluate the quality, we take three matrices from the University of Florida Sparse Matrix Collection (UF Collection) and measure the `Recall` and `Precision` of P_0 of FNNLSi.

The results are given in Table 15.1. It can be seen that the quality of P_0 in FNNLSi is not satisfactory. For example, `Precision` for the matrix HB/illc1850 is as low as 57.71%, which means lots of computation is required to remove the elements in the equality set from the initial passive set. While `Recall` of matrix HB/well1850 is as low as 58.76%, which means large among of slack set elements are not in the initial passive set, so large number of iteration steps may still be needed.

Table 15.1 Recall and Precision of P_0 in FNNLSi(%)

Matrix	Recall	Precision
HB/illc1850	60.84	57.71
HB/well1850	58.76	72.89
LPnetlib/lp_degen3	82.24	67.83

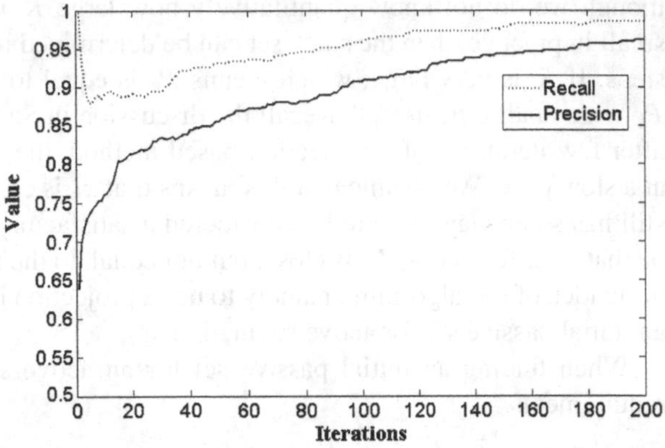

Fig. 15.1 Recall and Precision at each iteration (test matrix: illc1850)

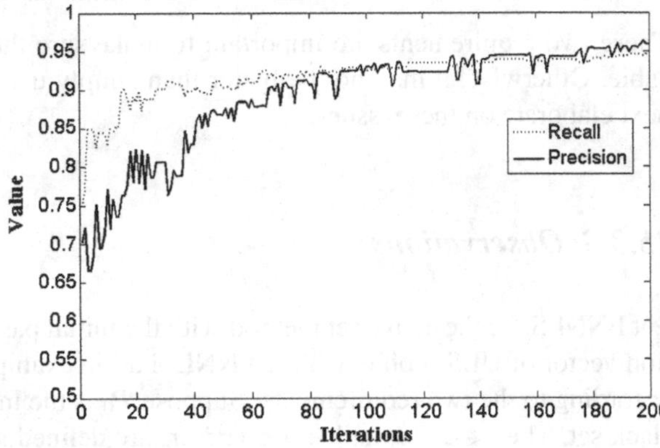

Fig. 15.2 Recall and Precision at each iteration (test matrix: lp_degen3)

To investigate the quality of projection based initialization method, suppose x_k and P_k are defined as in the proof of Theorem 15.3.1, we examined the evolution of Recall and Precision in the course of the iterations. Two representative cases are given in Figs. 15.1 and 15.2. From the figures, we have two observations:

First, the starting values of both Recall and Precision for the two test matrices are higher than before. For Recall, it is above 75%, and for Precision it is about 65%. Subsequently, both Recall and Precision generally increase as the iterations proceed. Even when there is an occasional decrease,. e.g., in the first several steps for matrix HB/illc1850, Recall decreases from 100% to about 87%, it does not alter much the general increasing trend. We thus hope that the initial passive set obtained from a few steps of the projection based algorithm is close to the slack set.

Second, we can see that the increase of `Recall` and `Precision` is much more in the earlier steps than in the later steps. For example, after 20 iteration steps, `Recall` and `Precision` are about 90% and 80% respectively, but after 200 iterations steps, they are just about 95%. This means that if we use projection based algorithm as initialization for active set methods, it should not take many iteration steps, because the main benefits are obtained in the course of the early steps.

Based on the these observations, we conclude that it is applicable to use projection based methods to initialize the passive set for active set methods. More discussion and the description of an implementation, PiNNLS, are given in the next subsection.

15.3.4 PiNNLS

The above observations suggest that projection based methods could be used as initialization for active set methods. We name this approach as *projection based initialization*. We next need to discuss some issues pertaining to such an initialization.

First, the active set methods need to be modified in order to handle the case of a non-empty initial passive set. In active set methods, a valid passive set is one for which the solution to the corresponding ULS problem does not contain zero or negative elements. If the initial passive set is not valid, it may cause the active set method generate incorrect result or ignore the initialization. To ensure correctness there should be a validation loop to validate the initial passive set; cf. [5]. The validation loop is exactly the same as the inner loop in active set methods.

Second, unlike in FNNLSi, where the initialization finishes after the solve of the ULS problem, it needs a stopping condition for projection based initialization. Because the slack set S is unknown until the final solution is obtained, we cannot simply compare P_k to S to decide whether we should stop the initialization. A simple way to control the number of iterations is to set the number of iteration steps λ. If $\lambda = 1$, that means the initial passive set comes directly from the gradient. In general, λ should not be large because we know from the observations in Sect. 15.3.3 that prolonging the initialization will not bring much additional benefit.

In general, any projection based method that satisfies the conditions in Theorem 15.3.1 could be used for initialization, and any active set method could be used after the initialization. In this paper, we choose Projected Quasi-Newton (PQN) method [11] for the initialization and FNNLS [5] as the active set method to form an algorithm of solving NNLS problem. We name this specific algorithm PiNNLS (Projection initialized NNLS solver).

In PiNNLS, the PQN method is implemented with some specializations and adjustments for the initialization. Some details are specified as the following:

1. The gradient scaling step is omitted because it is not expected to improve the results significantly while taking additional storage and computation. We directly use the gradient as the "search direction" (Line 5).

2. We use both the APA rule [1] (Line 6 to Line 12) and line-search techniques (Line 13 to Line 14) to determine how far the current solution vector x should go along the "search direction" (Line 15). The parameters σ, s and τ are introduced by the APA rule. In our experiment reported in the next section, these parameters are set as $\sigma = 1$, $s = 0.3$, $\tau = 0.35$ respectively.
3. The δ is the same as the numerical tolerance and we have discussed the reasons in Sect. 15.3.2.
4. The iteration terminates when the number of steps reaches λ (Line 18).
5. An initial passive set P is determined (Line 19). The validation loop is followed to make sure the passive set is valid (Line 20 to Line 24). After the validation, the results are the input for the active set method FNNLS (Line 25), on which our algorithm relies to get the final solution.

The description of the PiNNLS is given in the following. The description of FNNLS part is ignored. Refer to [5] for details.

PiNNLS:
Inputs($A \in R^{m\times n}, b \in R^{m\times 1}, x \in R^{n\times 1}$)
Parameters($\lambda, \delta, \sigma, s, \tau$)

1 repeat
2 $\quad \omega = A^\top b - A^\top Ax$
3 $\quad I = \{i \mid x^i = 0 \text{ and } \omega^i < 0\}, J = \{i \mid x^i > 0 \text{ or } \omega^i \geq 0\}$
4 $\quad y = x_J,\ \forall i \in J, y^i = x^i \text{ and } \forall i \in I, y^i = 0$
5 $\quad d = \omega_J,\ \forall i \in J, d^i = \omega^i \text{ and } \forall i \in I, d^i = 0$
6 $\quad \bar{f} = \frac{1}{2}\|Ay - b\|_2, \bar{\omega} = A^\top b - A^\top Ay, \beta = \sigma$
7 $\quad$ repeat
8 $\quad\quad \beta = \beta s, \gamma = \gamma + \beta d$
9 $\quad\quad \gamma = \frac{(\gamma + |\gamma|)}{2}$
10 $\quad\quad f = \frac{1}{2}\|A\gamma - b\|_2$
11 $\quad\quad g = \tau \times \bar{\omega}^T(y - \gamma)$
12 $\quad$ until $(\bar{f} - f) > g$
13 $\quad u = \gamma - y$
14 $\quad \alpha = \frac{u^\top (A^\top b - A^\top Ay)}{u^\top A^\top Au}, \alpha = \mathtt{mid}\{0, 1, \alpha\}$
15 $\quad x = y + \alpha(\gamma - y)$
16 $\quad k = k + 1$
17 $\quad P_k = \{i \mid x^i > \delta\}$
18 until $(k = \lambda)$
19 $P = P_k, z = \mathtt{argmin}\|A_P z - b\|^2, z^i = 0, \forall i \notin P$
20 while$(\exists i, z^i < 0)$
21 $\quad \alpha = \min\{\frac{x^j}{x^i - z^i} : z^i < 0, i \in P\}$
22 $\quad x = x + \alpha(z - x)$
23 $\quad P = \mathtt{find}(x > 0), z = \mathtt{argmin}\{\|A_P z - b\|^2, z^i = 0, \forall i \notin P\}$
24 end
25 $x = \mathtt{FNNLS}(A, b, x, P)$

Table 15.2 The synthetic dense matrices

Name	D1	D2	D3	D4	D5	D6
m	4000	4000	4000	6000	6000	6000
n	2000	2000	2000	3000	3000	3000
cond	1.61×10^3	6.71×10^4	3.66×10^7	3.95×10^3	2.19×10^4	4.63×10^6

15.4 Numerical Experiments

The experiments are designed and conducted to evaluate the effectiveness of the proposed initialization method as a jump-start for the active set methods and to test the performance of PiNNLS as a solver for the NNLS problem. In this section, we explain some details of our experiments and discuss the numerical results.

15.4.1 Experiments Design and Implementation

Experiments are designed to verify two assumptions. First, that the initialization used in PiNNLS is of good quality. When verifying this, we use FNNLSi as the reference for the comparison because it has the similar structure as PiNNLS. Second, due to the initialization, the overall computation is reduced for active set methods compare to the same method without initialization.

In the experiment, the methods are implemented in MATLAB. As active set algorithm we used the FNNLS code by Bro in [5]. The same algorithm was used in FNNLSi and PiNNLS. It is also used in FNNLSi and PiNNLS as the active set method to be initialized. In FNNLSi, the initialization is implemented by solving the ULS problem followed by the validation of initial passive set, which is the same as in PiNNLS Line 20 to 24. In PiNNLS, the initialization is implemented as shown in Sect. 15.3.4. The MATLAB version used is 2009a for Windows XP. The test are performed on a computer with an Intel E5400@2.7GHz CPU and 4 GB RAM.

Due to the limitation of the computation resources, the size of test problems is limited in the scale of thousands. The matrices of test problems include both dense and sparse matrices.

The first set of test problems consist of synthetic dense matrices. For each test problem, the matrix A and the corresponding vector b are generated by MATLAB `rand` function. We take the singular value decomposition (SVD) of a randomly generated matrix and manually scale the singular values to get matrices with different condition numbers. The size and condition number of the matrices are given in Table 15.2.

The second set of test matrices is sparse matrices from University of Florida (UF) collection. The information of the matrices are given in Table 15.3.

Table 15.3 The matrices from the UF collection

Name	S1	S2	S3	S4	S5	S6
matrix	illc1850	well1850	blckhole	lp_nug12	lp_pilot87	lp_ degen3
m	1850	1850	2132	8856	6680	2604
n	712	712	2132	3192	2030	1503
cond	1.40×10^3	1.11×10^2	4.17×10^3	1.22×10^{17}	8.15×10^3	1.76×10^{16}
sparsity	99.34%	99.34%	99.67%	99.86%	99.45%	99.35%

Table 15.4 `Recall` and `Precision` of initial guess in FNNLSi and PiNNLS (%)

		D1	D2	D3	D4	D5	D6	mean
FNNLSi	recall	51.54	47.42	48.77	52.72	50.00	50.00	49.96
	precision	52.17	46.79	48.52	52.61	52.21	52.21	50.10
PiNNLS	recall	97.11	99.07	98.26	97.97	98.31	95.70	97.64
	precision	87.06	96.71	96.87	97.20	96.91	94.58	95.37

Table 15.5 `Recall` and `Precision` of initial guess in FNNLSi and PiNNLS (%)

		S1	S2	S3	S4	S5	S6	mean
FNNLSi	recall	60.84	58.76	51.83	50.11	75.71	82.24	63.20
	precision	57.71	72.89	17.77	29.83	92.99	67.83	56.50
PiNNLS	recall	89.16	93.60	99.73	99.89	75.88	85.01	90.55
	precision	76.21	86.28	87.80	99.15	87.87	71.92	84.87

15.4.2 Quality of Initialization

Our first set of experiments is designed to compare the quality of the initial passive sets generated by PiNNLS and FNNLSi. For the results reported here, the parameters in PQN based initialization are set as described in Sect. 15.3.4 while t the projection iteration steps is set to $\lambda = 10$. The `Recall` and `Precision` of initial guesses of both FNNLSi and PiNNLS methods are measured for each test problem. Results are shown in Tables 15.4 and 15.5.

From Table 15.4 we can see that 10 steps sufficed for PiNNLS to get an initial passive set with much higher `Recall` and `Precision` than by solving the ULS problem. For FNNLSi, the `Recall` and `Precision` of the initial passive set are all around 50%, while for PiNNLS the `Recall` and `Precision` are all above 90% except the *precision* = 87.06% for the test problem D1 which is also close to 90%. The last column of Table 15.5 shows the mean of the quantity. It tells that in average `Recall` and `Precision` are in about 50% in FNNLSi. This agrees with the intuition that the positive and negative elements would each take about

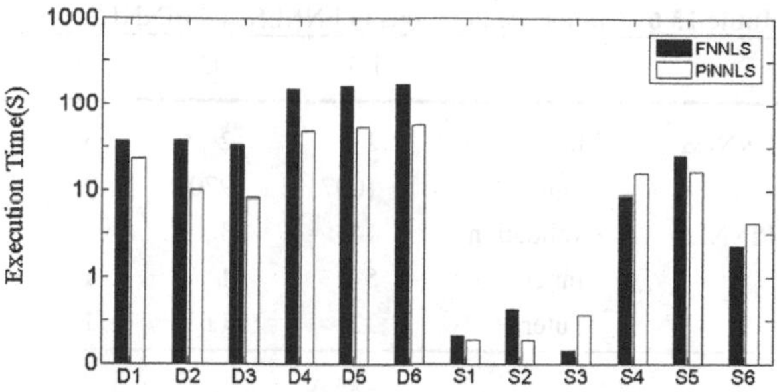

Fig. 15.3 Execution time of FNNLS and PiNNLS on test problems

50% among the total elements of the solution vector if the test problem matrices are randomly generated with the uniform distribution.

The test results are similar for the sparse test problems although the data are not as consistent as in dense cases. As in Table 15.4, Table 15.5 also shows that the initial passive set in PiNNLS also has higher `Recall` and `Precision` than that guessed by solving the associated ULS problem. On average, PiNNLS has 27% higher `Recall` and about 28% higher `Precision`. It shows in the table that the `Precision` of FNNLSi vary greatly from 17.77% in S3 to 92.99% in S5. This indicates that the initial guess by using the ULS solution may not be a reliable general method. Its performance may be problem dependent, e.g., for S6, the initial passive by FNNLSi is with higher `Recall` and `Precision` than by PiNNLS.

15.4.3 Performance of PiNNLS

In this section, we compare the performance of PiNNLS to that of FNNLS. Two metrics, execution time and iteration count, are used for the evaluation. It should be pointed out that the size of the operands in the operations may vary with the iterations. The size of the vector or matrix operands is directly proportional to the size of passive set. The parameters are set as mentioned in Sect. 15.3.4 and $\lambda = 10$. Due to the widely different scales in the execution times, results in Fig. 15.3 are plotted in logarithmic scale. A lower bar means less execution time. The two bars in one group represent the execution time of the two methods respectively on one test problem. The results of iteration count are shown in Table 15.6 for dense test problems and Table 15.7 for sparse test problems. The number of iterations of the validation, inner and outer loops are counted separately. No validation is needed for FNNLS.

For dense test problems, PiNNLS outperforms FNNLS in both execution time and iteration count. The data in Table 15.4 could explain these results. High `Precision` leads to less iterations of the validation loop and high `Recall` leads to less iterations in the inner and outer loop in the active set method. The results in Table 15.7 also testify the relation. For PiNNLS, due to a good initial passive set, the subsequent active set method only needs an order of ten iterations to determine the slack set. If instead we start with an empty passive set, as in FNNLS, the number of

Table 15.6 Number of iterations in FNNLS and PiNNLS

		D1	D2	D3	D4	D5	D6
FNNLS	inner	2	2	3	4	7	1
	outer	1007	970	981	1529	1541	1476
PiNNLS	validation	143	36	14	34	48	3
	inner	5	1	1	5	3	0
	outer	32	11	18	35	29	91

Table 15.7 Number of iterations in FNNLS and PiNNLS

		S1	S2	S3	S4	S5	S6
FNNLS	inner	16	52	0	0	46	2
	outer	422	583	368	940	1240	689
PiNNLS	validation	83	78	50	8	153	231
	inner	41	11	0	0	29	27
	outer	55	44	1	1	350	133

iterations is of the order of hundreds and sometimes thousands. Since only $\lambda = 10$ iteration steps are used in projection based initialization, the overall computation is greatly reduced with low cost initialization.

For sparse test problems, the situation appears a little more complicated. Overall, PiNNLS is faster than FNNLS except for problem S6. In terms of iteration count, PiNNLS outperforms FNNLS in all cases (see Table 15.7). We use Table 15.5 to explain. It shows that S6 has the lowest `Precision`. The reason that PiNNLS cannot be faster than FNNLS even though it takes fewer iterations is that the computational cost for each iteration is not the same as we have already pointed out. For FNNLS, the passive set starts empty, so that the operand size starts from zero and then grows. In PiNNLS, on the other hand, the low `Precision` and high `Recall` lead to a large but inaccurate initial passive set, so that the size of the operands in the validation loop start large (corresponding to the number of elements in the initial passive set) and then decrease with the removal of false elements. Therefore, although the number of iterations for PiNNLS is smaller than that for FNNLS, the overall computational cost may not. In other cases, due to the high `Precision`, their validation loop is not iterated many times, so the PiNNLS returns better performance. On the other hand, S5 has higher `Precision` and lower `Recall` than S6. This means the initial passive set is smaller but more accurate. It may need more iterations in the subsequent active set method, but the cost of the validation loops would be lower. From these two cases we conclude that the performance of PiNNLS is more sensitive to `Precision`, especially when `Recall` is high because the validation loop is more costly than the outer loop when the initial passive set is large.

Finally, we show by means of an example how the parameter setting could effect the performance of PiNNLS. It is seen in the previous results that for S6, the

Table 15.8 Parameter tuning for S6

λ	10	30
recall(%)	85.01	89.96
precision(%)	71.92	80.89
FNNLS execution time (S)	4.36	4.36
PiNNLS execution time (S)	5.70	3.81
PiNNLS initialization (S)	0.25	0.63
PiNNLS validation loop (S)	3.59	1.97
PiNNLS active set (S)	1.85	1.21

initial passive set is not good enough if $\lambda = 10$. We also know that more iteration steps can lead to better initial guess of slack set. So we try to increase the number of iterations in initialization to gain better result. As a test, we set the iteration steps as $\lambda = 30$, then PiNNLS could outperform FNNLS. The details are shown in Table 15.8. The 20 more iteration steps enhance the `Recall` from 85.01% to 89.89% and the `Precision` from 71.92% to 80.89% with the cost of 0.38 additional seconds. As expected, higher `Precision` leads to the reduction in execution time of the validation loop from 3.59 to 1.97 seconds. The higher `Recall` leads to the reduction of the number of iterations, as well as the execution time, of the active set algorithm FNNLS. The total runtime is also reduced from 5.70 to 3.81 seconds.

15.5 Conclusion and Future Work

In this paper, a projection based initialization algorithm for active set methods for solving NNLS problems is proposed. Experiments indicate that this algorithm can speed-up active set methods. This is work in progress and we point to some open questions and research issues. First, in our implementation, PiNNLS, the iteration steps λ in initialization is predetermined as an input parameter. Its best value is obviously problem dependent. A better choice for implementation should be adaptive. Secondly, in large scale parallel computing, global reductions are costly. Therefore we expect that PiNNLS would be more suitable than FNNLS because there is only one global reduction operation in each inner or outer loop, and the experimental results suggest that PiNNLS can lower the number of loops so as the number of global reduction operations. This needs to be validated in a parallel environment. Finally, in this study of projection based initializations of active set methods we focused on one special method of this kind. An extensive study for evaluating several projection based methods with different parameters remains to be done.

Acknowledgements The work is supported by National High-Tech Research and Development Plan of China under Grant No. 2010AA012302, and National Basic Research Program of China (973) under Grant No.2006CB605102. The authors would like to thank Professor Ahmed Sameh at Purdue University for the helpful discussion and support from the beginning of this study. Thanks also go to Professor Stratis Gallopoulos at University of Patras for his comments and pointers to helpful information during the study of the method and the writing of the article. The authors would also like to thank Professor Guangwen Yang and Wei Xue for providing a great supportive environment for this study.

References

1. Bertsekas, D.P.: On the Goldstein-Levitin-Poljak gradient projection method. IEEE Trans. Autom. Control **21**, 174–184 (1976)
2. Bertsekas, D.P.: Projected Newton methods for optimization problems with simple constraints. SIAM J. Control Optim. **20**(2), 221–246 (1982)
3. Bierlaire, M., Toint, P.L., Tuyttens, D.: On iterative algorithms for linear least squares problems with bound constraints. Linear Algebra Appl. **143**, 111–143 (1991)
4. Boutsidis, C., Gallopoulos, E.: SVD based initialization:ahead start for nonnegative matrix factorization. Pattern Recognit. **41**(4), 1350–1362 (2009)
5. Bro, R., Jong, S.D.: A fast non-negativity-constrained least squares algorithm. J. Chem. **11**(5), 393–401 (1997)
6. Chen, D., Plemmons, R.J.: Nonnegativity constraints in numerical analysis. In: Bultheel, A., Cools, R. (eds.) Symposium on the Birth of Numerical Analysis, pp. 109–140. Katholieke Universiteit Leuven, Belgium (2007)
7. Drineas, P., Boutsidis, C.: Random projections for the nonnegative least-squares problem. Linear Algebra Appl. **431**(5–7), 760–771 (2009)
8. Franc, V., Hlavc, V., Navara, M.: Sequential coordinate-wise algorithm for non-negative least squares problem. Research report ctu-cmp-2005-06, Center for Machine Perception, Czech Technical University (2005)
9. Guerrero-García, P., Santos-Palomo, A.: A sparse implementation of Lawson and Hanson's convex NNLS method. Technical report ma-05-03, University of Málaga (2005)
10. Kim, D., Sra, S., Dhillon, I.S.: Fast Newton-type methods for the least squares nonnegative matrix approximation problem. In: Proceedings of the 2007 SIAM International Conference on Data Mining, pp. 343–354 (2007)
11. Kim, D., Sra, S., Dhillon, I.S.: Fast projection-based methods for the least squares nonnegative matrix approximation problem. Stat. Anal. Data Min. **1**(1), 38–51 (2008)
12. Kim, J., Park, H.: Toward faster nonnegative matrix factorization: A new algorithm and comparisons. In: Proceedings of the IEEE International Conference on Data Mining, pp. 353–362 (2008)
13. Kuhn, H.W., Tucker, A.W.: Nonlinear programming. In: Neyman, J. (ed.) Proceedings of the second Berkeley symposium on mathematical statistics and probability, pp. 481–492. University of California Press, Berkeley (1951)
14. Lawson, C., Hanson, R.: Solving Least Squares Problems. Prentice-Hall, New York (1974)
15. Lin, C.J.: Projected gradient methods for nonnegative matrix factorization. Neural Comput. **19**(10), 2756–2779 (2007)
16. Luo, Y., Duraiswami, R.: Efficient parallel nonnegative least squares on multicore architectures. SIAM J. Sci. Comput. **33**, 2848 (2011)
17. Portugal, L.F., Judice, J.J., Vicente, L.N.: A comparison of block pivoting and interior-point algorithms for linear least squares problems with nonnegative variables. Math. Comput. **63**, 625–643 (1994)
18. Timotheou, S.: Nonnegative least squares learning for the random neural network. In: ICANN, pp. 195–204 (2008)
19. Van Benthem, M.H., Keenan, M.R.: Fast algorithm for the solution of large-scale non-negativity-constrained least squares problems. J. Chem. **18**, 441–450 (2004)
20. Vo, N., Moran, B., Challa, S.: Nonnegative-least-square classifier for face recognition. In: Advances in Neural Networks, pp. 449–456 (2009)

Chapter 16
Fast Nonnegative Tensor Factorization with an Active-Set-Like Method

Jingu Kim and Haesun Park

Abstract We introduce an efficient algorithm for computing a low-rank nonnegative CANDECOMP/PARAFAC (NNCP) decomposition. In text mining, signal processing, and computer vision among other areas, imposing nonnegativity constraints to the low-rank factors of matrices and tensors has been shown an effective technique providing physically meaningful interpretation. A principled methodology for computing NNCP is alternating nonnegative least squares, in which the nonnegativity-constrained least squares (NNLS) problems are solved in each iteration. In this chapter, we propose to solve the NNLS problems using the block principal pivoting method. The block principal pivoting method overcomes some difficulties of the classical active method for the NNLS problems with a large number of variables. We introduce techniques to accelerate the block principal pivoting method for multiple right-hand sides, which is typical in NNCP computation. Computational experiments show the state-of-the-art performance of the proposed method.

16.1 Introduction

Tensors are mathematical objects for representing multidimensional arrays. Tensors include vectors and matrices as first-order and second-order special cases, respectively, and more generally, tensors of Nth-order can represent an outer product of N vector spaces. Recently, decompositions and low-rank approximations of tensors have been actively studied and applied in numerous areas including signal processing, image processing, data mining, and neuroscience. Several different decomposition models, their algorithms, and applications are summarized in recent reviews by Kolda and Bader [19] and Acar and Yener [1].

J. Kim (✉) · H. Park
School of Computational Science and Engineering, College of Computing, Georgia Institute of Technology, Atlanta, USA
e-mail: jingu@cc.gatech.edu

H. Park
e-mail: hpark@cc.gatech.edu

M.W. Berry et al. (eds.), *High-Performance Scientific Computing*,
DOI 10.1007/978-1-4471-2437-5_16,

In this chapter, we discuss tensors with nonnegative elements and their low-rank approximations. In particular, we are interested in computing a CANDECOMP/PARAFAC decomposition [5, 11] with nonnegativity constraints on factors. In the context of matrices, when data or signals are inherently represented by nonnegative numbers, imposing nonnegativity constraints to low-rank factors was shown to provide physically meaningful interpretation [21, 26]. Widely known as nonnegative matrix factorization (NMF), it has been extensively investigated and utilized in areas of computer vision, text mining, and bioinformatics. In higher-order tensors with nonnegative elements, tensor factorizations with nonnegativity constraints on factors have been developed in several papers [4, 6, 24, 29]. Interestingly, some method for finding nonnegative factors of higher-order tensors, such as [6], were introduced even before NMF. Recent work dealt with properties such as degeneracy [23] and applications such as sound source separation [9], text mining [2], and computer vision [27].

Suppose a tensor of order three, $\mathcal{X} \in \mathbb{R}^{M_1 \times M_2 \times M_3}$, is given. We will introduce main concepts using this third-order tensor for the sake of simplicity, and will deal with a tensor with a general order later. A canonical decomposition (CANDECOMP) [5], or equivalently the parallel factor analysis (PARAFAC) [11], of $\mathcal{X}$ can be written as

$$\mathcal{X} = \sum_{k=1}^{K} \mathbf{a}_k \circ \mathbf{b}_k \circ \mathbf{c}_k, \tag{16.1}$$

where $\mathbf{a}_k \in \mathbb{R}^{M_1}$, $\mathbf{b}_k \in \mathbb{R}^{M_2}$, $\mathbf{c}_k \in \mathbb{R}^{M_3}$, and "$\circ$" represents an outer product of vectors. Following [19], we will call a decomposition in the form of Eq. (16.1) the CP (CANDECOMP/PARAFAC) decomposition. A tensor in a form of $\mathbf{a} \circ \mathbf{b} \circ \mathbf{c}$ is called a *rank-one* tensor: In the CP decomposition, tensor $\mathcal{X}$ is represented as a sum of K rank-one tensors. A smallest integer K for which Eq. (16.1) holds with some vectors $\mathbf{a}_k$, $\mathbf{b}_k$, and $\mathbf{c}_k$ for $k \in \{1, \ldots, K\}$ is called the *rank* of tensor $\mathcal{X}$. The CP decomposition can be more compactly represented with *factor matrices* (or *loading matrices*), $\mathbf{A} = [\mathbf{a}_1 \cdots \mathbf{a}_K]$, $\mathbf{B} = [\mathbf{b}_1 \cdots \mathbf{b}_K]$, and $\mathbf{C} = [\mathbf{c}_1 \cdots \mathbf{c}_K]$, as follows:

$$\mathcal{X} = [\![\mathbf{A}, \mathbf{B}, \mathbf{C}]\!],$$

where $[\![\mathbf{A}, \mathbf{B}, \mathbf{C}]\!] = \sum_{k=1}^{K} \mathbf{a}_k \circ \mathbf{b}_k \circ \mathbf{c}_k$ (see [19]). With a tensor $\mathcal{X}$ of rank R, given an integer $K \leq R$, the computational problem of the CP decomposition is finding factor matrices $\mathbf{A}$, $\mathbf{B}$, and $\mathbf{C}$ that best approximates $\mathcal{X}$.

Now, for a tensor $\mathcal{X}$ with only nonnegative elements, we are interested in recovering factor matrices $\mathbf{A}$, $\mathbf{B}$, and $\mathbf{C}$ that also contain only nonnegative components. Using the Frobenius norm as a criterion for approximation, the factor matrices can be found by solving an optimization problem:

$$\min_{\mathbf{A},\mathbf{B},\mathbf{C}} \left\| \mathcal{X} - [\![\mathbf{A}, \mathbf{B}, \mathbf{C}]\!] \right\|_F^2 \quad \text{s.t.} \quad \mathbf{A}, \mathbf{B}, \mathbf{C} \geq 0. \tag{16.2}$$

Inequalities $\mathbf{A}, \mathbf{B}, \mathbf{C} \geq 0$ denote that all the elements of $\mathbf{A}$, $\mathbf{B}$, and $\mathbf{C}$ are nonnegative. The factorization problem in Eq. (16.2) is known as nonnegative CP (NNCP). The computation of NNCP is demanding not only because many variables are involved in optimization but also because nonnegativity constraints are imposed on the factors. A number of algorithms have been developed for NNCP [4, 10, 15, 29], and we will review them in Sect. 16.2.

In this chapter, extending our prior work on NMF [17], we present a new and efficient algorithm for computing NNCP. Our algorithm is based on alternating nonnegativity-constrained least squares (ANLS) framework, where in each iteration the nonnegativity-constrained least squares (NNLS) subproblems are solved. We propose to solve the NNLS problems based on the block principal pivoting method [12]. The block principal pivoting method accelerates the traditional active-set method [20] by allowing exchanges of multiple variables between index groups per iteration. We adopt ideas that improve the block principal pivoting method in multiple right-hand sides [17].

The remaining of this chapter is organized as follows. In Sect. 16.2, related work is reviewed. In Sect. 16.3, the ANLS framework is described, and in Sect. 16.4, the block principal pivoting method is introduced as well as ideas for improvements for multiple right-hand sides. In Sect. 16.5, we describe how the proposed method can be used to solve regularized and sparse formulations. In Sect. 16.6, experimentation settings and results are shown. We conclude this chapter in Sect. 16.7.

Notations Let us summarize some notations used in this chapter. A lowercase or an uppercase letter, such as x or X, is used to denote a scalar; a boldface lowercase letter, such as $\mathbf{x}$, is used to denote a vector; a boldface uppercase letter, such as $\mathbf{X}$, is used to denote a matrix; and a boldface Euler script letter, such as $\boldsymbol{\mathscr{X}}$, is used to denote a tensor of order three or higher. Indices typically grow from 1 to its uppercase letter: For example, $n \in \{1, \dots, N\}$. Elements of a sequence of vectors, matrices, or tensors are denoted by superscripts within parentheses: $\mathbf{X}^{(1)}, \dots, \mathbf{X}^{(N)}$. For a matrix $\mathbf{X}$, $\mathbf{x}_i$ denotes its ith column, and x_{ij} denotes its (i, j) component.

16.2 Related Work

Several computational methods have been developed for solving NNCP. Within the ANLS framework, different methods for solving the NNLS subproblems have been proposed. A classical method for solving the NNLS problem is the active set method of Lawson and Hanson [20]; however, applying Lawson and Hanson's method directly to NNCP is extremely slow. Bro and De Jong [4] suggested an improved active-set method to solve the NNLS problems, and Ven Benthem and Keenan [28] further accelerated the active-set method, which was later utilized in NMF [14] and NNCP [15]. In Friedlander and Hatz [10], the NNCP subproblems are solved by a two-metric projected gradient descent method.

In our work of this chapter, we solve the NNLS subproblems using the block principal pivoting method [12, 17]. The block principal pivoting method is similar to the active set method in that (1) the groups of zero and nonzero variables are explicitly kept track of, and (2) a system of linear equations is solved at each iteration. However, unlike the active set method, the objective function value in the block principal pivoting method does not monotonically decrease. Instead, by exchanging multiple variables between variable groups after each iteration, the block principal pivoting method is much faster than the active set method. Due the relationship with the active set method, we note the block principal pivoting method as an *active-set-like* method.

Numerous other algorithms that are not based on the ANLS framework were suggested. Paatero discussed a Gauss-Newton method [24] and a conjugate gradient method [25], but nonnegativity constraints were not rigorously handled in those work. Extending the multiplicative updating rule of Lee and Seung [22], Welling and Weber [29] proposed a multiplicative updating method for NNCP. Earlier in [6], Carroll et al. proposed a simple procedure that focuses on a rank-one approximation conditioned that other variables are fixed. Recently, Cichocki et al. proposed a similar algorithm, called hierarchical alternating least squares (HALS), which updates each column of factor matrices at a time [8].

16.3 ANLS Framework

We describe the ANLS framework for solving NNCP. Let us consider the a Nth-order tensor $\mathcal{X} \in \mathbb{R}^{M_1 \times \cdots \times M_N}$ and a corresponding factorization problem

$$\begin{aligned} \min_{\mathbf{A}^{(1)},\ldots,\mathbf{A}^{(N)}} \quad & f(\mathbf{A}^{(1)},\ldots,\mathbf{A}^{(N)}) = \left\| \mathcal{X} - [\![\mathbf{A}^{(1)},\ldots,\mathbf{A}^{(N)}]\!] \right\|_F^2 \\ \text{s.t.} \quad & \mathbf{A}^{(n)} \geq 0 \quad \text{for } n = 1,\ldots,N, \end{aligned} \tag{16.3}$$

where $\mathbf{A}^{(n)} \in \mathbb{R}^{M_n \times K}$ for $n = 1,\ldots,N$, and

$$[\![\mathbf{A}^{(1)},\ldots,\mathbf{A}^{(N)}]\!] = \sum_{k=1}^{K} \mathbf{a}_k^{(1)} \circ \cdots \circ \mathbf{a}_k^{(N)}.$$

In order to introduce the ANLS framework, we need definitions of some tensor operations. See Kolda and Bader [19] and references therein for more details of these operations.

Mode-n matricization The mode-n matricization of a tensor $\mathcal{X}$, denoted by $\mathbf{X}_{(n)}$, is a matrix obtained by linearizing all indices except n. More formally, $\mathbf{X}_{(n)}$ is a matrix of size $M_n \times \prod_{k=1,k\neq n}^{N} M_k$, and the $(m_1,\ldots,m_N)$th element of $\mathcal{X}$ is mapped to the (m_n, I)th element of $\mathbf{X}_{(n)}$ where

$$I = 1 + \sum_{k=1}^{N} (m_k - 1) I_k, \quad \text{and} \quad I_k = \prod_{j=1,j\neq n}^{k-1} M_j.$$

Khatri–Rao product The Khatri–Rao product of two matrices $\mathbf{A} \in \mathbb{R}^{J_1 \times L}$ and $\mathbf{B} \in \mathbb{R}^{J_2 \times L}$, denoted by $\mathbf{A} \odot \mathbf{B}$, is defined as

$$\mathbf{A} \odot \mathbf{B} = \begin{bmatrix} a_{11}\mathbf{b}_1 & a_{12}\mathbf{b}_2 & \cdots & a_{1L}\mathbf{b}_L \\ a_{21}\mathbf{b}_1 & a_{22}\mathbf{b}_2 & \cdots & a_{2L}\mathbf{b}_L \\ \vdots & \vdots & \ddots & \vdots \\ a_{J_1 1}\mathbf{b}_1 & a_{J_1 2}\mathbf{b}_2 & \cdots & a_{J_1 L}\mathbf{b}_L \end{bmatrix}.$$

Using above notations, the approximation model

$$\mathcal{X} \approx [\![\mathbf{A}^{(1)}, \ldots, \mathbf{A}^{(N)}]\!]$$

can be written as, for any $n \in \{1, \ldots, N\}$,

$$\mathbf{X}^{(n)} \approx \mathbf{A}^{(n)} \times \left(\mathbf{B}^{(n)}\right)^T, \tag{16.4}$$

where

$$\mathbf{B}^{(n)} = \mathbf{A}^{(N)} \odot \cdots \odot \mathbf{A}^{(n+1)} \odot \mathbf{A}^{(n-1)} \odot \cdots \odot \mathbf{A}^{(1)} \in \mathbb{R}^{(\prod_{k=1,k\neq n}^{N} M_k) \times K}. \tag{16.5}$$

Equation (16.4) is a key relationship that is utilized in the ANLS framework. The ANLS framework is a block-coordinate-descent method applied to Eq. (16.3). First, $\mathbf{A}^{(2)}, \ldots, \mathbf{A}^{(N)}$ are initialized with nonnegative components. Then, for $n = 1, \ldots, N$, the following subproblem is solved iteratively:

$$\begin{aligned} \min_{\mathbf{A}^{(n)}} \quad & \left\| \mathbf{B}^{(n)} \times \left(\mathbf{A}^{(n)}\right)^T - \left(\mathbf{X}^{(n)}\right)^T \right\|_F^2 \\ \text{s.t.} \quad & \mathbf{A}^{(n)} \geq 0. \end{aligned} \tag{16.6}$$

The convergence property of a block-coordinate-descent method [3] states that if each subproblem in the form of Eq. (16.6) has a unique solution, then every limit point produced by the ANLS framework is a stationary point. In particular, if matrices $\mathbf{B}^{(n)}$ are of full column rank, each subproblem has a unique solution.

The problem in Eq. (16.6) is in the form of the nonnegativity-constrained least squares (NNLS) problems, and an efficient algorithm to solve the problem will be the subject of next section. For now, typical characteristics of the subproblem in Eq. (16.6) deserves to be noted. Due to the flattening by the Khatri–Rao product, matrix $\mathbf{B}^{(n)}$ in Eq. (16.6) is typically long and thin. Also, as NNCP is often used for low-rank approximation, matrix $(\mathbf{A}^{(n)})^T$ in Eq. (16.6) is typically flat and wide. These properties will be important in designing efficient algorithms for solving Eq. (16.6), which we now describe.

16.4 Block Principal Pivoting Method

The block principal pivoting method, which we adopt in this work to solve Eq. (16.6), was earlier proposed by Judice and Pires [12] for a single right-hand side case. We will first explain this method and then explain efficient ways to accelerate the multiple right-hand side case as proposed in [17].

The motivation of the block principal pivoting method comes from the difficulty of conventional active set algorithms which occur when the number of variables increases. In the active set method, because typically only one variable is exchanged per iteration between working sets, the number of iterations until termination heavily depends on the number of variables. To accelerate computation, an algorithm whose iteration count does not depend on the number of variables is desirable. The block principal pivoting method manages to do so by exchanging multiple variables at a time.

For the moment, consider an NNLS problem with a single right-hand side vector:

$$\min_{\mathbf{x}\geq 0} \|\mathbf{V}\mathbf{x}-\mathbf{w}\|_2^2, \tag{16.7}$$

where $\mathbf{V}\in\mathbb{R}^{P\times Q}$, $\mathbf{x}\in\mathbb{R}^{Q\times 1}$, and $\mathbf{w}\in\mathbb{R}^{P\times 1}$. The subproblems in Eq. (16.6) are decomposed to independent instances of Eq. (16.7) with respect to each column vector of $(\mathbf{A}^{(n)})^T$. Hence, an algorithm for Eq. (16.7) is a basic building block of an algorithm for Eq. (16.6).

The Karush–Kuhn–Tucker (KKT) optimality conditions for Eq. (16.7) are given as

$$\mathbf{y}=\mathbf{V}^T\mathbf{V}\mathbf{x}-\mathbf{V}^T\mathbf{w}, \tag{16.8a}$$

$$\mathbf{y}\geq 0,\quad \mathbf{x}\geq 0, \tag{16.8b}$$

$$x_q y_q = 0,\quad q=1,\dots,Q. \tag{16.8c}$$

We assume that the matrix $\mathbf{V}$ has full column rank. In this case, a solution $\mathbf{x}$ that satisfies the conditions in Eqs. (16.8a)–(16.8c) is the optimal solution of Eq. (16.7).

We divide the index set $\{1,\dots,Q\}$ into two subgroups $\mathcal{F}$ and $\mathcal{G}$ where $\mathcal{F}\cup\mathcal{G}=\{1,\dots,Q\}$ and $\mathcal{F}\cap\mathcal{G}=\emptyset$. Let $\mathbf{x}_{\mathcal{F}}$, $\mathbf{x}_{\mathcal{G}}$, $\mathbf{y}_{\mathcal{F}}$, and $\mathbf{y}_{\mathcal{G}}$ denote the subsets of variables with corresponding indices, and let $\mathbf{V}_{\mathcal{F}}$ and $\mathbf{V}_{\mathcal{G}}$ denote the submatrices of $\mathbf{V}$ with corresponding column indices. Initially, we assign zeros to $\mathbf{x}_{\mathcal{G}}$ and $\mathbf{y}_{\mathcal{F}}$. Then, by construction, $\mathbf{x}=(\mathbf{x}_{\mathcal{F}},\mathbf{x}_{\mathcal{G}})$ and $\mathbf{y}=(\mathbf{y}_{\mathcal{F}},\mathbf{y}_{\mathcal{G}})$ always satisfy Eq. (16.8c) for any $\mathbf{x}_{\mathcal{F}}$ and $\mathbf{y}_{\mathcal{G}}$. Now, we compute $\mathbf{x}_{\mathcal{F}}$ and $\mathbf{y}_{\mathcal{G}}$ using Eq. (16.8a) and check whether the computed values of $\mathbf{x}_{\mathcal{F}}$ and $\mathbf{y}_{\mathcal{G}}$ satisfy Eq. (16.8b). Computation of $\mathbf{x}_{\mathcal{F}}$ and $\mathbf{y}_{\mathcal{G}}$ is done as follows:

$$\mathbf{V}_{\mathcal{F}}^T\mathbf{V}_{\mathcal{F}}\mathbf{x}_{\mathcal{F}}=\mathbf{V}_{\mathcal{F}}^T\mathbf{w}, \tag{16.9a}$$

$$\mathbf{y}_{\mathcal{G}}=\mathbf{V}_{\mathcal{G}}^T(\mathbf{V}_{\mathcal{F}}\mathbf{x}_{\mathcal{F}}-\mathbf{w}). \tag{16.9b}$$

One can first solve for $\mathbf{x}_{\mathcal{F}}$ in Eq. (16.9a) and use it to compute $\mathbf{y}_{\mathcal{G}}$ in Eq. (16.9b). We call the computed pair $(\mathbf{x}_{\mathcal{F}},\mathbf{y}_{\mathcal{G}})$ a complementary basic solution.

If a complementary basic solution $(\mathbf{x}_{\mathcal{F}},\mathbf{y}_{\mathcal{G}})$ satisfies $\mathbf{x}_{\mathcal{F}}\geq 0$ and $\mathbf{y}_{\mathcal{G}}\geq 0$, then it is called *feasible*. In this case, $\mathbf{x}=(\mathbf{x}_{\mathcal{F}},0)$ is the optimal solution of Eq. (16.7), and the algorithm terminates. Otherwise, a complementary basic solution $(\mathbf{x}_{\mathcal{F}},\mathbf{y}_{\mathcal{G}})$

is *infeasible*, and we need to update $\mathcal{F}$ and $\mathcal{G}$ by exchanging variables for which Eq. (16.8b) does not hold. Formally, we define the following index set:

$$\mathcal{H} = \{q \in \mathcal{F} : \mathbf{x}_q < 0\} \cup \{q \in \mathcal{G} : \mathbf{y}_q < 0\} \tag{16.10}$$

and choose a nonempty subset $\hat{\mathcal{H}} \subset \mathcal{H}$. Then, $\mathcal{F}$ and $\mathcal{G}$ are updated by the following rules:

$$\mathcal{F} = (\mathcal{F} - \hat{\mathcal{H}}) \cup (\hat{\mathcal{H}} \cap \mathcal{G}), \tag{16.11a}$$

$$\mathcal{G} = (\mathcal{G} - \hat{\mathcal{H}}) \cup (\hat{\mathcal{H}} \cap \mathcal{F}). \tag{16.11b}$$

The number of elements in set $\hat{\mathcal{H}}$, which we denote by $|\hat{\mathcal{H}}|$, represents how many variables are exchanged per iteration between $\mathcal{F}$ and $\mathcal{G}$. If $|\hat{\mathcal{H}}| > 1$, then an algorithm is called a block principal pivoting algorithm; if $|\hat{\mathcal{H}}| = 1$, then an algorithm is called a single principal pivoting algorithm. The active set algorithm can be understood as an instance of single principal pivoting algorithms. An algorithm repeats this procedure until the number of infeasible variables (i.e., $|\hat{\mathcal{H}}|$) becomes zero.

In order to speed up the search procedure, one usually uses $\hat{\mathcal{H}} = \mathcal{H}$, which we call the *full exchange rule*. The full exchange rule means that we exchange all variables of $\mathcal{F}$ and $\mathcal{G}$ that do not satisfy Eqs. (16.8a)–(16.8b), and the rule accelerates computation by reducing the number of iterations. However, contrary to the active set algorithm in which the variable to exchange is carefully selected to reduce the residual, the full exchange rule may lead to a cycle and fail to find an optimal solution although it occurs rarely. To ensure finite termination, we need to employ a backup rule, which uses the following exchange set for Eqs. (16.11a) and (16.11b):

$$\hat{\mathcal{H}} = \{q : q = \max\{q \in \mathcal{H}\}\}. \tag{16.12}$$

The backup rule, where only the infeasible variable with the largest index is exchanged, is a single principal pivoting rule. This simple exchange rule guarantees a finite termination: Assuming that matrix $\mathbf{V}$ has full column rank, the exchange rule in Eq. (16.12) returns the solution of Eqs. (16.8a)–(16.8c) in a finite number of iterations [12]. Combining the full exchange rule and the backup rule, the block principal pivoting method for Eq. (16.7) that terminates within a finite number of iterations is summarized in [12].

Now, let us move on to the multiple right-hand side case:

$$\min_{\mathbf{X} \geq 0} \|\mathbf{V}\mathbf{X} - \mathbf{W}\|_F^2, \tag{16.13}$$

where $\mathbf{V} \in \mathbb{R}^{P \times Q}$, $\mathbf{X} \in \mathbb{R}^{Q \times L}$ and $\mathbf{W} \in \mathbb{R}^{P \times L}$. One can solve Eq. (16.13) by separately solving NNLS problems for each right-hand side vector. Although this approach is possible, we will see that there exist efficient ways to accelerate the multiple right-hand side case employing two important improvements suggested in [17].

Observe that the sets $\mathcal{F}$ and $\mathcal{G}$ change over iterations, and Eqs. (16.9a) and (16.9b) has to be solved for varying $\mathcal{F}$ and $\mathcal{G}$ every time. The first improvement is based on the observation that matrix $\mathbf{V}$, which corresponds to $\mathbf{B}^{(n)}$ of Eq. (16.6),

Algorithm 16.1: Block principal pivoting algorithm for the NNLS with multiple right-hand side vectors. $\mathbf{x}_{\mathcal{F}_l}$ and $\mathbf{y}_{\mathcal{G}_l}$ represents the subsets of lth column of $\mathbf{X}$ and $\mathbf{Y}$ indexed by $\mathcal{F}_l$ and $\mathcal{G}_l$, respectively

Input: $\mathbf{V} \in \mathbb{R}^{P\times Q}, \mathbf{W} \in \mathbb{R}^{Q\times L}$
Output: $\mathbf{X}(\in \mathbb{R}^{Q\times L}) = \arg\min_{\mathbf{X}\geq 0} \|\mathbf{V}\mathbf{X} - \mathbf{W}\|_F^2$
Compute $\mathbf{V}^T\mathbf{V}$ and $\mathbf{V}^T\mathbf{W}$.
Initialize $\mathcal{F}_l = \emptyset$ and $\mathcal{G}_l = \{1, \ldots, q\}$ for all $l \in \{1, \ldots, L\}$. Set $\mathbf{X} = 0$, $Y = -\mathbf{V}^T\mathbf{W}$, $\boldsymbol{\alpha}(\in \mathbb{R}^r) = 3$, and $\boldsymbol{\beta}(\in \mathbb{R}^r) = q + 1$.
Compute $\mathbf{x}_{\mathcal{F}_l}$ and $\mathbf{y}_{\mathcal{G}_l}$ for all $l \in \{1, \ldots, L\}$ by Eqs. (16.9a) and (16.9b) using column grouping.
while any $(\mathbf{x}_{\mathcal{F}_l}, \mathbf{y}_{\mathcal{G}_l})$ is infeasible **do**
  Find the indices of columns in which the solution is infeasible: $I = \{j : (\mathbf{x}_{\mathcal{F}_j}, \mathbf{y}_{\mathcal{G}_j})$ is infeasible$\}$.
  Compute $\mathcal{H}_l$ for all $l \in I$ by Eq. (16.10).
  For all $l \in I$ with $|\mathcal{H}_l| < \beta_l$, set $\beta_l = |\mathcal{H}_l|$, $\alpha_l = 3$ and $\hat{\mathcal{H}}_l = \mathcal{H}_l$.
  For all $l \in I$ with $|\mathcal{H}_l| \geq \beta_l$ and $\alpha_l \geq 1$, set $\alpha_l = \alpha_l - 1$ and $\hat{\mathcal{H}}_l = \mathcal{H}_l$.
  For all $l \in I$ with $|\mathcal{H}_l| \geq \beta_l$ and $\alpha_l = 0$, set $\hat{\mathcal{H}}_l$ by Eq. (16.12).
  Update $\mathcal{F}_l$ and $\mathcal{G}_l$ for all $l \in I$ by Eqs. (16.11a)–(16.11b).
  Update $\mathbf{x}_{\mathcal{F}_l}$ and $\mathbf{y}_{\mathcal{G}_l}$ for all $l \in I$ by Eqs. (16.9a) and (16.9b) using column grouping.
end while

is typically very long and thin. In this case, constructing matrices $\mathbf{V}_{\mathcal{F}}^T\mathbf{V}_{\mathcal{F}}$, $\mathbf{V}_{\mathcal{F}}^T\mathbf{w}$, $\mathbf{V}_{\mathcal{G}}^T\mathbf{V}_{\mathcal{F}}$, and $\mathbf{V}_{\mathcal{G}}^T\mathbf{w}$ before solving Eqs. (16.9a) and (16.9b) is computationally very expensive. To ease this difficulty, $\mathbf{V}^T\mathbf{V}$ and $\mathbf{V}^T\mathbf{W}$ can be computed in the beginning and reused in later iterations. One can easily see that $\mathbf{V}_{\mathcal{F}}^T\mathbf{V}_{\mathcal{F}}$, $\mathbf{V}_{\mathcal{F}}^T\mathbf{w}_l$, $\mathbf{V}_{\mathcal{G}}^T\mathbf{V}_{\mathcal{F}}$, and $\mathbf{V}_{\mathcal{G}}^T\mathbf{w}_l$, $l \in \{1, \ldots, L\}$, can be directly retrieved as a submatrix of $\mathbf{V}^T\mathbf{V}$ or $\mathbf{V}^T\mathbf{W}$. Because the column size of $\mathbf{V}$ is small, storage needed for $\mathbf{V}^T\mathbf{V}$ and $\mathbf{V}^T\mathbf{W}$ is also small.

The second improvement involves exploiting common computations in solving Eq. (16.9a). Here we simultaneously run the block principal pivoting algorithm for multiple right-hand side vectors. At each iteration, we have index sets $\mathcal{F}_l$ and $\mathcal{G}_l$ for each column $l \in \{1, \ldots, L\}$, and we must compute $\mathbf{x}_{\mathcal{F}_l}$ and $\mathbf{y}_{\mathcal{G}_l}$ using Eqs. (16.9a) and (16.9b). The idea is to find groups of columns that share the same index sets $\mathcal{F}_l$ and $\mathcal{G}_l$. We reorder the columns with respect to these groups and solve Eqs. (16.9a) and (16.9b) for the columns in the same group. By doing so, we avoid repeated Cholesky factorization computations required for solving Eq. (16.9a). When matrix $\mathbf{X}$ is flat and wide, which is typically the case for $(\mathbf{A}^{(n)})^T$ in Eq. (16.6), more columns are likely to share their index sets $\mathcal{F}_l$ and $\mathcal{G}_l$, allowing bigger speed-up.

Incorporating these improvements, a full description of the block principal pivoting method for Eq. (16.13) is shown in Algorithm 16.1. Finite termination of Algorithm 16.1 is achieved by controlling the number of infeasible variables using $\boldsymbol{\alpha}$ and $\boldsymbol{\beta}$. For more details of how it is controlled, see [17, 18].

16.5 Regularized and Sparse NNCP

The ANLS framework described in Sect. 16.3 can be easily extended to formulations with regularization. In a general form, a regularized formulation appears as

$$\begin{aligned} \min_{\mathbf{A}^{(1)},\ldots,\mathbf{A}^{(N)}} \quad & \left\| \mathcal{X} - [\![\mathbf{A}^{(1)},\ldots,\mathbf{A}^{(N)}]\!] \right\|_F^2 + \sum_{n=1}^{N} \lambda_n \phi_n(\mathbf{A}^{(n)}), \\ \text{s.t.} \quad & \mathbf{A}^{(n)} \geq 0 \text{ for } n = 1, \cdots, N, \end{aligned} \tag{16.14}$$

where $\phi_n(\mathbf{A}^{(n)})$ represents a regularization term and $\lambda_n \geq 0$ is a parameter to be chosen. A commonly used regularization term is the Frobenius norm:

$$\phi_n(\mathbf{A}^{(n)}) = \left\| \mathbf{A}^{(n)} \right\|_F^2.$$

In this case, the subproblem for finding $\mathbf{A}^{(n)}$ is modified as

$$\begin{aligned} \min_{\mathbf{A}^{(n)}} \quad & \left\| \begin{pmatrix} \mathbf{B}^{(n)} \\ \sqrt{\lambda_n}\mathbf{I}_{K\times K} \end{pmatrix} \times (\mathbf{A}^{(n)})^T - (\mathbf{X}^{(n)})^T \right\|_F^2 \\ \text{s.t.} \quad & \mathbf{A}^{(n)} \geq 0, \end{aligned} \tag{16.15}$$

where $\mathbf{I}_{K\times K}$ is a $K \times K$ identity matrix. Observe that matrix $\begin{pmatrix} \mathbf{B}^{(n)} \\ \sqrt{\lambda_n}\mathbf{I}_{K\times K} \end{pmatrix}$ is always of full column rank; hence, when $\mathbf{B}^{(n)}$ is not necessarily of full column rank, the Frobenius norm regularization can be adopted to ensure that the NNLS subproblem is of full column rank, satisfying the requirement of the convergence property of a block-coordinate-descent method, mentioned in Sect. 16.3. In addition, the block principal pivoting method assumes that the matrix $\mathbf{V}$ in Eq. (16.13) is of full column rank, and the Frobenius norm regularization automatically satisfies this condition.

If it is desired to promote sparsity on factor matrix $\mathbf{A}^{(n)}$, l_1-norm regularization can be used:

$$\phi_n(\mathbf{A}^{(n)}) = \sum_{j=1}^{M_n} \left\| (\mathbf{A}^{(n)})^T (:, j) \right\|_1^2,$$

where $(\mathbf{A}^{(n)})^T(:, j)$ represents the jth column of $(\mathbf{A}^{(n)})^T$. See [13, 16] for applications of this l_1-norm regularization in microarray data analysis and clustering. In this case, the subproblem for finding $\mathbf{A}^{(n)}$ is modified as

$$\begin{aligned} \min_{\mathbf{A}^{(n)}} \quad & \left\| \begin{pmatrix} \mathbf{B}^{(n)} \\ \sqrt{\lambda_n}\mathbf{1}_{1\times K} \end{pmatrix} \times (\mathbf{A}^{(n)})^T - (\mathbf{X}^{(n)})^T \right\|_F^2 \\ \text{s.t.} \quad & \mathbf{A}^{(n)} \geq 0, \end{aligned} \tag{16.16}$$

where $\mathbf{1}_{1\times K}$ is a row vector of ones. Regularization term $\phi_n(\cdot)$ can be separately chosen for each factor $\mathbf{A}^{(n)}$, and if necessary, both of the Frobenius norm and the l_1-norm may be used.

16.6 Implementation and Results

In this section, we describe the details of our implementation, data sets used, and comparison results. All experiments were executed in MATLAB on a Linux machine with a 2.66 GHz Intel Quad-core processor and 6 GB memory. The multi-threading option of MATLAB was disabled. In all the executions, all the algorithms were provided with the same initial values.

16.6.1 Algorithms for NNCP Used for Comparisons

The following algorithms for NNCP were included in our comparison.

1. (ANLS-BPP) ANLS with the block principal pivoting method proposed in this chapter
2. (ANLS-AS) ANLS with H. Kim and Park's active set method [15]
3. (HALS) Cichocki and Phan's hierarchical alternating least squares algorithm [7, 8]
4. (MU) Welling and Weber's multiplicative updating algorithm [29].

We implemented all algorithms in MATLAB. Besides above methods, we also have tested Friedlander and Hatz's two-metric projected gradient method [10] using their MATLAB code;[1] however, not only it was much slower than methods listed above, but it also required so much memory that we could not execute all comparison cases. We hence do not include the results of Friedlander and Hatz's method here. In all the algorithms, once we obtain factors $\{\mathbf{A}^{(1)}, \ldots, \mathbf{A}^{(N)}\}$, they are used as initial values of the next iteration.

16.6.2 Data Sets

We have used three data sets for comparisons. The first data set include dense tensors using synthetically generated factors. For each of $K = 10$, 20, 60, and 120, we constructed $\mathbf{A}^{(1)}$, $\mathbf{A}^{(2)}$, and $\mathbf{A}^{(3)}$ of size $300 \times K$ using random numbers from the uniform distribution over [0, 1]. Then, we randomly selected 50 percent of elements

[1] http://www.cs.ubc.ca/~mpf/2008-computing-nntf.html.

in $\mathbf{A}^{(1)}$, $\mathbf{A}^{(2)}$, and $\mathbf{A}^{(3)}$ to make them zero. Finally, a three way tensor of size $300 \times 300 \times 300$ is constructed by $[\![\mathbf{A}^{(1)}, \mathbf{A}^{(2)}, \mathbf{A}^{(3)}]\!]$. Different tensors were created for different K values.

The second data set is a dense tensor obtained from Extended Yale Face Database B[2]. We used aligned and cropped images of size 168×192. From total 2424 images, we obtained a three-way tensor of size $168 \times 192 \times 2424$.

The third data set is a sparse tensor from NIPS conference papers.[3] This data set contains NIPS papers volume 0 to 12, and a tensor is constructed as a four-way tensor representing author×documents×term×year. By counting the occurrence of each entry, a sparse tensor of size $2037 \times 1740 \times 13649 \times 13$ was created.

16.6.3 Experimental Results

To observe the performance of several algorithms, at the end of each iteration we have recorded the relative objective value, $\|\mathcal{X} - [\![\mathbf{A}^{(1)}, \ldots, \mathbf{A}^{(N)}]\!]\|_F / \|\mathcal{X}\|_F$. Time spent to compute the objective value is excluded from the execution time. One execution result involves relative objective values measured at discrete time points and appears as a piecewise-linear function. We averaged piecewise-linear functions from different random initializations to plot figures.

Results on the synthetic data set are shown in Fig. 16.1. This data set was synthetically created, and the value of global optimum is zero. From Fig. 16.1, it can be seen that ANLS-AS and ANLS-BPP performed the best among the algorithms we tested. The HALS method showed convergence within the time window we have observed, but the MU method was too slow to show convergence. ANLS-AS and ANLS-BPP showed almost the same performance although ANLS-BPP was slightly faster when $k = 120$. The difference between these two methods are better shown in next results.

Results on YaleB and NIPS data sets are shown in Fig. 16.2. Similarly to the results in Fig. 16.1, ANLS-AS and ANLS-BPP showed the best performance. In Fig. 16.2, it can be clearly observed that ANLS-BPP outperforms ANLS-AS for $k = 60$ and $k = 120$ cases. Such a difference demonstrates a difficulty of the active-set method: Since typically only one variable is exchanged between working sets, the active-set method is slow for a problem with a large number of variables. On the other hand, the block principal pivoting method quickly solves large problems by allowing exchanges of multiple variables between $\mathcal{F}$ and $\mathcal{G}$. The convergence of HALS and MU was slower than ANLS-AS and ANLS-BPP. Although the convergence of HALS was faster than MU in the YaleB data set, the initial convergence of MU was faster than HALS in the NIPS data set.

---

[2]http://vision.ucsd.edu/~leekc/ExtYaleDatabase/ExtYaleB.html

[3]http://www.cs.nyu.edu/~roweis/data.html.

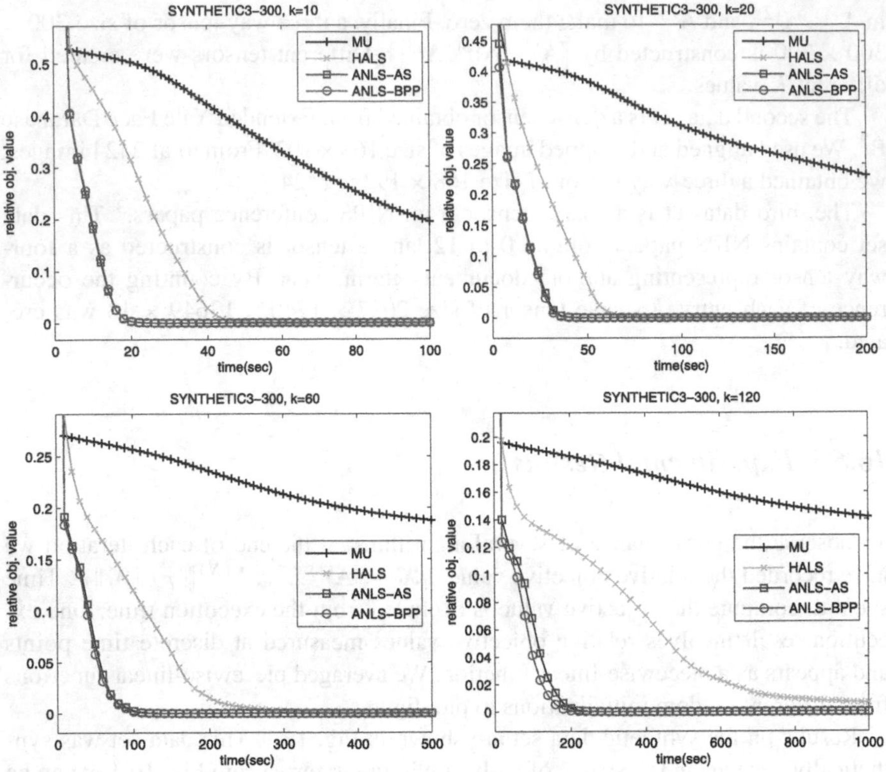

Fig. 16.1 Relative objective value ($\|\mathcal{X} - [\![\mathbf{A}^{(1)}, \ldots, \mathbf{A}^{(N)}]\!]\|_F / \|\mathcal{X}\|_F$) vs. execution time on the synthetic tensors. Average results of 5 different random initializations are shown. *Top row*: $k = 10$ and $k = 20$, *bottom row*: $k = 60$ and $k = 120$

Lastly, we present more detailed information regarding the executions of ANLS-AS and ANLS-BPP in Fig. 16.3. In Fig. 16.1 and Fig. 16.2, we have observed that ANLS-BPP clearly outperforms ANLS-AS for large k's. Because both of the methods solve each NNLS subproblem exactly, solutions after each iteration from the two methods are the same up to numerical rounding errors. Hence, it suffices to compare the amount of time spent at each iteration. In Fig 16.3, we showed average execution time of each iteration of the two methods. It can be seen that the time required for ANLS-BPP is significantly shorter than the time required for ANLS-AS in early iterations, and their time requirements became gradually closer to each other. The types of NNLS problem in which ANLS-BPP accelerates ANLS-AS is the case that there is much difference in the zero and nonzero pattern between the initial value and the final solution of the NNLS problem. As iteration goes on, factors $\{\mathbf{A}^{(1)}, \ldots, \mathbf{A}^{(N)}\}$ do not change much from one iteration to the next; hence there are little differences between the computational costs of the two methods.

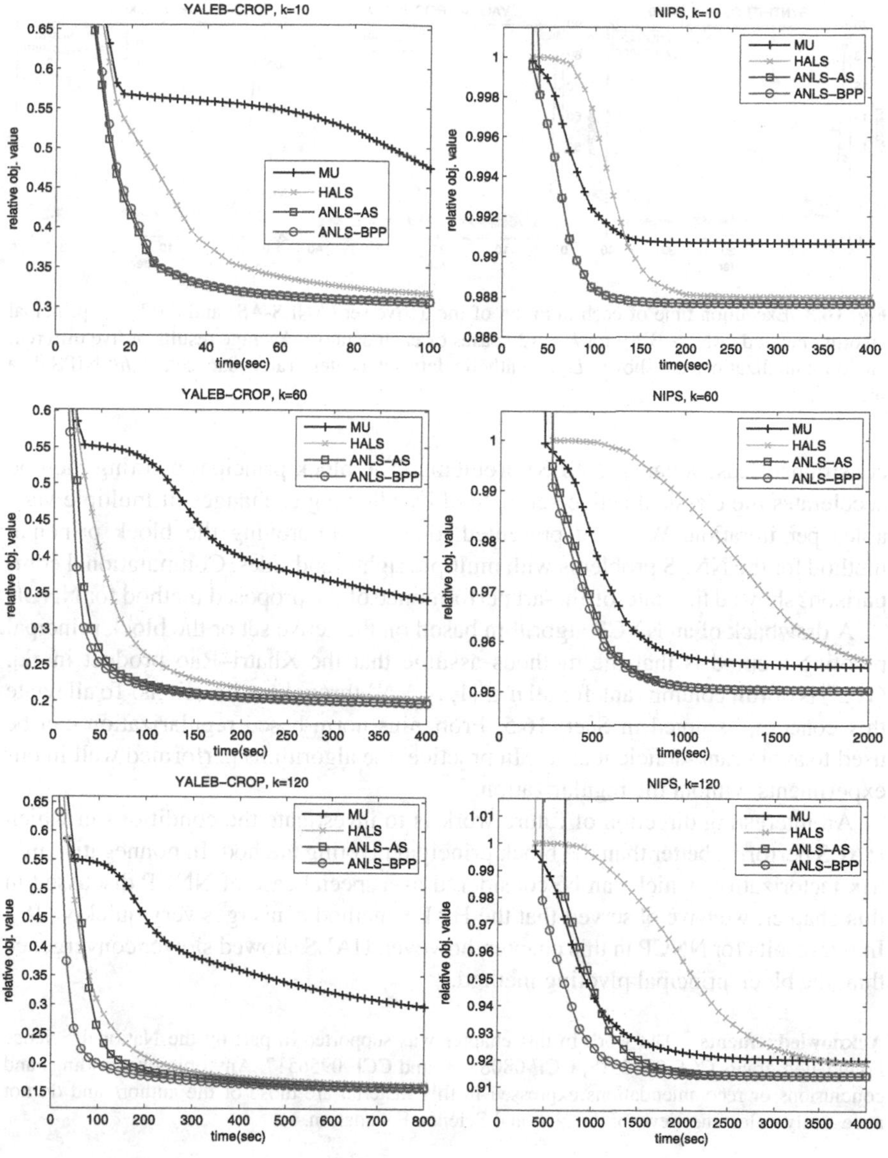

Fig. 16.2 Relative objective value ($\|\mathcal{X} - [\![\mathbf{A}^{(1)}, \ldots, \mathbf{A}^{(N)}]\!]\|_F / \|\mathcal{X}\|_F$) vs. execution time on the YaleB and NIPS data sets. Average results of 5 different random initializations are shown. *Left*: NIPS data set, *right*: YaleB data set, *top row*: $k = 10$, *middle row*: $k = 60$, and *bottom row*: $k = 120$

16.7 Conclusions and Discussion

We have introduced an efficient algorithm for nonnegative CP (NNCP). The new method is based on the block principal pivoting method for the nonnegativity-

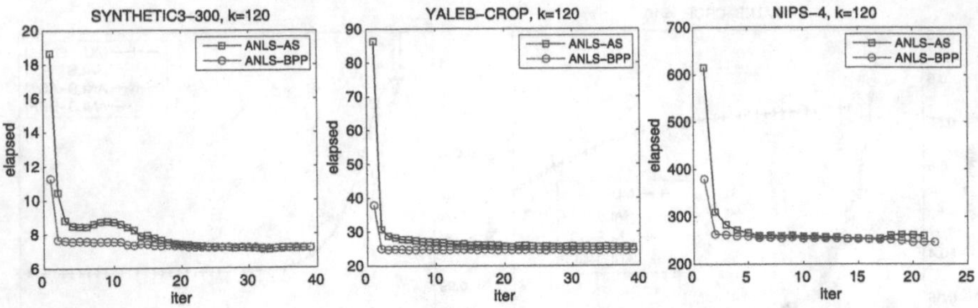

Fig. 16.3 Execution time of each iteration of the active set (ANLS-AS) and the block principal pivoting method (ANLS-BPP) for $k = 120$ cases of each data set. Average results of five different random initializations are shown. *Left*: synthetic data set, center: YaleB data set, *right*: NIPS data set

constrained least squares (NNLS) problems. The block principal pivoting method accelerates the classical active-set method by allowing exchanges of multiple variables per iteration. We have presented ideas for improving the block principal method for the NNLS problems with multiple right-hand sides. Computational comparisons showed the state-of-the-art performance of the proposed method for NNCP.

A drawback of an NNCP algorithm based on the active set or the block principal pivoting method is that the methods assume that the Khatri–Rao product in Eq. (16.5) is of full column rank for all $n \in \{1, \dots, N\}$ throughout iterations. To alleviate this concern, as noted in Sect. 16.5, Frobenius norm-based regularization can be used to avoid rank-deficient cases. In practice, the algorithms performed well in our experiments without the regularization.

An interesting direction of future work is to investigate the conditions in which HALS performs better than the block principal pivoting method. In nonnegative matrix factorization, which can be considered as a special case of NNCP discussed in this chapter, we have observed that the HALS method converges very quickly [18]. In our results for NNCP in this chapter, however, HALS showed slower convergence than the block principal pivoting method.

Acknowledgements The work in this chapter was supported in part by the National Science Foundation grants CCF-0732318, CCF-0808863, and CCF-0956517. Any opinions, findings and conclusions or recommendations expressed in this material are those of the authors and do not necessarily reflect the views of the National Science Foundation.

References

1. Acar, E., Yener, B.: Unsupervised multiway data analysis: A literature survey. IEEE Trans. Knowl. Data Eng. **21**(1), 6–20 (2009)
2. Bader, B.W., Berry, M.W., Browne, M.: Discussion tracking in Enron email using PARAFAC. In: Survey of Text Mining II: Clustering, Classification, and Retrieval, pp. 147–163. Springer, Berlin (2008)
3. Bertsekas, D.P.: Nonlinear Programming. Scientific, Athena (1999)

4. Bro, R., De Jong, S.: A fast non-negativity-constrained least squares algorithm. J. Chem. **11**, 393–401 (1997)
5. Carroll, J.D., Chang, J.J.: Analysis of individual differences in multidimensional scaling via an N-way generalization of "Eckart-Young" decomposition. Psychometrika **35**(3), 283–319 (1970)
6. Carroll, J.D., Soete, G.D., Pruzansky, S.: Fitting of the latent class model via iteratively reweighted least squares CANDECOMP with nonnegativity constraints. In: Multiway data analysis, pp. 463–472. North-Holland, Amsterdam (1989). http://portal.acm.org/citation.cfm?id=120565.120614
7. Cichocki, A., Phan, A.H.: Fast local algorithms for large scale nonnegative matrix and tensor factorizations. IEICE Trans. Fundam. Electron. Commun. Comput. Sci. **E92-A**(3), 708–721 (2009)
8. Cichocki, A., Zdunek, R., Amari, S.I.: Hierarchical ALS algorithms for nonnegative matrix and 3D tensor factorization. In: Lecture Notes in Computer Science, vol. 4666, pp. 169–176. Springer, Berlin (2007)
9. FitzGerald, D., Cranitch, M., Coyle, E.: Non-negative tensor factorisation for sound source separation. In: Proceedings of the Irish Signals and Systems Conference (2005)
10. Friedlander, M.P., Hatz, K.: Computing nonnegative tensor factorizations. Comput. Optim. Appl. **23**(4), 631–647 (2008). doi:10.1080/10556780801996244
11. Harshman, R.A.: Foundations of the PARAFAC procedure: Models and conditions for an "explanatory" multi-modal factor analysis. In: UCLA Working Papers in Phonetics, vol. 16, pp. 1–84 (1970)
12. Júdice, J.J., Pires, F.M.: A block principal pivoting algorithm for large-scale strictly monotone linear complementarity problems. Comput. Oper. Res. **21**(5), 587–596 (1994)
13. Kim, H., Park, H.: Sparse non-negative matrix factorizations via alternating non-negativity-constrained least squares for microarray data analysis. Bioinformatics **23**(12), 1495–1502 (2007)
14. Kim, H., Park, H.: Nonnegative matrix factorization based on alternating nonnegativity constrained least squares and active set method. SIAM J. Matrix Anal. Appl. **30**(2), 713–730 (2008). doi:10.1137/07069239X
15. Kim, H., Park, H., Eldén, L.: Non-negative tensor factorization based on alternating large-scale non-negativity-constrained least squares. In: Proceedings of IEEE 7th International Conference on Bioinformatics and Bioengineering (BIBE07), vol. 2, pp. 1147–1151 (2007)
16. Kim, J., Park, H.: Sparse nonnegative matrix factorization for clustering. Tech. rep., Georgia Institute of Technology Technical Report GT-CSE-08-01 (2008)
17. Kim, J., Park, H.: Toward faster nonnegative matrix factorization: A new algorithm and comparisons. In: Proceedings of the 2008 Eighth IEEE International Conference on Data Mining (ICDM), pp. 353–362 (2008)
18. Kim, J., Park, H.: Fast nonnegative matrix factorization: An active-set-like method and comparisons. SIAM J. Sci. Comput. **33**, 3261 (2011)
19. Kolda, T.G., Bader, B.W.: Tensor decompositions and applications. SIAM Rev. **51**(3), 455–500 (2009)
20. Lawson, C.L., Hanson, R.J.: Solving Least Squares Problems. Prentice Hall, New York (1974)
21. Lee, D.D., Seung, H.S.: Learning the parts of objects by non-negative matrix factorization. Nature **401**(6755), 788–791 (1999)
22. Lee, D.D., Seung, H.S.: Algorithms for non-negative matrix factorization. In: Advances in Neural Information Processing Systems, vol. 13, pp. 556–562. MIT Press, Cambridge (2001)
23. Lim, L.H., Comon, P.: Nonnegative approximations of nonnegative tensors. J. Chem., **23**(7–8), 432–441 (2009)
24. Paatero, P.: A weighted non-negative least squares algorithm for three-way PARAFAC factor analysis. Chemom. Intell. Lab. Syst. **38**(2), 223–242 (1997)
25. Paatero, P.: The multilinear engine: A table-driven, least squares program for solving multilinear problems, including the n-way parallel factor analysis model. J. Comput. Graph. Stat. **8**, 854–888 (1999)

26. Paatero, P., Tapper, U.: Positive matrix factorization: A non-negative factor model with optimal utilization of error estimates of data values. EnvironMetrics **5**(1), 111–126 (1994)
27. Shashua, A., Hazan, T.: Non-negative tensor factorization with applications to statistics and computer vision. In: ICML '05: Proceedings of the 22nd International Conference on Machine Learning, pp. 792–799. ACM, New York (2005). doi: http://doi.acm.org/10.1145/1102351.1102451
28. Van Benthem, M.H., Keenan, M.R.: Fast algorithm for the solution of large-scale non-negativity-constrained least squares problems. J. Chem. **18**, 441–450 (2004). doi:10.1002/cem.889
29. Welling, M., Weber, M.: Positive tensor factorization. Pattern Recognit. Lett. **22**(12), 1255–1261 (2001). doi:10.1016/S0167-8655(01)00070-8

Chapter 17
Knowledge Discovery Using Nonnegative Tensor Factorization with Visual Analytics

Andrey A. Puretskiy and Michael W. Berry

Abstract Non-negative tensor factorization (NTF) is a technique that has been used effectively for the purposes of analyzing large textual datasets. This article describes the improvements achieved by creating a Python implementation of the NTF algorithm, and by integrating it with several pre-processing and post-processing functions within a single Python-based analysis environment. The improved implementation allows the user to construct and modify the contents of the tensor, experiment with relative term weights and trust measures, and experiment with the total number of algorithm output features. Non-negative tensor factorization output feature production is closely integrated with a visual post-processing tool, FutureLens, that allows the user to perform in-depth analysis of textual data, facilitating scenario extraction and knowledge discovery.

17.1 Background

A wide variety of fields, such as biology, medical science, various social sciences, the legal field, and business have the potential to greatly benefit from an ability to analyze vast amounts of textual data. The digitalization trend of recent decades, combined with the readily available and increasingly inexpensive digital storage capabilities, has resulted in a newfound ability to gather, organize, store, and analyze vast repositories of knowledge in all of those fields and many others. As computing and digitalization increasingly permeate virtually every aspect of society, researchers and analysts sometimes find themselves overwhelmed with enormous quantities of information. The fields of data mining and visual analytics developed alongside the

A.A. Puretskiy (✉)
Department of Electrical Engineering and Computer Science, University of Tennessee,
203 Claxton Complex, Knoxville, TN 37996-3450, USA
e-mail: puretski@eecs.utk.edu

M.W. Berry
Center for Intelligent Systems and Machine Learning (CISML), University of Tennessee,
203 Claxton Complex, Knoxville, TN 37996-3450, USA
e-mail: berry@eecs.utk.edu

M.W. Berry et al. (eds.), *High-Performance Scientific Computing*,
DOI 10.1007/978-1-4471-2437-5_17,

ever-increasing information stores in order to provide analytical knowledge discovery capabilities.

17.1.1 NTF-PARAFAC: Background

There exists a plethora of approaches to analyzing large amounts of textual information. The exact nature of the dataset and the goals of the analysis process influence which approach has the potential to be most effective in each particular case. For cases where the dataset contains tagged entities and a clearly defined time-line, nonnegative tensor factorization (NTF) techniques have been shown to be highly effective. NTF allows the analyst to extract term-by-entity associations from the data. With the addition of a visual post-processing tool (FutureLens), it becomes possible to trace the progression of term-entity, term-term, and entity-entity relationships through the data space over time. One example of such a study involved scenario discovery using the fictional news article dataset from the IEEE VAST-2007 contest [2, 7]. As shown by this example, NTF based on the well-known PARAFAC [4] model for multidimensional data can be highly effective in extracting important features from a large textual dataset.

17.1.2 NTF-PARAFAC: The Algorithm

The Parallel Factors (PARAFAC) model, also known as Canonical Decomposition, was proposed by Harshman in 1970 [3, 4]. Given a third-order tensor $\mathcal{X}$ of size $m \times n \times p$ and a desired approximation rank r, the PARAFAC model approximates $\mathcal{X}$ as a sum of r rank-1 tensors formed by the outer products of three vectors, i.e.,

$$\mathcal{X} \approx \sum_{i=1}^{r} a_i \circ b_i \circ c_i, \tag{17.1}$$

where the symbol $\circ$ denotes the outer (tensor) product.

The goal of NTF is to find best fitting nonnegative matrices, A, B, and C, that fit the data in $\mathcal{X}$. That is,

$$\min_{A,B,C} = \left\| \mathcal{X} - \sum_{i=1}^{r} a_i \circ b_i \circ c_i \right\|. \tag{17.2}$$

Graphically, this process may be illustrated by as a decomposition of a datacube into component features. For example, for an email-based dataset, the corresponding datacube may be term-by-author-by-time. The term-by-author-by-time decomposition [6] is illustrated in Fig. 17.1. Each of the groups resulting from the decomposition can be said to represent a feature of the data—some group of interrelated components found within the dataset.

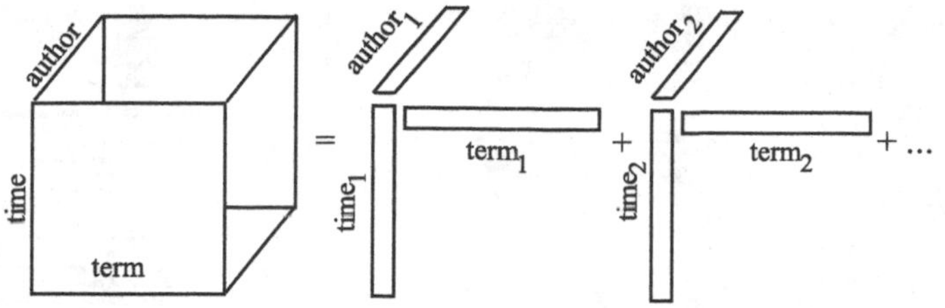

Fig. 17.1 The 3-way NTF PARAFAC decomposition model produces a number of data features, each one corresponding to a potentially significant underlying theme or scenario contained in the email dataset

17.1.3 NTF-PARAFAC: Example of Effective Use

In the study described in [2], NTF-PARAFAC was applied to a $12{,}121 \times 7{,}141 \times 15$ sparse tensor that contained 1,142,077 nonzeros. This resulted in twenty-five total output groups, each described by fifteen interrelated entities and thirty-five interrelated terms.

The groups corresponding to the two fictional *hidden* scenarios were correctly identified, although the identification process required a significant time commitment and several post-processing steps. The study was subsequently replicated using a visual post-processing software tool, FutureLens, in order to improve the effectiveness and efficiency of processing NTF output group results [8]. The two figures below illustrate how FutureLens was used to identify and gather evidence for the IEEE VAST-2007 scenario involving a bioterrorism-induced monkeypox outbreak.

Figure 17.2 hows one of the NTF output files ("Tensor Group 15") loaded into FutureLens. This group is described by a list of top 15 most relevant entities and 35 most relevant terms. In this figure, the user has selected two of the top terms (*monkeypox* and *outbreak*), and then combined them into a collection of terms (*monkeypox, outbreak*). FutureLens located two articles containing both terms. The first article describes much of the bioterrorism scenario, however, a few crucial details regarding the perpetrator are missing in this article.

In order to locate addition information pertaining to this scenario, the user adds the entity corresponding to the suspect's name (*Cesar Gil*), and a collection of terms (*chinchilla, Gil*), to the FutureLens display. Doing so allows the user to locate an article where Cesar Gil explains his philosophy regarding the trade in exotic animals (he states that breaking a few laws is an acceptable tactic in stopping such trade). In addition, as shown in Fig. 17.3 below, the user is also able to locate an article corresponding to an advertisement of chinchillas for sale by a business called "Gil Breeders". A complete storyline corresponding to this NTF output feature now emerges.

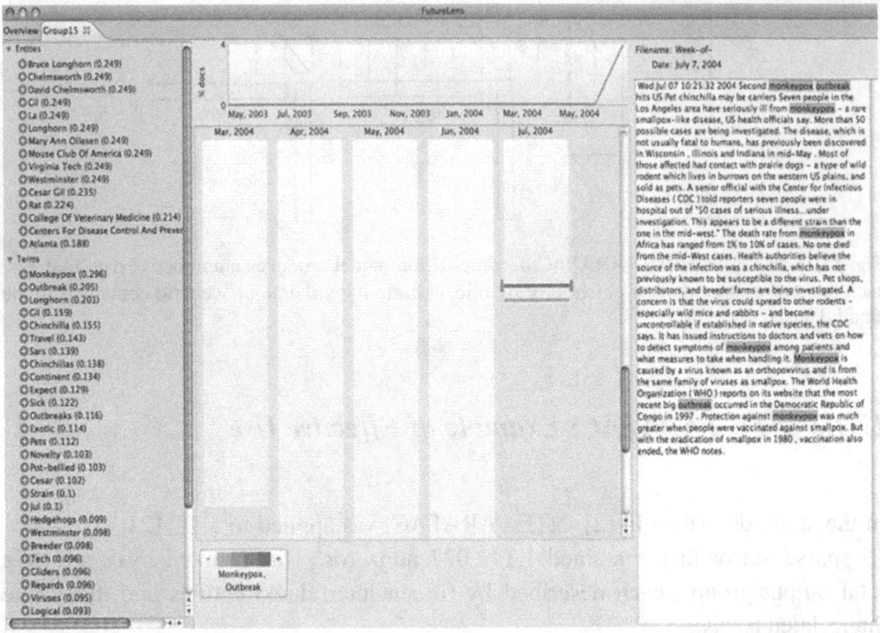

Fig. 17.2 FutureLens can greatly facilitate the interpretation of an NTF output file. Here, a collection including two top terms (*monkeypox* and *outbreak*) has been created by the user and relevant articles located within the data

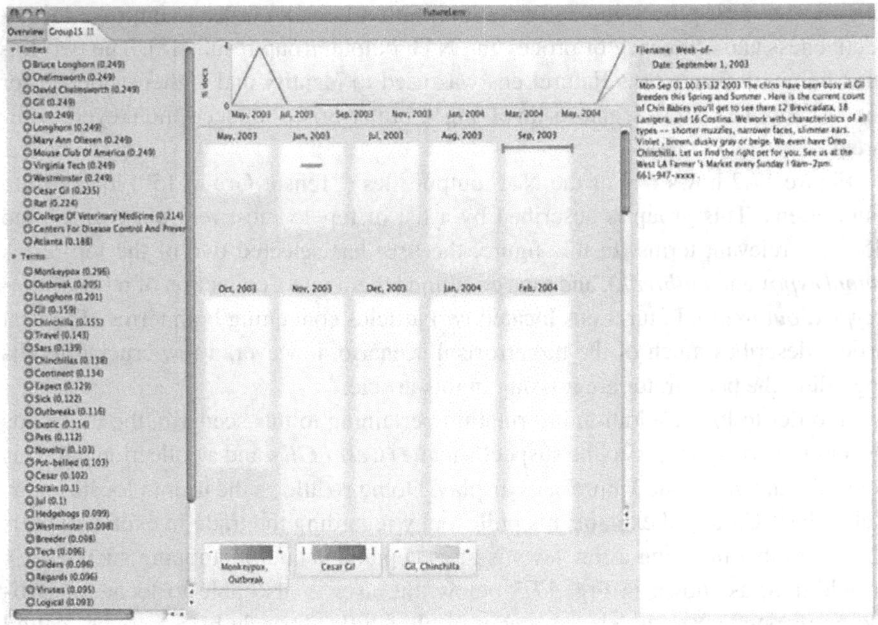

Fig. 17.3 A more complete description of the scenario corresponding to this NTF output feature can be obtained using FutureLens

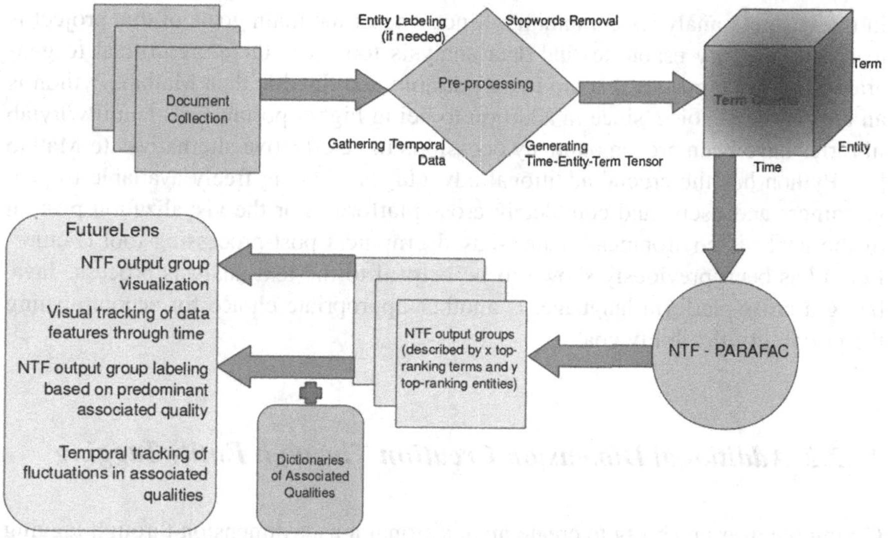

Fig. 17.4 Summary of the design of the NTF-based text analysis environment. This design allows the user to easily perform a number of operations on an input dataset, such as entity tagging, timestamp insertion, and tensor term weight adjustment. The environment also allows the user to easily execute the NTF algorithm, and analyze the results. Among the most important aids in results analysis is the environment's capability to automatically label the resulting NTF output features in accordance with a user-defined categorization scheme

17.2 Python Implementation: Goals and Purpose

While the Matlab-based NTF algorithm has been shown effective, a recent Python conversion adds a number of significant improvements that greatly facilitate the analysis process. One of the main goals of the conversion process was to create a single, unified, user-friendly textual dataset analysis environment. Most significantly, the conversion attempts a novel integration of techniques from such vastly different fields as text mining, visual analytics, and sentiment tracking into a single, convenient, portable, and highly usable text analysis environment. The overall design of the integrated analysis environment is summarized in Fig. 17.4.

17.2.1 Portability, Flexibility, Cost of Use

One of the most significant goals of the integrated analysis environment project is portability and flexibility. A great amount of NTF-related work has in the past been performed using Matlab. The Matlab Tensor Toolbox that was created at Sandia National Laboratory is a great example of such work [1]. However, experience suggests that even though Matlab is a powerful programming environment for scientific applications, code written in it does not transition well into general usage, particularly in the business analytics community. Since one of the main goals of this project is to create a highly usable textual data analysis tool, it is therefore critical to generate code in languages that are more portable and flexible than Matlab. Python is an appropriate choice, since in addition to being highly portable, its NumPy/Pylab libraries have been proven on many occasions to be effective alternatives to Matlab [5]. Python has the crucial additional advantages of being freely available to programmers and users, and completely cross-platform. For the visualization portion of the analysis environment, a Java-based graphical post-processing tool (FutureLens) has been previously shown to be helpful to the text analysis process. Java, being a cross-platform language, is another appropriate choice for accomplishing the portability/flexibility goal.

17.2.2 Additional Dimension Creation Through Entity Tagging

Giving the user an ability to create an additional tensor dimension through tagging a subset of significant terms or entities is one of the major NTF improvements included in the integrated analysis environment [6]. This is distinct from the trust measures described in the subsequent section, because relative significance in the case of entities is the result of their type, rather than of the nature of the specific terms. For example, *Person*-type entities could include all the people's names found in the dataset. *Location*-type entities could include a wide variety of geographical labels: city names, state/province names, countries, mountain ranges, lakes, etc. In other words, a user could emphasize an entire group of terms (created because of common type), without having to consider each individual term's potential significance.

17.2.3 Significance or Trust Measure Integration into NTF

Under some circumstances, it could be greatly helpful to the analysis process for the environment to include an integrated significance or trust measures capability. It is possible, indeed likely, that a knowledgeable user will have access to potentially important information which normally would be inaccessible to the NTF algorithm. In other words, different elements of the data may have different levels of significance to the user because of the user's prior knowledge about the data. Alternatively, this may be viewed as a trustworthiness issue-meaning, for example, that the user may consider certain sources as inherently worthy of trust, while others may be entirely untrustworthy in the user's mind. The Python NTF implementation includes the ability to alter the tensor values in accordance with a user-supplied trust list. The trust list is simply a list of terms and corresponding weights. Terms that are more worthy of consideration may be assigned a higher weight by the user, while some other terms may be assigned a lower weight. The NTF-PARAFAC approach then integrates these significance/trust measures into the factorization process. Incorporation of different term weighting schemes could also be included as part of this user-influenced NTF approach. The integrated analysis environment provides the user with significance/trust controls that do not requiring the user to be exposed to the underlying NTF code.

17.3 Integrated Analysis Environment Capabilities

The following sections describe the various capabilities of the analysis environment. The required input formats and pre-processing steps needed to build an NTF model are well described in [6]. Here, we focus on how the NTF can be used within FutureLens to facilitate knowledge discovery.

17.3.1 Deployment of NTF Algorithm (in Python)

While the features of analysis environment described in [6] are important and enhance the potential effectiveness of the environment as it relates to knowledge discovery, the NTF step is by far the most significant. In order to utilize this feature, the user will need to provide an NTF input file that may or may not contain tagged entities. The inclusion of tagged entities, however, may greatly enhance the analysis process. The additional dimension that can be constructed based on the tagged entities may allow for the establishment of connections that would not have otherwise been revealed.

The user chooses the number of desired NTF output features, and the NTF algorithm attempts to create that number of output groups, each described in a separate file and labeled *GroupX.txt*, where *X* is the arbitrarily assigned group number. It should be noted that the group number does not carry any significance. For example, Group1.txt does not necessarily describe a feature of the data that is more interesting or important than that described by *Group20.txt*. This is in large part due to the highly subjective and context-dependent nature of concepts such as "interesting" and "important". These concepts depend on the nature and the context of the analysis, the nature of the dataset and the problem, as well as the user's personal opinions and biases. It is impossible to quantify all of these highly subjective and unstable variables to incorporate them into a deterministic computer algorithm.

When entities are included in the dataset, each NTF output group file includes a list of top 15 most relevant entities and top 35 most relevant terms. The entities and terms are ranked in accordance with an internally generated relevance score. The score attempts to quantify the term's relative importance to this particular feature. As shown in Fig. 17.5, both the terms and the entities are listed in descending order of importance in an NTF output group file. However, it is again important to remember that this quantification is just an attempt at reflecting subjective, human judgment, and may not reflect the opinions of a human analyst precisely.

As demonstrated in Fig. 17.5, the output of the NTF algorithm is simply a series of lists of terms, each list describing some feature of the dataset. Further human

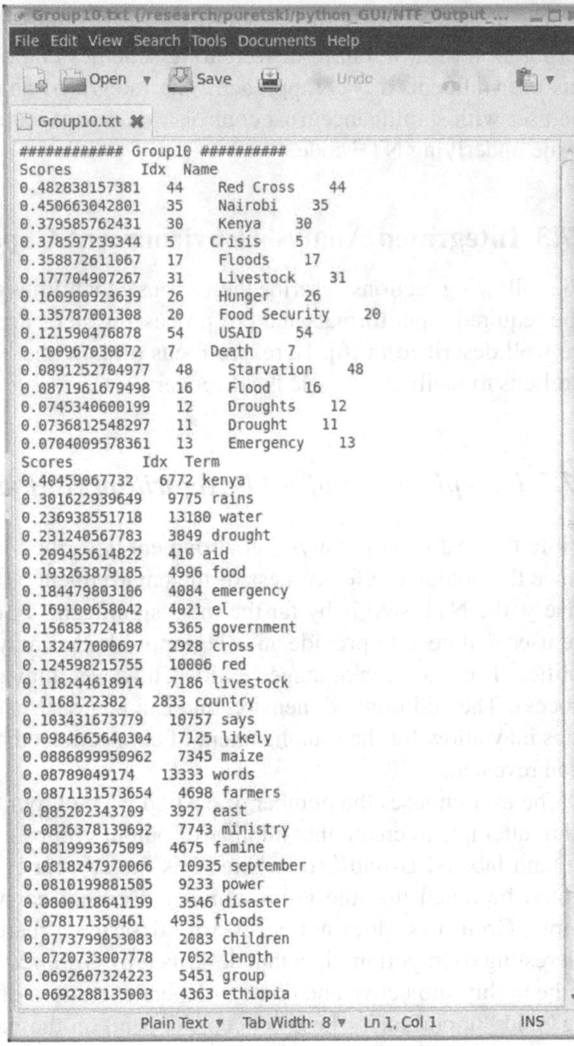

Fig. 17.5 A sample NTF output file. This file was generated by the Python-based analysis environment using the NTF-PARAFAC algorithm. The algorithm was applied to a dataset of news articles about Kenya, covering the years of 2001–2009. As can be seen in this figure, this NTF output feature describes a drought-related theme in the dataset. Terms such as rains, water, drought, emergency, and aid appear near the top of the terms list

analysis and knowledge discovery may be difficult to accomplish based on nothing more than a list of terms. This was the motivation for the creation of the visual NTF output analysis tool called FutureLens [8].

FutureLens allows the user to import the output of the NTF algorithm and analyze it further, while connecting it back to the original dataset. The user has the option of loading any number of NTF output groups at the same time, and in any combination. Each group is allocated its own separate tab in the graphical user interface. The button labeled with a "+" symbol that appears to the left of each term may be used to add that term to the main FutureLens display. Once a term has been added, FutureLens will plot that term's temporal distribution summary in the top-center display panel (see Fig. 17.6). This allows the user to get a quick impression of how the term

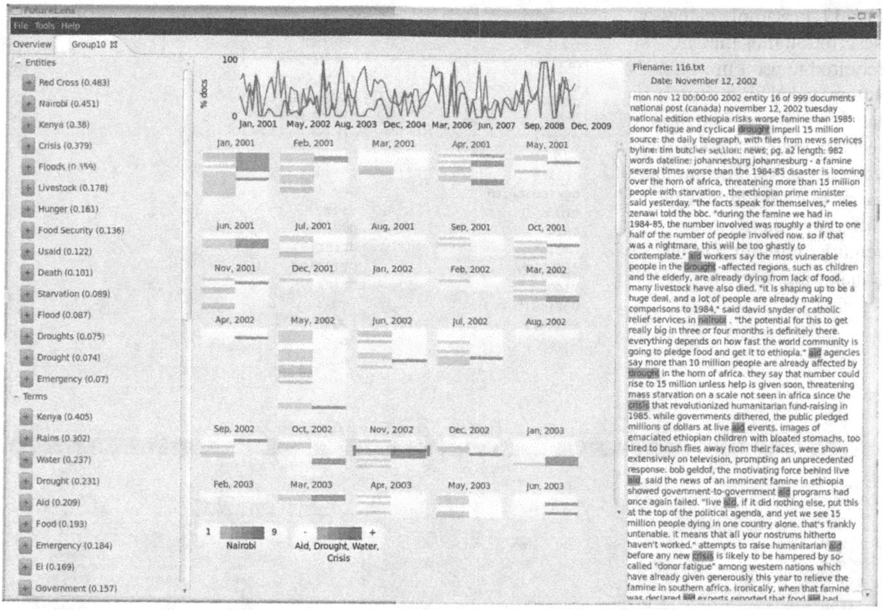

Fig. 17.6 FutureLens allows the user to analyze NTF output results in depth by tracking the constituent NTF group terms through the dataset

is used throughout the dataset, perhaps taking note of peak usage times. FutureLens also locates and color-codes the term within the dataset's document space. This is shown in the central display panel, where every line segment is clickable and corresponds to a single document within the dataset. If the user clicks on one of these line segments, the corresponding document will be displayed in the panel on the right.

It is important to note that FutureLens may be highly useful as a text analysis tool even without NTF output results, since it functions quite effectively as stand-alone software. For instance, the user has the ability to load a dataset into FutureLens independently of NTF output groups. Once a dataset is loaded, the user may search for particular terms and track their occurrence temporally through the dataset (if the dataset contains SGML-style date tags, which can be added using the feature of the analysis environment [6]). It is also possible to display all of the terms contained within the dataset (excluding the ones on a user-defined stop words list), sorted either alphabetically or by frequency. FutureLens displays the terms thirty at a time, providing the user with *Next Page* and *Previous Page* buttons.

Automated NTF output labeling is a significant addition to FutureLens that was made as part of its integration into the analysis environment. Automated NTF group labeling has the ability to speed up the analysis process by allowing the user to quickly focus attention of most relevant groups. Naturally, relevance and relative importance are highly subjective and depend on the exact nature of the user's particular research study. It is therefore highly beneficial to allow easily customizable, plain-text files to serve as category descriptors. The format of these files is extremely straightforward, as shown in Fig. 17.7.

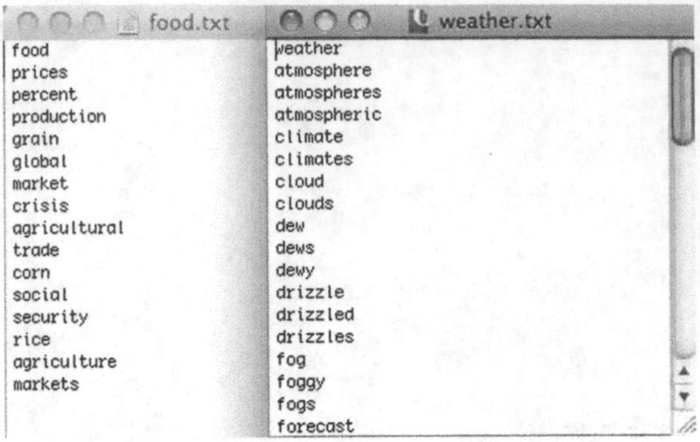

Fig. 17.7 Sample category description files that are required to use FutureLens's automated NTF output labeling feature. The first term in a file is used as the category label. The number of terms in each file may be different—there is no required minimum number of a maximum limit

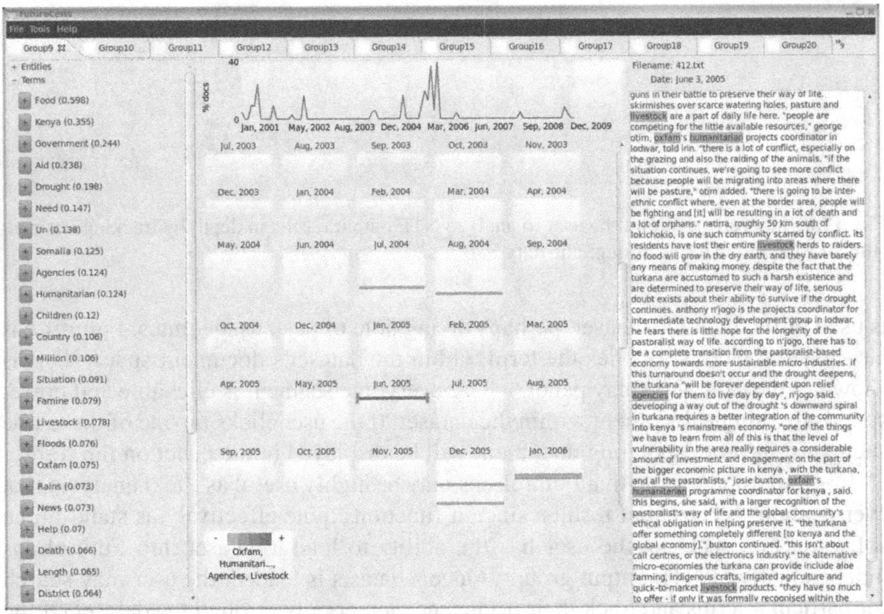

Fig. 17.8 After adjusting the NTF algorithm to have an agriculture focus, the user may utilize FutureLens for further visual analysis of the NTF results. Shown here, the discovery of the impact of a 2004–2005 drought on Kenyan agriculture and the corresponding social unrest it caused

The category descriptor files can be very easily created and/or modified by the user, in accordance with the exact nature of the goals and desired focus of each particular study or model. Any number of categories is possible, but experience has shown that it is generally more helpful to keep the number relatively small. After the categories have been loaded, FutureLens compares the terms constituting each NTF output group with the terms found in the category descriptor files. The category with the highest number of matches becomes the label for that NTF group. Figures 17.8 and 17.9 demonstrate how this feature may be highly useful

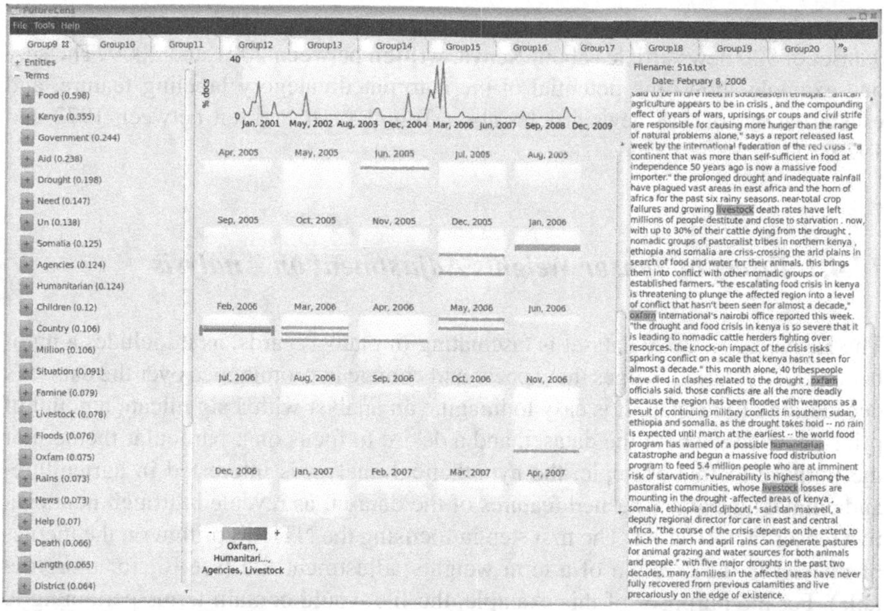

Fig. 17.9 The situation in Kenya's Rift Valley seems to have become even more dangerous by February of 2006. The articles corresponding to the spike in the selected term collection described a region "flooded" with weapons and on the brink of an outbreak of major violent conflict. This makes the subsequent leveling off in the frequency of this collection all the more mysterious

to furthering text analysis. In this example, the user can immediately see that of the ten NTF output groups loaded into FutureLens, five have been labeled as belonging to the *weather* category (light yellow), four have been labeled under the *water* category (dark green), and one has been labeled as belonging to the *food* category (dark red). It should be noted that the category labels also appear as a tool-tip if the user places the mouse cursor over GUI tab containing the NTF group file name.

As discussed in this section, the integrated analysis environment provides the analyst with a number of significant features, ranging from data pre-processing, to NTF execution, to deeper, post-processing NTF results analysis. The next section goes into greater detail in describing the potential effectiveness of this approach, focusing on two newly added features: term weight adjustment capability and automated NTF results labeling.

17.4 Examples of Knowledge Discovery

The two examples described in this section demonstrate the potential effectiveness of the integrated analysis environment and its potential for knowledge discovery. The first example focuses on demonstrating the potential effectiveness of adjusting term weights as it applies to knowledge discovery. This example utilizes a dataset of 900 news articles about Kenya, written between 2001 and 2009. The second example shows the potential of the automated category labeling feature, and uses a dataset of 818 news articles about Bangladesh, written between 1972 and 1976.

17.4.1 Effect of Tensor Weights Adjustment on Analysis

The Kenya 2001–2009 dataset is fascinating in many regards, as it includes a number of greatly varied themes that appear and change in prominence over the dataset's decade-long time span. It is easy to imagine an analyst with a significant amount of prior knowledge about the dataset, and a desire to focus on a particular theme. For the purpose of this example, the hypothetical analyst is interested in agriculture- and animal husbandry-related features of the dataset, as revealed through nonnegative tensor factorization. The first step in focusing the NTF algorithm on the themes of interest is the creation of a term weights adjustment file (see [6] for more details). For the purposes of this example, the file would contain terms pertaining to agriculture, giving them increased weight.

Figure 17.8 shows a significant spike in the user-created term collection (*Oxfam*, *Humanitarian*, *Agencies*, *Livestock*), which occurs starting in mid-2005 and levels off by mid-2006. Selecting one of the color-coded (blue) bars in the June 2005 box in the central panel causes the corresponding article to be displayed in the panel on the right. Here, the user quickly learns about a recent spike in conflict over limited resources and grazing rights in Kenya's Rift Valley, partly caused by a recent drought's wiping out of 70 percent of the livestock in the Turkana province.

The dataset, however, includes news articles from 2001 through 2009, and the peak in the selected term group levels off in mid-2006. It may be interesting to track this collection further temporally, in order to attempt to determine why its importance decreased toward the end of this time period. Taking a look at a strong February 2006 spike in this collection's frequency, one may note that matters have in fact gotten worse at this time. The article shown in Fig. 17.9 discusses escalating and increasingly violent conflict, made even worse by the fact that the region is "flooded" with weapons due to continuing military conflict in neighboring Sudan. This dire description of the situation makes the subsequent leveling off all the more mysterious.

To explore this mystery further, the user simply has to continue tracking the term collection temporally through the dataset, reading only a very small portion of the articles contained in the entire dataset. This is has the potential to greatly increase analyst efficiency, saving significant time and resources. The subsequent months' articles that were revealed by continued tracking of this term collection show the causes of the eventual sudden leveling off that indicates that the conflicts described in the previous articles may have been resolved. As shown in Fig. 17.10,

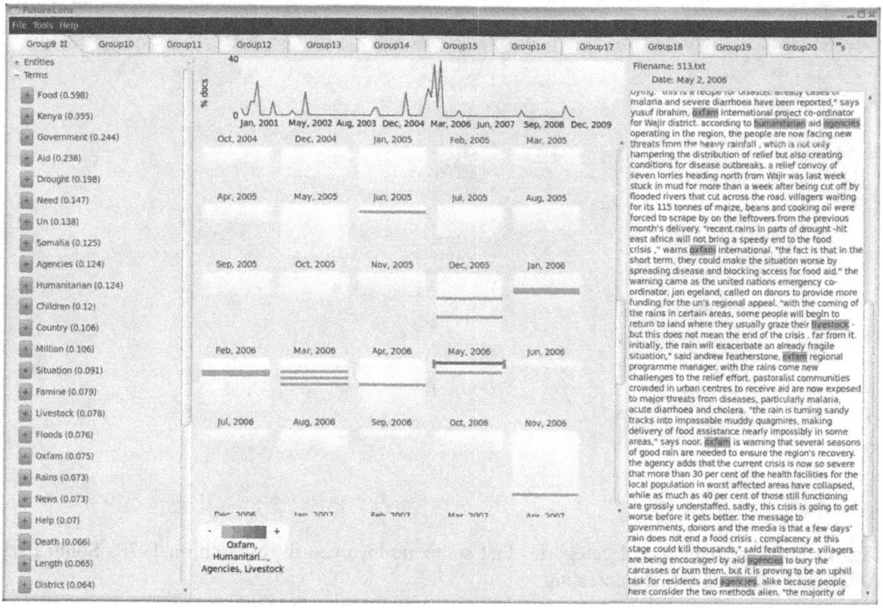

Fig. 17.10 Continuing to track the term collection further through the dataset reveals that the dangerous situation described in Figs. 17.8 and 17.9 had been resolved largely due to a high amount of rainfall that occurred in April and May of 2006

the growing conflict was alleviated by a significant amount of rainfall that occurred in April and May of 2006 in this area of Kenya. The rainfall amount was in fact so great that it even caused some additional danger through a risk of flooding. However, it did eventually stabilize the situation in the area by eliminating the drought. While the crisis had not been completely resolved, positive trends had began to emerge and cattle herders had began to return to previously abandoned land.

Thus, the use of a number of different features of the integrated analysis development environment has lead to significant knowledge discovery. Even an analyst who is completely new to this environment, having gone through the process described above, could learn a number of important pieces of information in just an hour or two. First, an agriculture-themed initial exploration had revealed serious and potentially critically important agriculture-based conflicts in the region of interest. Second, tracking the evolution of these conflicts through the dataset had revealed that these conflicts are by no means fully resolved. Even though they were alleviated before turning strongly violent, the alleviation was essentially just a lucky, weather-related break. The underlying risk factors and dangers, such as the *flood* of weapons and competition for scarce resources remain. And thus one might conclude that the situation in this region remains dangerous, though perhaps not immediately so.

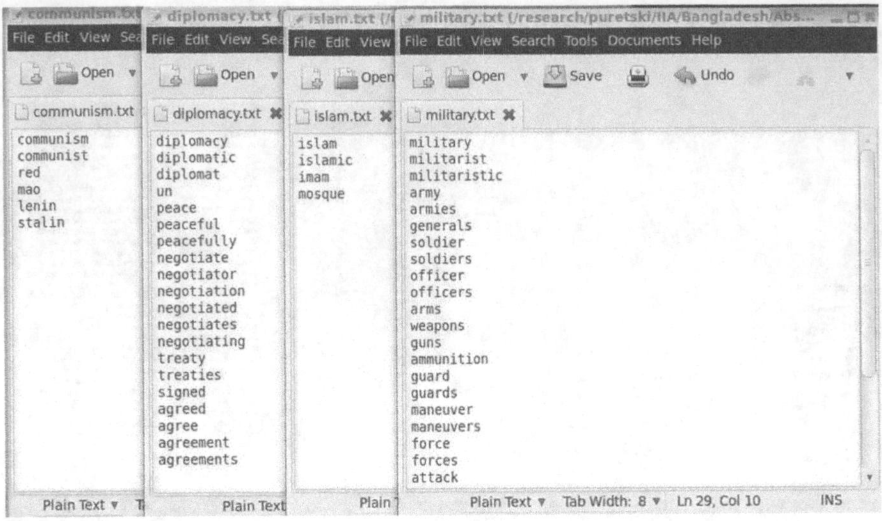

Fig. 17.11 A realistic set of categories that someone involved in research on 1970s South East Asia could potentially find interesting

17.4.2 Effect of Automated NTF Output Labeling on Analysis

The integrated analysis environment's automated NTF output labeling capability is one of its most important features. As will be shown in this section, it can enormously improve an analyst's efficiency by providing a quick automatic ability to sort NTF results in accordance with analyst-defined categories of interest.

For this example, the Bangladesh 1972–1976 dataset was processed using the analysis environment. As the first step, several category descriptor files were created. These categories represent realistic potential areas of interest to someone involved in research on 1970s South East Asia. However, for the purposes of this example, let us assume that the analyst is most interested in developments pertaining to Islam. The category described by the files shown in Fig. 17.11, include Communism, Diplomacy, Islam, and Military.

Following the creation of these category descriptors and the previously described process of execution of the NTF algorithm to generate NTF output group files, the user may utilize FutureLens's automated group labeling feature. Without the automated labeling feature, the analyst must focus in great detail on every single one of the NTF output groups (25 total, for this example). This could take a considerable amount of time, and the process would be prone to human error. Using the automated NTF group labeling feature of the analysis environment, however, takes just a few second. The results are shown in Fig. 17.12, where those groups that did not fit into any one of the four categories of interest have already been closed. Of the labeled groups, one fit into the Islam category, four were labeled as Military-related, ten had a Diplomacy theme, while the rest did not fit into any of the categories created by the user. There were no Communism-labeled groups in this set.

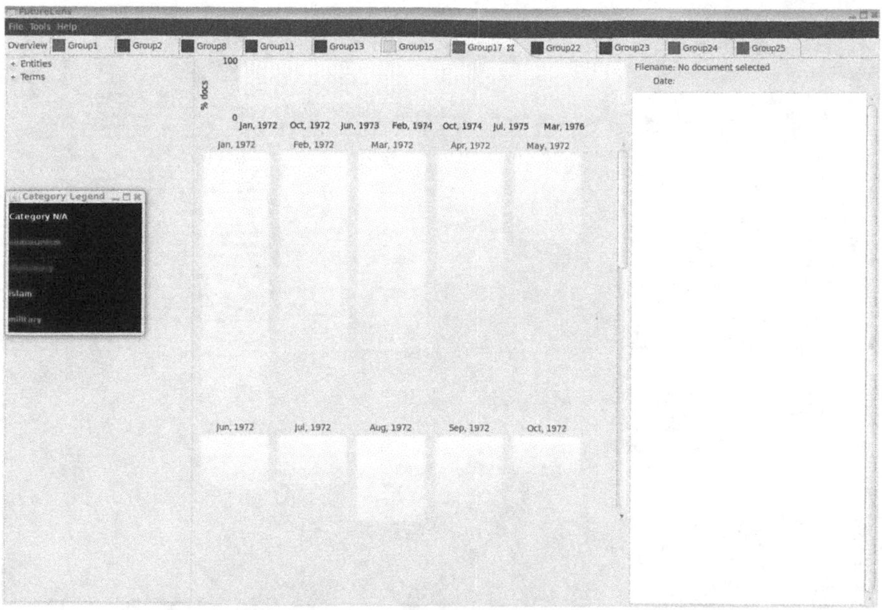

Fig. 17.12 NTF output groups have been automatically labeled in accordance with the categories loaded by the user (shown in the *legend window on the right*)

As one may recall, the hypothetical analyst in this scenario is most interested in developments pertaining to Islam. It just happens that only one of the NTF output features has been automatically labeled as belonging to the Islam category. This already provides the analyst with some important and potentially new knowledge, namely that Islam did not figure prominently into the news coming out of Bangladesh in the 1970s. Even more importantly, the analyst can save a great deal of time by focusing exclusively on just one of the twenty-five total NTF output groups. Shown in Fig. 17.13, the analyst performs a detailed analysis of Group 15, labeled as belonging to the Islam category. Quickly revealed in the articles belonging to this category are Pakistan's efforts to improve its diplomatic position by strengthening ties with Islamic countries inside and outside of the South East Asia region.

17.5 Conclusions and Future Work

In this paper, we have presented a new text analysis environment that effectively integrates nonnegative tensor factorization with visual post-processing tools. The integrated environment also provides effective pre-processing tools for the construction and evaluation of NTF-based models. Non-negative tensor factorization output feature production and interpretation is facilitated by a visual post-processing tool, FutureLens. This Java-based software allows the user to easily mine tensor factors for the purpose of discovering new, interesting patterns or communications from

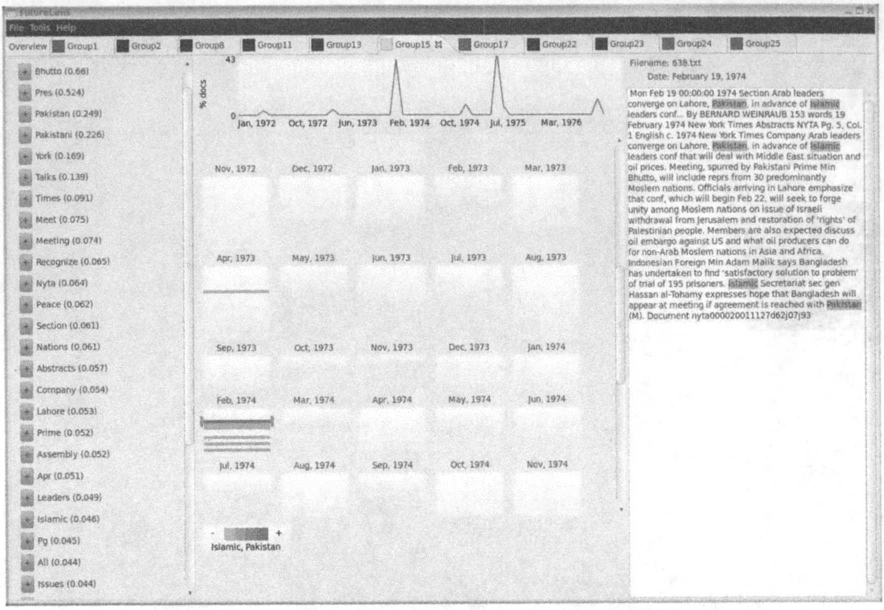

Fig. 17.13 The automated NTF group labeling feature allows the analyst to very quickly focus on the one most relevant group. Quickly revealed through deeper analysis of this group are Pakistan's efforts at diplomacy involving Islamic countries inside and outside of the South East Asia region

large text-based corpora. Customizing FutureLens and NTF for applications such as bioinformatics and spatial-temporal data mining with geocoding (addition of geographic descriptors) is planned.

References

1. Bader, B., Kolda, T.: MATLAB Tensor Toolbox, version 2.4 (2010). http://csmr.ca.sandia.gov/~tgkolda/TensorToolbox/
2. Bader, B., Puretskiy, A., Berry, M.: Scenario discovery using nonnegative tensor factorization. In: Ruiz-Schulcloper, J., Kropatsch, W. (eds.) Proceedings of the Thirteenth Iberoamerican Congress on Pattern Recognition, CIARP 2008. LNCS, vol. 5197, pp. 791–805. Springer, Berlin (2008)
3. Carroll, J., Chang, J.: Analysis of individual differences in multidimensional scaling via an N-way generalization of 'Eckart-Young' decomposition. Psychometrika **35**, 283–319 (1970)
4. Harshman, R.: Foundations of the PARAFAC procedure: models and conditions for an explanatory multi-modal factor analysis. UCLA Work. Pap. Phon. **16**, 1–84 (1970)
5. Numpy documentation (2010). http://docs.scipy.org/doc/numpy
6. Puretskiy, A.: A visual approach to automated text mining and knowledge discovery. Ph.D. thesis, Department of Electrical Engineering and Computer Science, University of Tennessee, Knoxville (2010)
7. Scholtz, J., Plaisant, C., Grinstein, G.: IEEE VAST 2007 Contest (2007). http://www.cs.umd.edu/hcil/VASTcontest07
8. Shutt, G., Puretskiy, A., Berry, M.: Futurelens: software for text visualization and tracking. In: Proc. Ninth SIAM International Conference on Data Mining, Sparks, NV (2009)

Index

DOI 10.1007/978-1-4471-2437-5,